Microarrays in Clinical Diagnostics

METHODS IN MOLECULAR MEDICINE™

John M. Walker, Series Editor

METHODS IN MOLECULAR MEDICINE™

Microarrays in Clinical Diagnostics

Edited by

Thomas O. Joos, PhD

Biochemistry Department
NMI Natural and Medical Sciences Institute
at the University of Tuebingen, Reutlingen, Germany

Paolo Fortina, MD, PhD

Center for Translational Medicine, Department of Medicine
Jefferson Medical College, Philadelphia, PA

HUMANA PRESS ✳ TOTOWA, NEW JERSEY

Cover design by Patricia F. Cleary.

Cover illustration: LMPC of cultured HepG2 cells in a PALM DuplexDish (Chapter 1, Fig. 7, *see* pp. 16, 17 and Color Plate 7).

For additional copies, pricing for bulk purchases, and/or information about other Humana titles, contact Humana at the above address or at any of the following numbers: Tel.: 973-256-1699; Fax: 973-256-8341; E-mail: orders@humanapr.com; or visit our Website: www.humanapress.com

Printed in the United States of America. 10 9 8 7 6 5 4 3 2 1

eISBN 1-59259-923-0
ISSN 1543-1894

Library of Congress Cataloging-in-Publication Data

Microarrays in clinical diagnostics / edited by Thomas O. Joos, Paolo Fortina.
    p. ; cm. -- (Methods in molecular medicine ; 114)
 Includes bibliographical references and index.
 ISBN 1-58829-394-7 (alk. paper)
 1. DNA microarrays--Diagnostic use.
 [DNLM: 1. Oligonucleotide Array Sequence Analysis. 2. Diagnostic
Techniques and Procedures. QZ 52 M6262 2005] I. Joos, Thomas. II. Fortina,
Paolo. III. Series.
 RB43.8.D62M53 2005
 616.07'58--dc22
                                                    2004026935

Preface

Within the last decade, microarray technology has evolved from an emerging technology developed and used by a few laboratories into a well-established technology used in laboratories all over the world. In fact, the need to characterize genetic alterations is one of the highest priorities for the future of medicine and the clinical management of disease. This technology allows the rapid detection of point mutations, insertions or deletions, loss of hetero-zygosity, and gene amplification, which constitute the major nucleic acid variations associated with human disease. Additional disease-causing changes may involve DNA methylation and microsatellite instability for which automatable methods to detect instability in as few as 100 cells at multiple loci are required. Furthermore, in some instances it may be necessary to detect one tumor cell among a large number of normal cells, as well as to profile differentially expressed genes. Eventually, the ultimate goals are to characterize the entire genome rapidly and inexpensively, ideally using a single cell in order to survey the whole genome for any nucleic acid variation.

Within the field of proteome research, microarray technology has been adapted to the protein arena. Although DNA microarrays are quite popular and in vogue, proteins (not genes) are the targets for drugs; therefore, there is an increasing need to develop protein chips. Specifically, tools and methods are needed for the identification and quantification of proteins, and for the study of protein–protein interactions, enzyme–substrate interactions, and small-molecule interactions. Enormous efforts have been undertaken to transfer standard sandwich immunoassays in miniaturized and parallel formats to analyze simultaneously the expression of a large number of proteins, e.g., serum or tumor biomarkers.

It is becoming clear that microarray technology is capable of fulfilling these needs. Although some technologies are still confined to research laboratories, such as those aimed at performing resequencing of known genes and protein identification, rapid and robust methods are becoming available to address each of these needs. However, some general goals for diagnostics including sensitivity, specificity, high throughput, cost effectiveness, and turnaround time still need improvement.

Finally, it is also clear that new tools in the nanoscale format are on the horizon: quantum dots, nanoparticles, carbon nanotubes, and atomic force microscopes are now being used to directly probe DNA structure. These technologies represent an emerging approach promising increased throughput, sensitivity, and sample processing, as well as facilitating single-cell and single-molecule detection.

Microarrays in Clinical Diagnostics offers an overview of the world of microarray technology. Because it is not clear which technology will eventually prevail, we have tried to assemble a comprehensive survey of the varied technologies now in use and to provide detailed methods sections in order to support scientists who design and perform microarray experiments.

Thomas O. Joos, PhD
Paolo Fortina, MD, PhD

Contents

Contributors

ANDREA ARDIZZONI • *Department of Biological Sciences, Imperial College London, London, UK*

TITO BACARESE-HAMILTON • *Department of Biological Sciences, Imperial College London, London, UK*

RHONDA BANGHAM • *Research and Development, Protometrix Inc., Branford, CT*

FRANCIS BARANY • *Department of Microbiology and Immunology, Weill Medical College of Cornell University, New York, NY*

PIERANGELO BONINI • *Diagnostica e Ricerca San Raffaele S.p.A., Università Vita-Salute S. Raffaele, Milan, Italy*

RENATE BURGEMEISTER • *P.A.L.M. Microlaser Technologies AG, Bernried, Germany*

JING CHENG • *State Key Laboratory of Biomembrane and Membrane Biotechnology, Department of Biological Sciences and Biotechnology, Tsinghua University; National Engineering Research Center for Beijing Biochip Technology; CapitalBio Corporation, Beijing, China*

YU-WEI CHENG • *Department of Microbiology and Immunology, Weill Medical College of Cornell University, New York, NY*

LAURA CREMONESI • *Unit of Genomics for Diagnosis of Human Pathologies, Istituto di Ricovero e Cura a Carattere Scientifico Ospedale San Raffaele, Milan, Italy*

ANDREA CRISANTI • *Department of Biological Sciences, Imperial College London, London, UK*

HUGH M. DAVIS • *Department of Clinical Pharmacology, Centocor Inc., Malvern, PA*

SHERRY A. DUNBAR • *Research and Development, Luminex Corporation, Austin, TX*

EDWARD EIRIKIS • *Department of Clinical Pharmacology, Centocor Inc., Malvern, PA*

REYNA FAVIS • *Department of Microbiology and Immunology, Weill Medical College of Cornell University, New York, NY*

MAURIZIO FERRARI • *Unit of Genomics for Diagnosis of Human Pathologies, Istituto di Ricovero e Cura a Carattere Scientifico Ospedale San Raffaele; Diagnostica e Ricerca San Raffaele S.p.A, Milan, Italy*

BARBARA FOGLIENI • *Unit of Genomics for Diagnosis of Human Pathologies, Istituto di Ricovero e Cura a Carattere Scientifico Ospedale San Raffaele, Milan, Italy*

PAOLO FORTINA • *Center for Translational Medicine, Department of Medicine, Jefferson Medical College, Thomas Jefferson University, Philadelphia, PA*

GABRIELE FRIEDEMANN • *P.A.L.M. Microlaser Technologies AG, Bernried, Germany*

RAINER GANGNUS • *P.A.L.M. Microlaser Technologies AG, Bernried, Germany*

NORMAN P. GERRY • *Department of Genetics and Genomics, Boston University School of Medicine, Boston, MA*

JULIAN GRAY • *Department of Biological Sciences, Imperial College London, London, UK*

YONG GUO • *State Key Laboratory of Biomembrane and Membrane Biotechnology, Department of Biological Sciences and Biotechnology, Tsinghua University, Beijing, China*

BRIAN B. HAAB • *Laboratory of Cancer Immunodiagnostics, Van Andel Research Institute, Grand Rapids, MI*

RUOCHUN HUANG • *Department of Gynecology and Obstetrics, Emory University School of Medicine, Atlanta, GA*

RUO-PAN HUANG • *Department of Gynecology and Obstetrics, Emory University School of Medicine, Atlanta, GA*

JAMES W. JACOBSON • *Research and Development, Luminex Corporation, Austin, TX*

THOMAS O. JOOS • *Head, Biochemistry Department, NMI Natural and Medical Sciences Institute at the University of Tuebingen, Reutlingen, Germany*

MARGARET A. KELLER • *Cardeza Special Hemostasis Laboratory, Cardeza Foundation for Hematologic Research and Division of Hematology, Jefferson Medical College, Thomas Jefferson University, Philadelphia, PA*

LARRY J. KRICKA • *Department of Pathology and Laboratory Medicine, University of Pennsylvania, Philadelphia, PA*

ZE LI • *National Engineering Research Center for Beijing Biochip Technology; CapitalBio Corporation, Beijing, China*

LOVISA LOVMAR • *Department of Medical Sciences, Uppsala University, Uppsala, Sweden*

STEPHEN R. MASTER • *Department of Pathology and Laboratory Medicine, University of Pennsylvania, Philadelphia, PA*

GREGORY A. MICHAUD • *Research and Development, Protometrix Inc., Branford, CT*

BRUCE E. MILLER • *Department of Clinical Pharmacology, Centocor Inc., Malvern, PA*

MARTINA MIRLACHER • *Department of Pathology, University Medical Center Hamburg-Eppendorf, Hamburg, Germany*

YILMAZ NIYAZ • *P.A.L.M. Microlaser Technologies AG, Bernried, Germany*

UMA PRABHAKAR • *Department of Clinical Pharmacology, Centocor Inc., Malvern, PA*

PAUL F. PREDKI • *Research and Development, Protometrix Inc., Branford, CT*

BERND SÄGMÜLLER • *P.A.L.M. Microlaser Technologies AG, Bernried, Germany*

ULRICH SAUER • *P.A.L.M. Microlaser Technologies AG, Bernried, Germany*

GUIDO SAUTER • *Department of Pathology, University Medical Center Hamburg-Eppendorf, Hamburg, Germany*

KARIN SCHÜTZE • *P.A.L.M. Microlaser Technologies AG, Bernried, Germany*

BARRY SCHWEITZER • *Research and Development, Protometrix Inc., Branford, CT*

QIAN SHI • *Department of Gynecology and Obstetrics, Emory University School of Medicine, Atlanta, GA*

RONALD SIMON • *Department of Pathology, University Medical Center Hamburg-Eppendorf, Hamburg, Germany*

STEFANIA STENIRRI • *Unit of Genomics for Diagnosis of Human Pathologies, Istituto di Ricovero e Cura a Carattere Scientifico Ospedale San Raffaele, Milan, Italy*

MONIKA STICH • *P.A.L.M. Microlaser Technologies AG, Bernried, Germany*

ANN-CHRISTINE SYVÄNEN • *Department of Medical Sciences, Uppsala University, Uppsala, Sweden*

SHENG-CE TAO • *State Key Laboratory of Biomembrane and Membrane Biotechnology, Department of Biological Sciences and Biotechnology, Tsinghua University, Beijing, China*

ACHIM WIXFORTH • *Chair for Experimental Physics I, University of Augsburg, Augsburg, Germany*

WEIMIN YANG • *Department of Gynecology and Obstetrics, Emory University School of Medicine, Atlanta, GA*

QIONG ZHANG • *National Engineering Research Center for Beijing Biochip Technology; CapitalBio Corporation, Beijing, China*

YAN ZHANG • *State Key Laboratory of Biomembrane and Membrane Biotechnology, Department of Biological Sciences and Biotechnology, Tsinghua University, Beijing, China*

ZHI-WEI ZHANG • *State Key Laboratory of Biomembrane and Membrane Biotechnology, Department of Biological Sciences and Biotechnology, Tsinghua University, Beijing, China*

YI-MING ZHOU • *State Key Laboratory of Biomembrane and Membrane Biotechnology, Department of Biological Sciences and Biotechnology, Tsinghua University, Beijing, China*

Color Plates

Color plates 1–14 follow p. 18 and plates 15–19 follow p. 178.

xiii

1

Noncontact Laser Microdissection and Pressure Catapulting

Sample Preparation for Genomic, Transcriptomic, and Proteomic Analysis

Yilmaz Niyaz, Monika Stich, Bernd Sägmüller, Renate Burgemeister, Gabriele Friedemann, Ulrich Sauer, Rainer Gangnus, and Karin Schütze

Summary

The understanding of the molecular mechanisms of cellular metabolism and proliferation necessitates accurate identification, isolation, and finally characterization of a specific cell or a population of cells and subsequently their subsets of biomolecules.

For the simultaneous analysis of thousands of molecular parameters within a single experiment, as realized by DNA, RNA, and protein microarray technologies, a defined number of homogeneous cells derived from a distinct morphological origin is required. Sample preparation is therefore a very crucial step for high-resolution downstream applications.

Laser microdissection and laser pressure catapulting (LMPC) enables such pure and homogeneous sample preparation, resulting in an eminent increase in the specificity of molecular analyses. For microdissection, the force of focused laser light is used to excise selected cells or large tissue areas from object slides or from living cell culture down to a resolution of individual single cells and subcellular components like organelles or chromosomes, respectively. After microdissection this sample is directly catapulted into an appropriate collection device. As the entire process works without any mechanical contact, it enables pure sample retrieval from morphologically defined origin without cross contamination. Wherever homogenous samples are required for subsequent analysis of, e.g., cell areas, single cells, or chromosomes, the PALM® MicroBeam system is an indispensable tool. The integration of image analysis platforms fully automates screening, identification, and finally subsequent high-throughput sample handling. These samples can be directly linked into versatile downstream applications, such as single-cell mRNA-extraction, different PCR methods, microarray techniques, and many others. Acceleration in sample generation vastly increases the throughput in molecular laboratories and leads to an increasing knowledge about differentially regulated mRNAs and expressed proteins, providing new insights into cellular mechanisms and therefore enabling the development of systems for tumor biomarker identification, early detection of disease-causing alterations, therapeutic targeting and/or patient-tailored therapy.

From: *Methods in Molecular Medicine, Vol. 144, Microarrays in Clinical Diagnostics*
Edited by: T. Joos and P. Fortina © Humana Press Inc., Totowa, NJ

Key Words: Laser microdissection and pressure catapulting; high-resolution sample preparation; single cell analysis; microarray technology.

1. Introduction

Histopathological tissue sections are typically composed of a variety of different cell types in a complex 3D architecture. Each of these cell types shows a different expression pattern of mRNA and protein within their tissue context. Modern biomedical research is usually interested in the molecular expression profile of a distinct cell type within the specimen. Such studies rely on pure sample preparation, as the cells of interest constitute only a small proportion of the specimen and therefore have to be separated from other compounds of the tissue. Among various options for achieving homogeneous samples, only laser microdissection offers high-resolution control of sample composition by selecting or rejecting individual cells and tissue areas of interest, respectively. Therefore, this ability to manipulate cells individually has made a great impact on our ongoing understanding of cellular physiology and molecular pathology by means of genomic, transcriptomic, and proteomic research *(1–3)*. Recent technical advances have now opened new horizons for cellular investigations.

The PALM® MicroBeam (P.A.L.M. Microlaser Technologies, Bernried, Germany, http://www.palm-microlaser.com) allows completely noncontact sample preparation by the combined applications of laser microdissection and pressure catapulting (LMPC). In brief, a pulsed UV-A laser is coupled to a microscope and focused on the sample plane. Thus the microscope, known as an optoanalytical device, turns into a versatile micromanipulation tool: selected specimens of different origins can first be laser microdissected and then ejected against gravity. This patented laser pressure catapulting (LPC) technology drives the sample from the objective plane toward a collection device, e.g., a standard microfuge cap, solely by a laser-induced transportation process. Thus the PALM micromanipulation system has neither physical nor mechanical contact with the specimen, and the risk of contamination of the isolated samples is minimized. Moreover, this technology is a paramount prerequisite for homogeneous sample capture as the extracted samples are traceable from a morphologically defined origin. These samples can be used in a variety of downstream applications (**Fig. 1** and Color Plate 1 following p.18.), e.g., genome-wide expression profiling using cDNA or protein microarrays *(4–7)*.

The unique combination of noncontact microsurgery and sample preparation (from subcellular compounds to entire tissue areas) is performed in numerous research institutes or industrial laboratories throughout the world (cf. http://www.palm-microlaser.com/aboutus/PALM-Referenzen.pdf).

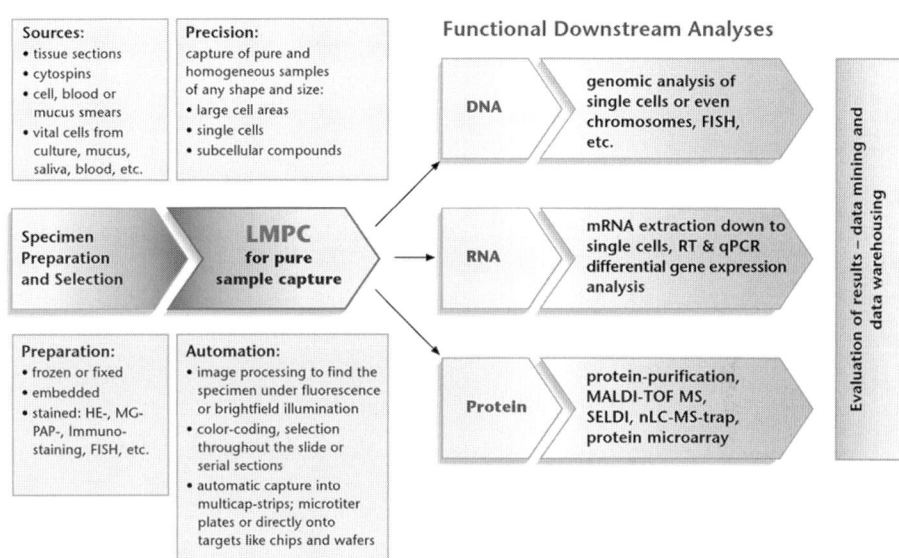

Fig. 1. Workflow for LMPC Applications. Abbreviations: FISH, fluorescence *in situ* hybridization; HE, Hematoxylin & Eosin; LMPC, laser microdissection and pressure catapulting; MALDI-TOF; MS, matrix-assisted laser desorption ionization time-of-flight mass spectrometry; nLC-MS, non-flow liquid chromatography mass spectrometry with ion trapping; PCR, quantitative polymerase chain reaction; RT, reverse transcreptase; SELDI, surface-entranced laser desorption ionization. (*See* Color Plate 1 following p.18.)

1.1. The Force of Focal Light

1.1.1. Laser Microdissection

The PALM MicroBeam (**Fig. 2;** *see* Color Plate 2 following p.18.) is equipped with a UV-A laser coupled through the epifluorescence path to an inverted research microscope and focused on a micron-sized spot on the sample via the objective lenses (**Fig. 3;** *See* Color Plate 3 following p.18.). The beam focus diameter results mainly from the wavelength and beam quality of the laser, the magnification and numerical aperture (NA) of the applied objective, and the specimen's absorbance behavior. Laser microdissection is possible with several objective magnifications from 5 to 100×. Higher aperture objectives, e.g., a 100× oil immersion objective (NA = 1.3), are necessary for a minimum cutting size of less than 700 nm *(8)*, allowing microdissection and microsurgery of single nuclei, filaments, chromosomes, or even chromosomal parts *(9)*. For best focusing results, a laser of high beam quality and an objective with an NA >1 is required.

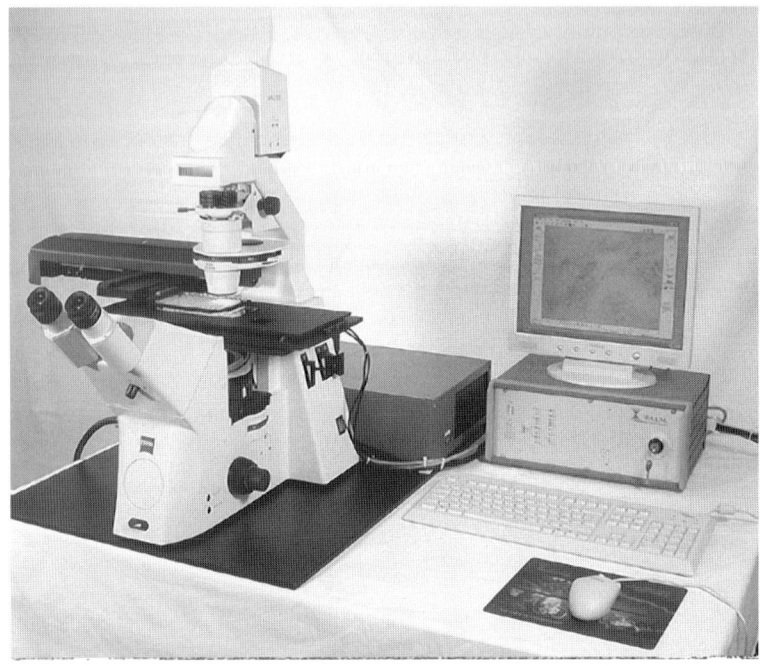

Fig. 2. The PALM® MicroBeam-HT. (*See* Color Plate 2 following p.18.)

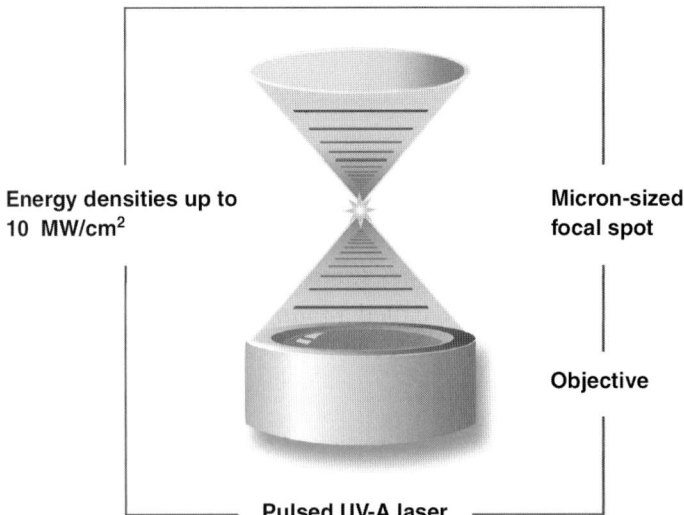

Energy densities up to
10 MW/cm²

Micron-sized
focal spot

Objective

Pulsed UV-A laser

Fig. 3. Scheme of laser ablation. A high-intensity beam of single wavelength radiation (laser) is focused through the objective. Depending on the numerical aperture of the objective, laser focus diameters of less than 1 µm can be achieved. Within this spot, material becomes ablated owing to a photofragmentation process. (*See* Color Plate 3 following p.18.)

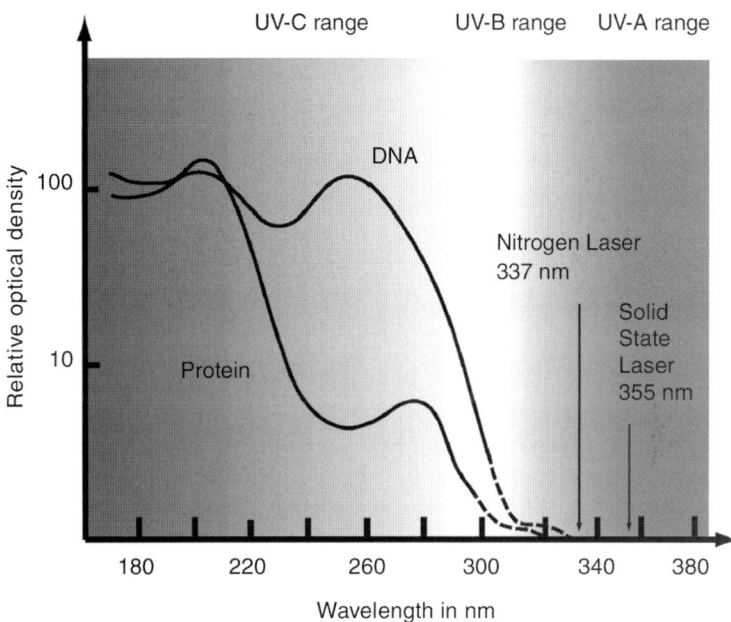

Fig. 4. Peak absorption wavelengths. Absorption maxima of DNA and proteins. The wavelengths of the laser used lie outside the local absorption maxima of these biomolecules and thus affect neither biomolecular information nor the viability of the microdissected specimen. (*See* Color Plate 3 following p.18.)

The pulsed UV-A laser beam used for LMPC is routinely focused to <1 µm, at which point it impacts the sample with high energy density (≤10 MW/cm^2). This condensed radiation within the focal spot leads to the formation of a high energetic microplasma, in which the energy transfer is sufficient to break the molecule bonds, resulting in photofragmentation of the radiated matter without any mechanical contact. Thus, at the focal point, unwanted material is disintegrated into atoms and small molecules, a not fully understood phenomenon called *ablative photo-decomposition*. However, as this cutting is a fast photochemical process devoid of heat transfer, adjacent material out of the focus is not affected *(10,11)*. The nonfocused laser light next to the focus becomes scattered and travels through adjacent areas without any impact on the specimen. Because of the reduced photon density and because it is out of range of the peak absorption wavelengths for DNA, RNA, or proteins (**Fig. 4**; *See* Color Plate 3 following p.18.), this radiation does not interfere with biological material *(12,13)*. Therefore biomolecules can routinely be isolated from the specimen for downstream analyses and applications *(14)*, and even living cells can be captured *(15,16)*. The viability of the treated cells is not constrained, as they readily enter the cell cycle and

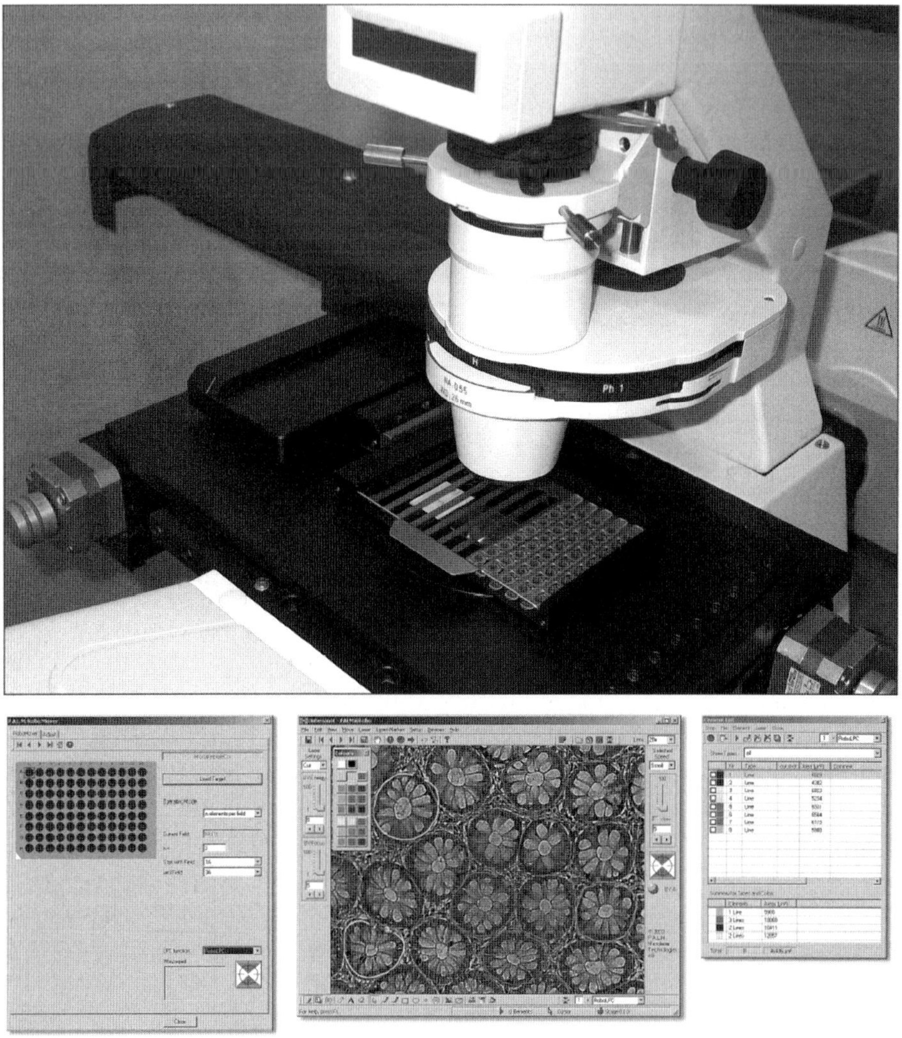

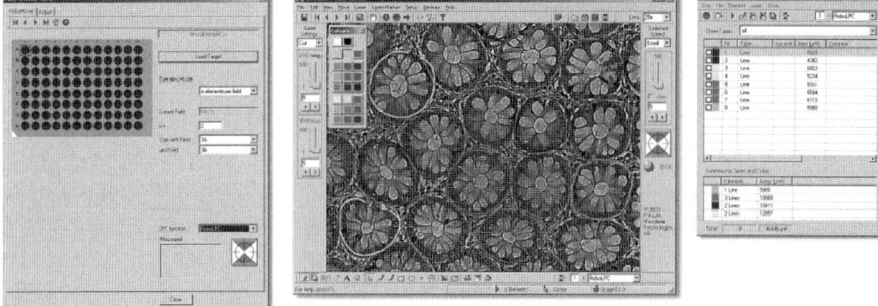

Fig. 5. PALM® RoboMover PALM® RoboSoftware. The RoboMover and the graphical user interface of the RoboSoftware. (*See* Color Plate 4 following p.18.)

proliferate after laser treatment *(16,17)*. As the effective laser energy is concentrated at the focal spot only, it is even possible to perform laser microsurgery or microinjection within living specimens. Numerous publications in the field of cell and developmental biology or from assisted human fertilization procedures have proved the safety of laser-based microsurgery and microdissection *(18–27)*. This technology has also been established in medical laser surgery as the only nonheating surgical process.

1.1.2. Laser Pressure Catapulting

The ablation process produces a clear gap between selected and unwanted material. Subsequently the selected and circumdissected area is ejected from the object plane with a single, slightly subfocal laser pulse. It has been shown that this pulse drives the isolated area out of the plane. The microdissected sample is transported with high speed (≈25 m/s) for several millimeters against gravity directly into a capture device mounted within the laser beam *(11)*. This patented LPC technology marks a breakthrough in modern laser capture methods and allows the entire noncontact preparation of pure and homogeneous samples in a fast and elegant manner *(28–30)*. The range of catapulted specimen spans subcellular organelles *(31)* up to entire small organisms, e.g., a juvenile *Caenorhabditis elegans,* which readily survives the catapult procedure *(15,17)*. At present LMPC is the only published technology able to microdissect and catapult viable cells and tissues (*see* **Subheading 3.3.**).

1.2. Technical Setup of the PALM® MicroBeam System

1.2.1. RoboMover and RoboStage II

A fully robotic unit called RoboMover functions as a multipurpose collection device, with adapters for routine microfuge tubes, multicap strips, and microtiter plate formats. Guided by the entries in the element list (**Fig. 5**; *see* Color Plate 4 following p.18.), complex experimental setups can be planned and processed automatically. The highly automated sample capture device is supported by the second generation of RoboStage. This newly developed microscope stage can travel large distances in *x/y* directions with high-accuracy relocation of selected areas, and allows collection of samples from various object dishes fitting into versatile customized holders like PALM DuplexDishes, Membrane-Slides, microtiter plate formats, and so on. Successful sample capture can easily be controlled using the so-called cap-check function of the RoboStage II. This fully automated generation of samples by the noncontact LMPC process for subsequent analyses vastly increases the throughput in routine and research laboratories.

A special software feature allows linking of similar regions of serial sections of a sample on different slides at the same time. Here, on the source slide (i.e., first slide), areas of interest can be outlined by the user and matched to the linked subset of the following serial sections (i.e., destination slides). The selected areas are individually addressed and adjusted to their actual shape, which may differ owing to torsions occurring during sectioning. By staining only the source slide and leaving the destination slides unstained, it becomes possible to recover unaltered proteins for further downstream proteomic analysis.

1.2.2. PALM RoboSoftware for Automated Microdissection and Catapulting

The controlling software of the PALM MicroBeam system is the so-called RoboSoftware, which manages the motion of the motorized microscope, the motorized microscope stage, the RoboMover capture device, and the optional fluorescence equipment. An intuitive graphical user interface for all software functions facilitates the use of this system. The RoboSoftware includes automated process routines as well as additional functions: a wide palette of drawing tools for marking the incision path in a preselected mode allows the outlining and color coding of independent target areas all over the entire slide or even from different slides after serial sectioning. These selected target areas are listed in an element list protocol, which allows target grouping and experimental scheduling. The list of elements (**Fig. 5**, lower right) is the main tool for summary and display of the outlined samples and corresponding area measures, the color-dependent sorting of the outlined areas, and laser activation. Choosing from the color chart, the computer will microdissect and/or catapult only elements with the designated color. Saving of the selected elements with respect to a reference position in personal files allows the relocation of the stored elements on each individual slide. Thus the slide can be taken out for later sample capture.

Noncontact laser microdissection and catapulting, as realized in PALM MicroBeam-HT, allows the largely automated and highly reliable capture of thousands of cells within a short time, thus allowing higher throughput specimen sampling, which is especially important for array techniques or proteomic studies (**Fig. 1**).

1.2.3. Automated Cell Recognition

Modern detection methods are often based on fluorescence techniques. Thus the PALM MicroBeam can optionally be equipped with features for fluorescence microscopy, allowing simultaneous fluorescence observation and LMPC. The high degree of automation realized in the latest generation of PALM systems (MicroBeam-HT) can be optionally augmented by image-analyzing software modules allowing automated fast scanning functions for specimen identification and image processing. Both fluorescence and bright field microscopy can be used for automated detection of cells or regions of interest. Coupled with any one of these software modules, the MicroBeam system is able to scan, detect, isolate, and finally capture the specimen of interest, e.g., immunostained areas (**Fig. 6A;** *See* Color Plate 6 following p.18.), metaphases, or fluorescent-labeled rare cells (**Fig. 6B**), in a fully automated manner. Recognized areas can subsequently be extracted automatically by the appropriate laser function. These versatile automated scanning software modules also have the advantage of fast and reliable detection and autoevaluation of particular cells, cell components or chromosomes, based on either optimized classifiers or rule sets by means of morphological

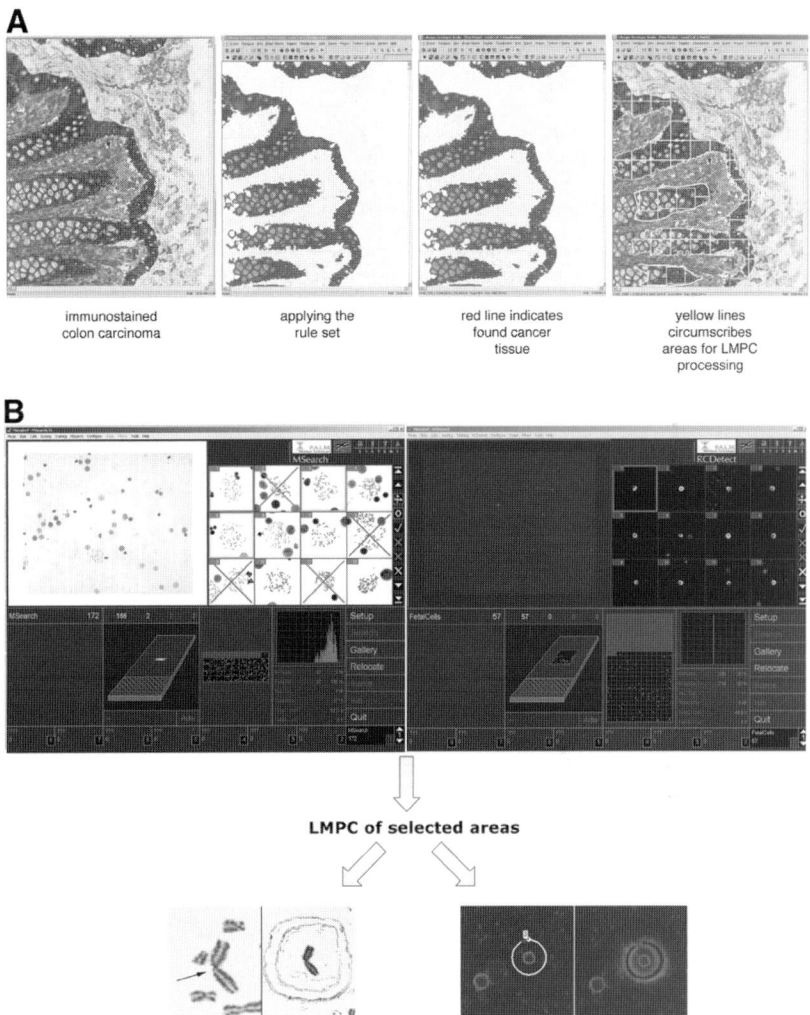

Fig. 6. (Top) Automatic recognition of defined sample. Left, Immunostained section of murine colon. Middle left, Autodetected crypt-containing areas. Middle right, Outlining of autodetected immunopositive tissue areas. Right, Generation of shape files for LMPC processing. **(Bottom)** Automatic detection of fluorescent-labeled samples. Integrated image analysis software provides unique high-performance slide scanning capacities with different modular concepts for the automated detection of, e.g., metaphases (left) and rare cells (right). LMPC, Laser microdissection and catapulting. (*See* Color Plate 5 and 6 following p.18.)

phenotypes. The efficient detection algorithms are trained by P.A.L.M. Microlaser Technologies to achieve integrated, interactive classifiers and rule sets for optimized recognition and accurate results.

1.3. Applications of PALM MicroBeam

In many biological applications the isolation of intact RNA is the most criti
cal step for subsequent analysis. Again, an important factor for good results in
molecular analysis is the homogeneity of the starting material. A selective pro-
curement of specific cells via LMPC technology leads to rapid and highly precise results. Furthermore, the amount of starting material can be minimized, as
only cells of interest are harvested. Direct catapulting with no mechanical interference will save time, prevent the danger of losing specimens during pipeting,
and minimize contamination with unwanted material.

LMPC will supplement functional genomic and proteomic studies correlating
gene or protein profiles with morphological relevant features. Laser catapulting is performed in either routine vials or microtiter plate formats, which allows experiments
with a higher throughput setup. From the respective capture device the specimens are
dissolved, and subsequently biomolecules of interest (i.e., DNA, RNA, or proteins)
are purified and committed to the corresponding downstream application such as
polymerase chain reaction (PCR), microarray hybridization chromosomal analysis,
matrix-assisted laser desorption ionization, or surface-enhanced laser desorption ionization (MALDI/SELDI) analysis, and others, **(Fig. 1)** *(3,14,32–40)*.

In summary, this versatile laser micromanipulation and microdissection system
is a state of the art tool; it is essential when pure sample generation is required
throughout the entire field of modern molecular research and medical analyses.

2. MATERIALS

2.1. Specimen Preparation

1. PALM MicroBeam-HT or PALM Combi System (P.A.L.M. Microlaser Technologies).
2. Standard tissue sectioning equipment.
3. PALM Liquid CoverGlass N (P.A.L.M., cat. no. 1440-0600).

2.2. Glass-Mounted Specimen vs Membrane-Mounted Specimen

1. PALM MembraneSlides (P.A.L.M., cat. no. 1440-1000).

2.3. LMPC of Living Cells

1. Standard cell culture equipment.
2. PALM DuplexDish (P.A.L.M., cat. no. 1440-0550).
3. PALM Tubes (P.A.L.M., cat. no. 1440-0200).

2.4. Downstream Applications

2.4.1. Isolation of DNA From Captured Samples

1. Catapult Buffer: 1 mM EDTA, pH 8.0, 20 mM Tris-HCl, pH 8.0, 0.5% Igepal CA-
   630 (Sigma cat. no. I-3021), optional: 0.2 mg/mL Proteinase K (Roche, Mannheim,
   Germany, cat. no. 1413783).

2. Standard DNA extraction kit.
3. Standard PCR equipment.

2.4.2. Isolation of RNA From Captured Samples

1. Qiagen RNeasy MicroKit, cat. no.74004.
2. Standard RNA evaluation equipment or Agilent Bioanalyzer RNA 6000 Pico LabChip® Kit (Agilent, Waldbronn, Germany).

2.4.3. Isolation of PolyA(+) mRNA From a Single Captured Cell

1. QuickPick™ mRNA nano kit (Bio-Nobile Oy, Turku, Finland).
2. LightCycler and LightCycler-FastStart DNA MasterPLUS SYBR Green I (Roche).

3. METHODS
3.1. Specimen Preparation

Nearly any biological sample is suitable for the noncontact LMPC technology. However, because tissue sections for microdissection cannot be routinely embedded and cover-slipped, their morphology sometimes appears quite poor compared with standard embedded tissue sections. Several approaches to improve the morphology of the specimen have been tried using various embedding materials. Of these tested materials, the PALM Liquid CoverGlass N has proved to be the most convenient solution for optimized morphological resolution. Moreover, this resin-based fluid cover slip protects the specimen from environmental influences, e.g., humidity. Thus it preserves RNA integrity in laser-microdissected tissue sections and allows prolonged sample capture time *(41)*.

3.1.1. Preparation of PALM Liquid CoverGlass N

The exact preparation depends on the type of the specimen used and its handling prior to the application of PALM Liquid CoverGlass N. It is advisable to start with a ratio of stock solution/thinning solution of 1:6.

For example, mix one part (e.g., 2 mL) of the stock solution and five parts (e.g., 10 mL) of the thinning solution. Pour this working solution mixture into the spray bottle provided.

3.1.2. Preparation of the Tissue Slides With PALM Liquid CoverGlass N

A good starting procedure for optimal preparation is to hold the microscope slide (*See* **Note 1**) with the histopathologic tissue highly inclined, to almost perpendicular to ground. Spray the liquid onto the slide once or twice, and allow a drying period of approx 5 min. The impregnated tissue section is now ready to be processed by LMPC.

A further improvement of visualization can be achieved with PALM AdhesiveCaps. In addition, these consumables allow the harvesting of laser-captured material without applying buffer to the collection vessel. This avoids evaporation, and the dangers of crystallization of the salty buffers and activation of RNases are strongly minimized owing to the dry environment.

3.2. Glass-Mounted Specimen vs Membrane-Mounted Specimen

Different sample sources can be used with LMPC. Numerous protocols for sample preparation and downstream application techniques have been published during the past few years for laser micromanipulation and microdissection *(1–3,5,15,16,33,40,42–44)*. Specimens can be prepared either directly onto routine glass slides or on special membrane-spanned slides (*See* **Note 2**).

Depending on the nature of the sample and the purpose of the experiment, catapulting may be performed directly without prior laser cutting, as is the case for glass-mounted specimens. However, to avoid contamination with adjacent material, it is advisable to perform laser circumdissection prior to catapulting to obtain pure samples. The laser removes a section of material between selected and nonselected regions. Within the narrow laser cut, all biological material is ablated, and the risk of contamination of the captured specimen is minimized. Direct catapulting from glass is recommended for cytocentrifuged specimens, small cell clusters, and pooled single cells, respectively.

The LMPC membrane (PALM MembraneSlides) serves as a backbone, which holds the selected tissue area close together and facilitates capture. This polyethylene naphthalate (PEN) membrane has a thickness of 1.35 µm and is highly absorptive in the UV-A range, which facilitates laser cutting. The laser first operates around the selected area following the preselected line and cuts the specimen together with the underlying membrane. Thus impulse catapulting of even large areas with one single laser pulse is enabled. With the catapult impulse of the laser, the entire selected area is ejected out of the object plane and catapulted directly into the collection device. With the supporting membrane, captured single cells or even large cell areas keep their morphology intact and can be visualized easily in the collection cap as cell-covered membrane islets. Catapulting from membrane-mounted slides is more suitable for large tissue areas, difficult specimens, or fragile samples such as cell smears, single individual cells, very small cells, cell nuclei, chromosomes, and especially living cells.

3.2.1. Preparation of Slides

1. *Poly-L-Lysine treatment:*
   a. Additional coating with poly-L-lysine (0.1 % w/v) will only be necessary for special materials (e.g., brain sections) and should be performed by distributing a drop of the solution on the glass or membrane slide.
   b. Let air-dry at room temperature for 30 min.

    c. Avoid any leakage underneath the membrane, as this might result in impairment of the LPC process.

2. *Removal of RNases:* To ensure RNase-free glass or membrane slides, dip them for a few seconds into RNase-ZAP, followed by two separate washings in diethyl pyrocarbonate (DEPC)-treated water and drying at 37°C for 30 min up to 2 h (*See* **Note 3**).

3. *Mounting sections onto slides:*

    a. Mount sections onto PALM MembraneSlides the same way as is routinely done with glass slides.

    b. For cutting and catapulting, a cover slip or standard mounting medium must not be applied.

4. *Paraffin-embedded sections:*

    a. Mount the sections (5–8 µm) onto the slide as is routinely done.

    b. Let dry at 37°C up to 56°C overnight in a drying oven.

    c. As paraffin will reduce the efficiency of laser cutting, it is advisable to remove the paraffin before LMPC. For deparaffinization, wash the slide twice in xylene (2 min). Transfer in successive ethanol washes (100% for 1 min, 96% for 1 min, and 70% for 1 min), and finally rinse with water.

5. *Frozen sections:* after mounting (8–15-µm sections), there are many possible ways to fix the material. If RNA preparations are intended with these frozen sections, an ethanol fixation is recommended. This is performed after 20 s of air-drying on ice by dipping the mounted sections for 1–5 min into ice-cold (−20°C) 70% ethanol.

    If OCT or another tissue freezing medium is used, it is important to remove these substances before laser microdissection, because these media will impact laser efficiency. For removing the freeze-supporting substances, the slides have to be washed gently in water (1 min). If the sections are to be stained, the supporting substance is removed in the aqueous solutions or the diluted ethanol series while staining is taking place. Frozen sections should always be allowed to dry for 5 up to 30 min at room temperature before use.

6. *Cytospins:*

    a. Cytospins can be prepared on glass slides or on MembraneSlides. After centrifugation with a cytocentrifuge, let the cells air-dry overnight.

    b. Then fix for 5 min in 100% methanol.

    c. Allow the cytospins to dry at room temperature before staining.

7. *Blood and tissue smear:* distribute a drop of (peripheral) blood or material of a swab smear over the slide. Let smears air-dry briefly, and fix them for 2 min or up to 5 min in 70% ethanol.

8. *UV Treatment:* to overcome the hydrophobic nature of the membrane, it is advisable to irradiate with UV light at 254 nm for 30 min. The membrane becomes more hydrophilic; therefore the sections (paraffin and cyrosections) will adhere better. Positive side effects are sterilization and destruction of potentially contaminating nucleic acids.

3.2.2. Staining of Sections

With frozen tissue, one has to be aware of endogeneous RNases that may still be active after short fixation steps. Therefore it is recommended to keep all incubation steps of histochemistry as short as possible. Please use RNase-free water and solutions.

3.2.2.1. DEPC TREATMENT OF SOLUTIONS

1. RNases in aqueous liquids can be destroyed with 0.1% DEPC.
2. To 1L of $H_2O_{bidest.}$ (or solution), add 1 mL DEPC and stir for 6–8 h at room temperature; then let incubate overnight in a fume hood.
3. The next day remove residual DEPC by autoclaving.
4. Store treated solutions at room temperature (*See* **Note 4**).

3.2.2.2. HISTOLOGICAL STAINING METHODS

Hematoxylin and eosin (H&E) staining is used routinely in most histological laboratories and does not interfere with DNA and RNA preparation. The nuclei are stained blue and the cytoplasm pink/red.

1. The slides are transferred directly from distilled water into Mayer's hematoxylin solution (3 min).
2. After rinsing in running RNase-free water (3 min), the mounted sections are stained in Eosin Y (0.5–3 min), washed quickly with increasing ethanol series, and air-dried.

Best results for RNA extraction are achieved with *cresyl violet acetate* staining.

1. Dissolve solid cresyl violet acetate at a concentration of 1% (w/v) in ACS-grade 100% ethanol at room temperature with agitation for several hours to overnight.
2. Filter the stain through a 0.2-μm filter unit prior to use.
3. After cutting and transfer of the cryosection to the slide, air-dry for 1 min on ice, and then incubate for 2 min in precooled (−20°C) 75% ethanol.
4. Gently tap the slide on an absorbent surface to remove excess ethanol.
5. Dip the slide in 1% cresyl violet acetate for 20 s (room temperature).
6. Gently tap the slide on an absorbent surface to remove excess stain before placing the slide into the next solution.
7. Wash the slide in 75% ethanol (only dipping briefly) and 100% ethanol (30 s).
8. Remove excess ethanol by gently tapping the slide on an absorbent surface.
9. Air-dry for about 10 min at room temperature. Slides can be used immediately or stored at −80°C before LMPC.

3.2.3. Specimen Isolation and Collection

Different laser functions are available to process the preselected and outlined specimens *(3,17)*.

3.2.3.1. For Glass-Mounted Preparations

1. ***AutoLPC*** allows automatic catapulting of larger areas from glass-mounted preparations, resulting in captured tissue flakes. With such preparations, only a small amount of cellular material can be transported with each single shot. Therefore, larger areas have to be catapulted with multiple shots. The distance between the single shots and the LPC-pattern depends on tissue characteristics and can be preselected in the setup menu.

2. In the ***LPC*** *mode*, only dot-marked specimens are catapulted. This function is of special benefit for cytocentrifuged specimens but is also used for individual isolated cells within a histological preparation or to catapult membrane-mounted specimens manually after microdissection.

3. With the use of ***CloseCut&AutoLPC***, critical preparations can be isolated prior to AutoLPC to avoid contamination with neighboring tissue. Thus pure sample preparations can be obtained.

4. The ***AutoCircle*** allows automatic circumcutting of a dot-marked specimen with a preset diameter and immediately catapulting of it from its center. This function is recommended only for single cells like, e.g., mucosa cells or sperm from smear tests.

3.2.3.2. For Membrane-Mounted Preparations: Cutting

1. The laser cuts precisely along the predrawn line. If ***CloseCut*** is selected, the user does not need to fulfill the drawing line around a specimen, as the computer will automatically do so, and cutting will be performed accordingly.

2. ***RoboLPC*** means cutting and catapulting in a single step. It is possible only for use with membrane-mounted specimens. The outlining is automatically cut up to a small connection bridge, from which the entire area is immediately catapulted with one single laser shot. The size of the connecting piece can be preselected from the laser setup menu. Within the cutting line, all biological material is eliminated owing to ablation. Thus pure sample preparation is possible without danger of contamination.

3.2.3.3. Laser Pressure Catapulting of the Samples

1. Pipete 3–5 µL $H_2O_{bidest.}$ or buffer into the inner ring of the cap or use original PALM AdhesiveCaps. Be aware that aqueous solutions will dry out after a while.

2. The catapulted cells or cell areas will stick onto the wet inner surface of the cap or the filling material of AdhesiveCaps and will not fall down after the catapulting procedure. When using glass-mounted samples, it may be better to put more liquid (up to 20 µL) into the cap, since the smaller "flakes" produced by AutoLPC deviate more strongly from the center of the cap during the catapult process than areas on membrane.

3. To control the efficiency of catapulting, it is possible to look into the collection device with the 5, 10, 40, and 63× objectives. Use the software function *go to checkpoint*; then the stage is moved to a position that makes it possible to look inside the cap.

4. After microdissection, close the microfuge tube with the sample-containing cap, and spin briefly (5–10 min, 13,000 rpm).

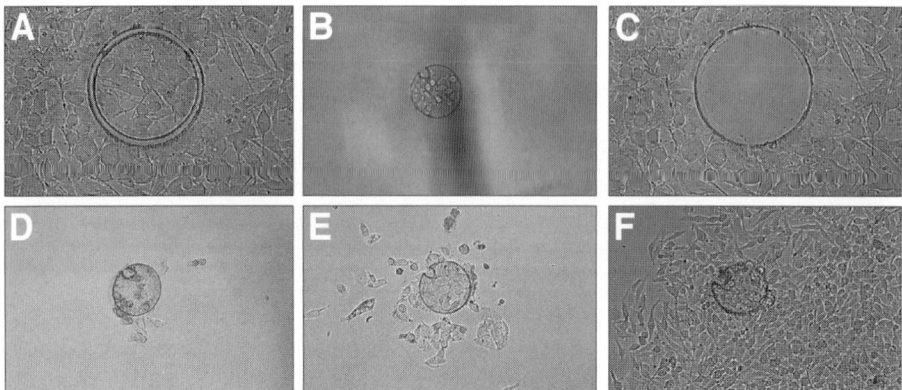

Fig. 7. LMPC of cultured HepG2 cells in a PALM DuplexDish. **(A)** Microdissection of an area of a confluent culture. **(B)** Catapulted cell-covered membrane islet in the collection device. **(C)** Remaining blank after catapulting in the originating culture dish. **(D)** Cell-covered membrane islet in a standard multiwell plate 2 d after initial transfer. **(E)** Proliferation of catapulted cells after 6 d. **(F)** Cell cluster grown around the membrane islet after 10 d of cultivation. Original magnification: (A, C) 200×; (B, D–F) 100×. (*See* Color Plate 7 following p.18.)

5. For RNA extraction first add the appropriate lysis buffer to the PCR tube, and after closure mix by inversion.
6. Spin the lysate down as in step 4.
7. For future RNA isolation/analysis, the tube can be placed on ice or stored at −80°C in a freezer.

3.3. LMPC of Living Cells

Common laser microdissection and capture techniques require a dry environment and are therefore restricted to fixed or frozen tissue sections or cell preparations *(30,33)*. In contrast, living cells need medium for growth and survival. Beyond the necessity of a humid environment, contamination is especially critical with living specimens, if the captured cells will be recultivated after microdissection. Therefore an entirely noncontact microdissection and capture method is needed for isolation and ongoing cultivation of cultured cells. This can be realized with specially developed culture dishes (PALM DuplexDish), which consist of two separate membranes. Cells are seeded out on top of the biocompatible upper thin PEN LMPC membrane. The lower membrane remains intact after LMPC and maintains a sterile environment. With these dishes, cells can be cultured, microdissected, and finally catapulted as cell-covered membrane islets into an appropriate collection device, although a thin layer of culture medium remains on the cells. Due to the precision of the LMPC method, even single cells can be isolated and recultivated to generate pure clones without the necessity of the time-consuming selection screenings.

3.3.1. Tissue Culture and LMPC

1. Seed out adherent growing cells into PALM DuplexDishes, and cultivate them under standard conditions in the recommended media to the desired confluency.
2. To perform microisolation and collection of the desired cells, remove the culture medium up to a humidity layer remaining on top of the cells.
3. Carry out microdissection of the desired cells, and catapult the isolated membrane islets with cells on top into a collection cap filled with 40 μL of culture medium. Depending on the objective applied, even large areas of more than 500 μm in diameter can be catapulted.

3.3.2. Recultivation

1. Centrifuge the collected cells at 900g for 1 min.
2. Gently resuspend the cell pellet in an additional 50 μL of culture medium, and transfer to a well of a Falcon 12-well plate.
3. Overlay the cell droplet carefully with 1–2 mL of culture medium, and cultivate under standard culture conditions.

The LMPC process has no detrimental effects on the isolated cells, as they routinely proliferate after variable reconvalescence periods *(16)*. These cells are still viable after several rounds of laser microdissection, capture, and recultivation (**Fig 7**; *See* Color Plate 7 following p.18). Ablative photodecomposition is the only way to perform microdissection or microsurgery within living cells or biological specimens without harming their viability. The same laser system can be used for microinjection, cellular fusion, or microdissection, and capture of selected biological specimens can be performed without impact on DNA, RNA, or protein integrity. Thus, new approaches toward establishing homogeneous cell populations from adherent cell cultures, fresh biopsies, or tissue cultures to characterize cell types or study differentiation processes in developmental biology are opened up with this new versatile tool. This allows microinjection (optoinjection) of drugs, particles, or genetic material, permitting genetic engineering without mechanical tools, disturbing chemical agents, or perturbing viral vectors, as well as directed membrane fusion of different cell types for cell fusion experiments *(17,45–49)*.

3.4. Downstream Applications

For all subsequent molecular analyses of LMPC-generated samples, the best results for the recovery of proteins, DNA, or RNA are achieved with freshly prepared cryosectioned specimens.

3.4.1. Isolation of DNA From Captured Samples

For DNA preparations from paraffin sections, it is recommended to use a Proteinase K-containing catapult buffer. For cryosections, it is not necessary to perform catapulting of the cells into a buffer with Proteinase K.

3.4.1.1. CAPTURE OF THE SAMPLES

1. The capture process may be performed in an AdhesiveCap.
2. If you are working with the AutoLPC mode, it is advisable to add 3–5 µL Catapult Buffer into the cap.
3. After sample collection, centrifuge the vial at full speed for 5 min.

3.4.1.2. PROTEINASE K DIGESTION

1. Add 10–15 µL Catapult Buffer containing Proteinase K onto the cells, and vortex gently.
2. Digest for 2–18 h (depending on the kind and number of catapulted cells) at 55°C followed by a heating step at 90°C for 10 min to inactivate Proteinase K.
3. Use a thermal cycler with a heating lid for the standard digestion.

3.4.1.3. EXTRACTION OF DNA

1. The DNA may be isolated by the instructions of the respective manufacturers of DNA extraction kits.

3.4.2. Isolation of RNA From Captured Samples

1. To ensure RNase-free slides, dip them for a few seconds into RNase-ZAP, followed by two separate washes in DEPC-treated water and let air-dry at 37–55°C for 30 min to 2 h.
2. Perform standard UV treatment as usual shortly before use.
3. For best RNA quality, use freshly cut specimens or cells from cell culture. Frozen sections should not be stored for more than a few days at –80°C. Freezing should be performed after ethanol fixation or after staining and drying (*See* **Note 5**).

3.4.2.1. CAPTURE OF THE SAMPLES

1. Again, the capture process is performed in an AdhesiveCap.
2. As Proteinase K digestion is required for paraffin sections or formalin-fixed samples and recommended for samples generated by AutoLPC, add Proteinase K to the Catapult buffer. If you are using cryosections, the Proteinase K digestion is not necessary.
3. Proceed with **Subheading 3.4.2.3.**, and extract RNA after the lysis.
4. If you are not proceeding immediately, store the samples at –80°C in lysis buffer.

3.4.2.2. PROTEINASE K DIGESTION

1. After centrifugation (5 min at full speed) add 20 µL of Catapult Buffer with Proteinase K to the tube.

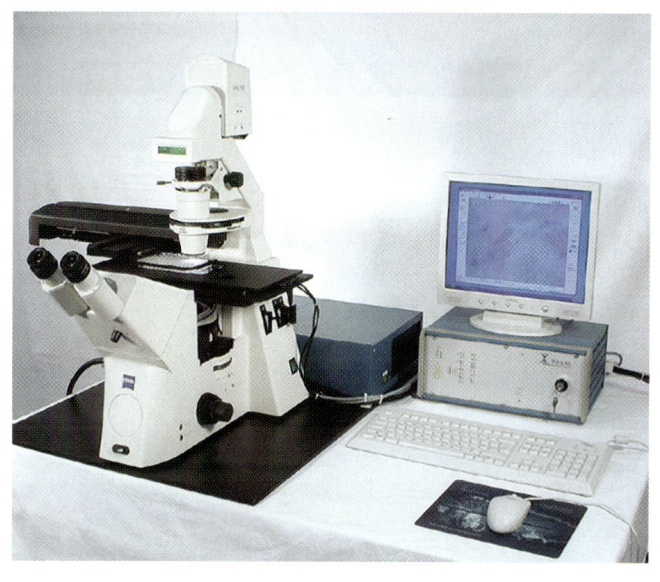

Color Plate 1. The PALM® MicroBeam-HT (Chapter 1, Fig. 2, *see* pp. 3, 4).

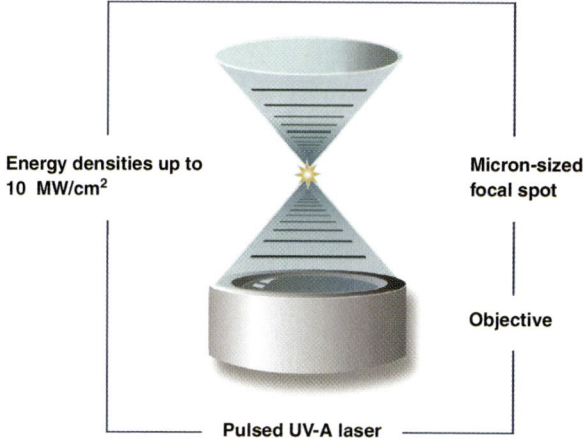

Energy densities up to 10 MW/cm²

Micron-sized focal spot

Objective

Pulsed UV-A laser

Color Plate 2. Scheme for laser ablation (Chapter 1, Fig. 3, *see* pp. 3, 4).

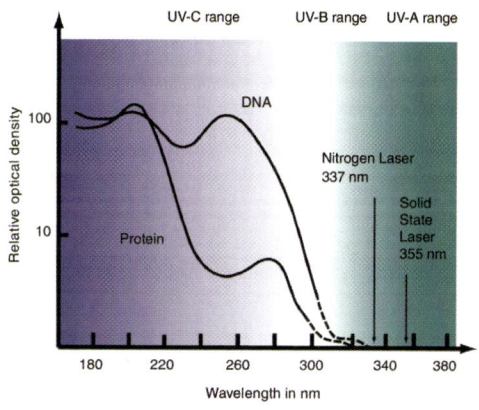

Color Plate 3. Peak absorption wavelengths; absorption maxima of DNA and proteins (Chapter 1, Fig. 4, *see* p. 5).

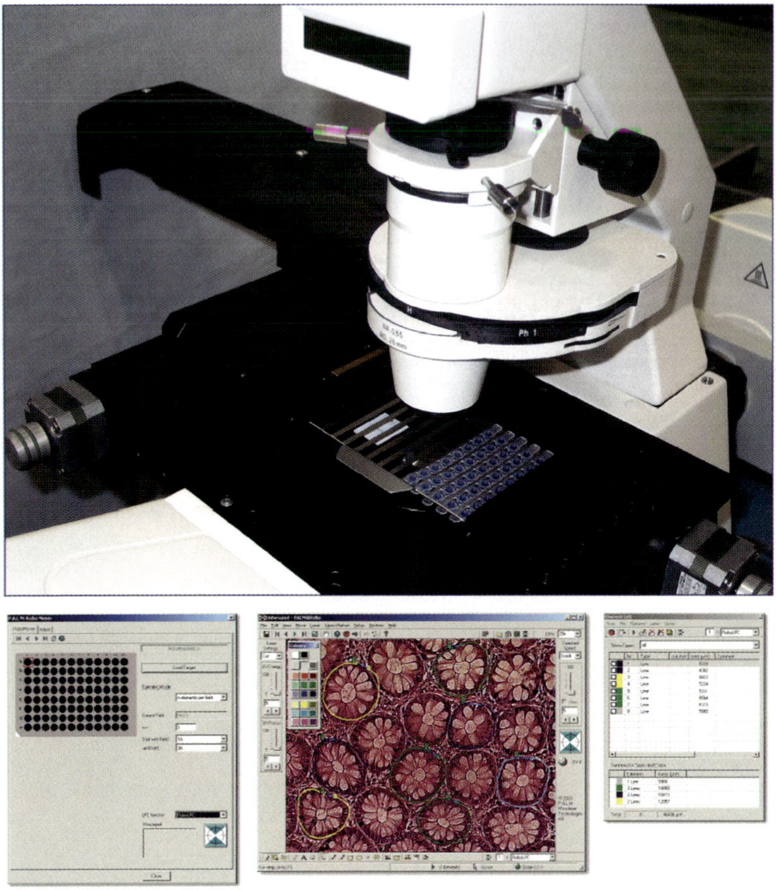

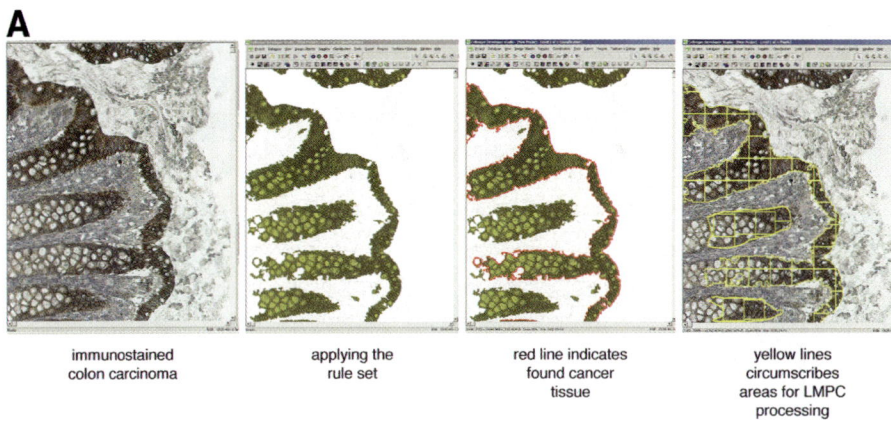

Color Plate 4. PALM® RoboMover and RoboSoftware (Chapter 1, Fig. 5, *see* pp. 6, 7).

A

| immunostained colon carcinoma | applying the rule set | red line indicates found cancer tissue | yellow lines circumscribes areas for LMPC processing |

Color Plate 5. MicroBeam-HT: automatic recognition of defined areas of samples (Chapter 1, Fig. 6A, *see* pp. 8, 9).

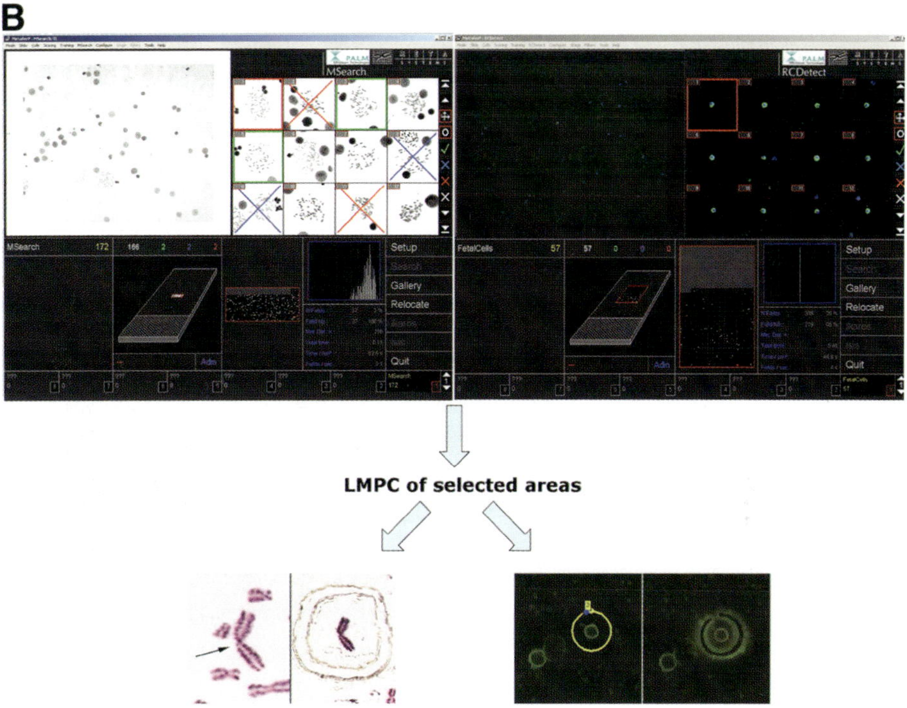

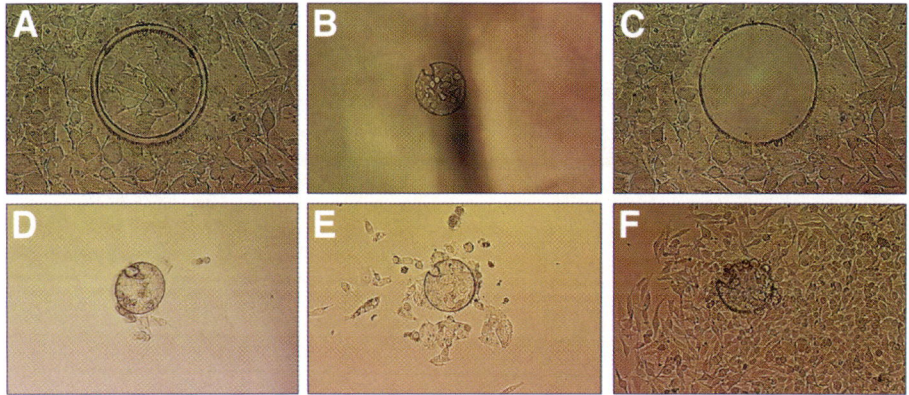

Color Plate 6. MicroBeam HT: automatic detection of labeled samples (Chapter 1, Fig. 6B, *see* pp. 8, 9).

Color Plate 7. LMPC of cultured HepG2 cells in a PALM DuplexDish (Chapter 1, Fig. 7, *see* pp. 16, 17).

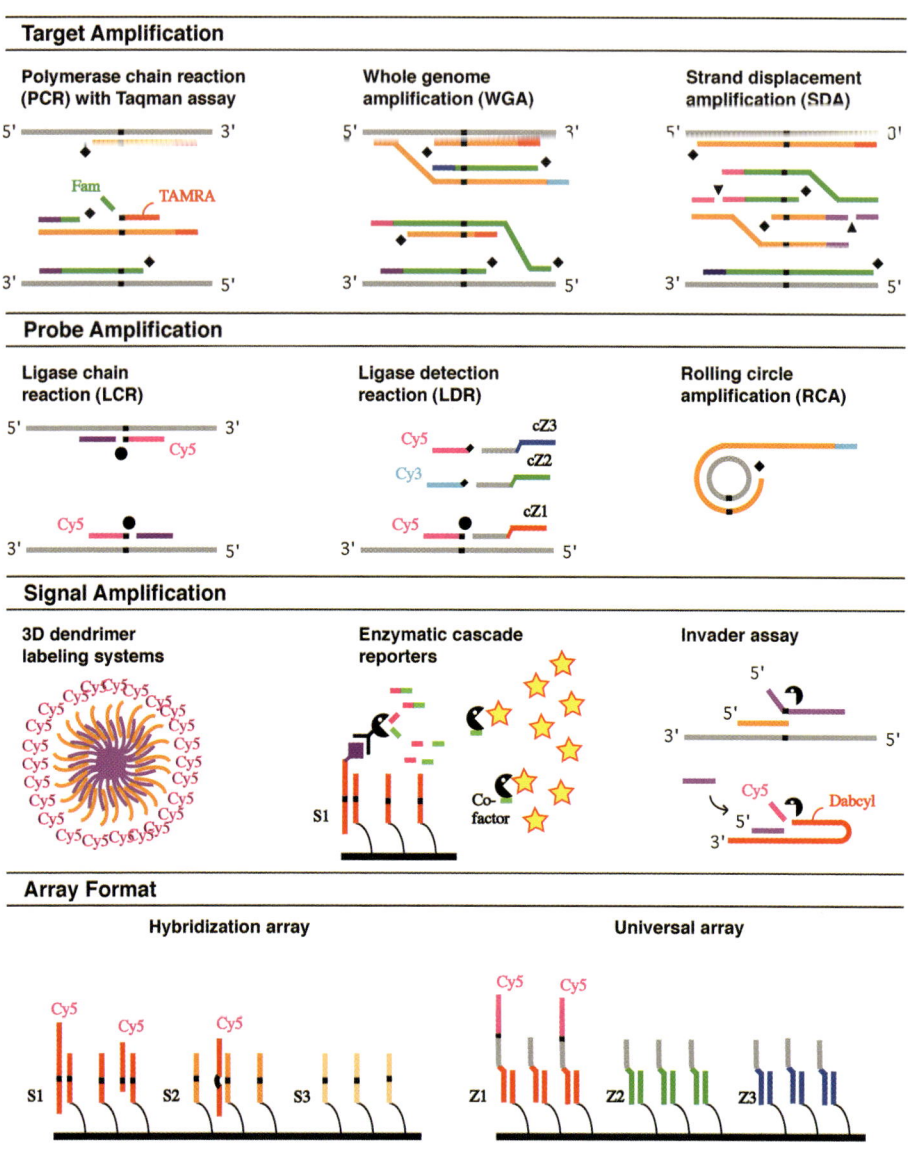

Color Plate 8. Overview of target amplification, probe amplification, signal amplification, and arrays used in molecular diagnostics (Chapter 2, Fig. 1, *see* pp. 26–29).

PCR / PCR / LDR / Universal Array

1. PCR amplify all p53 exons using gene-specific/universal primers and Taq polymerase. ◆

2. PCR amplify all primary products using universal primers and Taq polymerase. ◆

3. Perform LDR using mutation-specific LDR primers, common primers containing complementary zip code sequences, and thermostable ligase. ●

4. Capture fluorescent products on addressable array and score for presence of mutation.

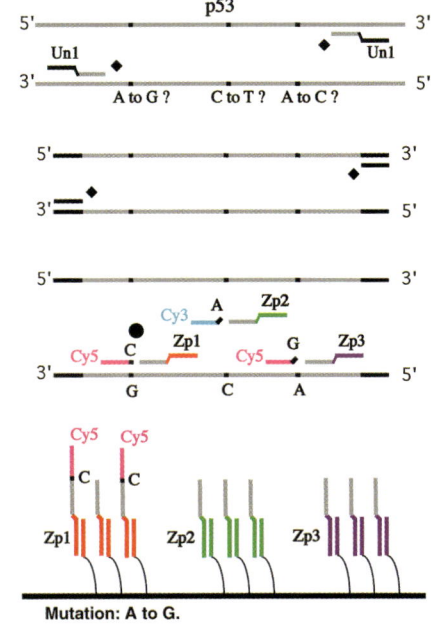

Mutation: A to G.

| Upper Strand | Lower Strand |

FIDUCIALS

AMPLICON CONTROLS

Exon 6   Exon 8 Exon 4

Exon 5   Exon 7

R72R

R273 C-T

| Cy 3 | Cy 5 | Bodipy | Alexa |

Color Plate 9. Schematic diagram and results for the detection of multiple mutations using PCR/PCR/LDR/Universal Array (Chapter 2, Fig. 2, *see* pp. 29–31).

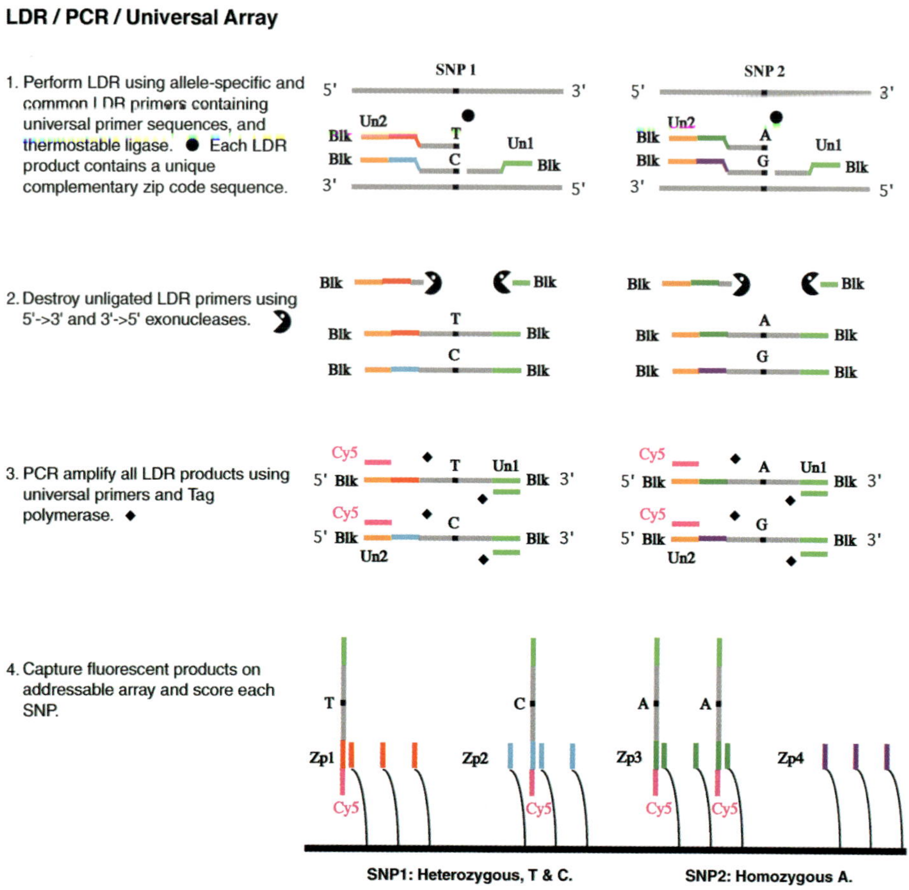

LDR / PCR / Universal Array

1. Perform LDR using allele-specific and common LDR primers containing universal primer sequences, and thermostable ligase. ● Each LDR product contains a unique complementary zip code sequence.

2. Destroy unligated LDR primers using 5'->3' and 3'->5' exonucleases. ❱

3. PCR amplify all LDR products using universal primers and Taq polymerase. ◆

4. Capture fluorescent products on addressable array and score each SNP.

SNP1: Heterozygous, T & C. **SNP2: Homozygous A.**

Color Plate 10. Schematic diagram for the detection of multiple single-nucleotide polymorphisms using LDR/PCR/Universal Array (Chapter 2, Fig. 3, *see* pp. 33, 34).

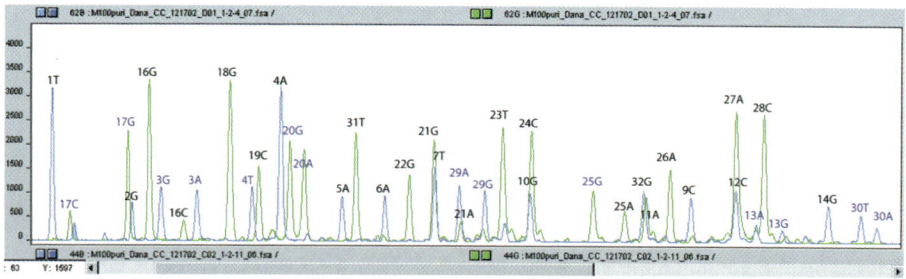

Color Plate 11. Electrophoretogram of 30-plex detection in genomic DNA (Chapter 2, Fig. 4, *see* pp. 33, 35).

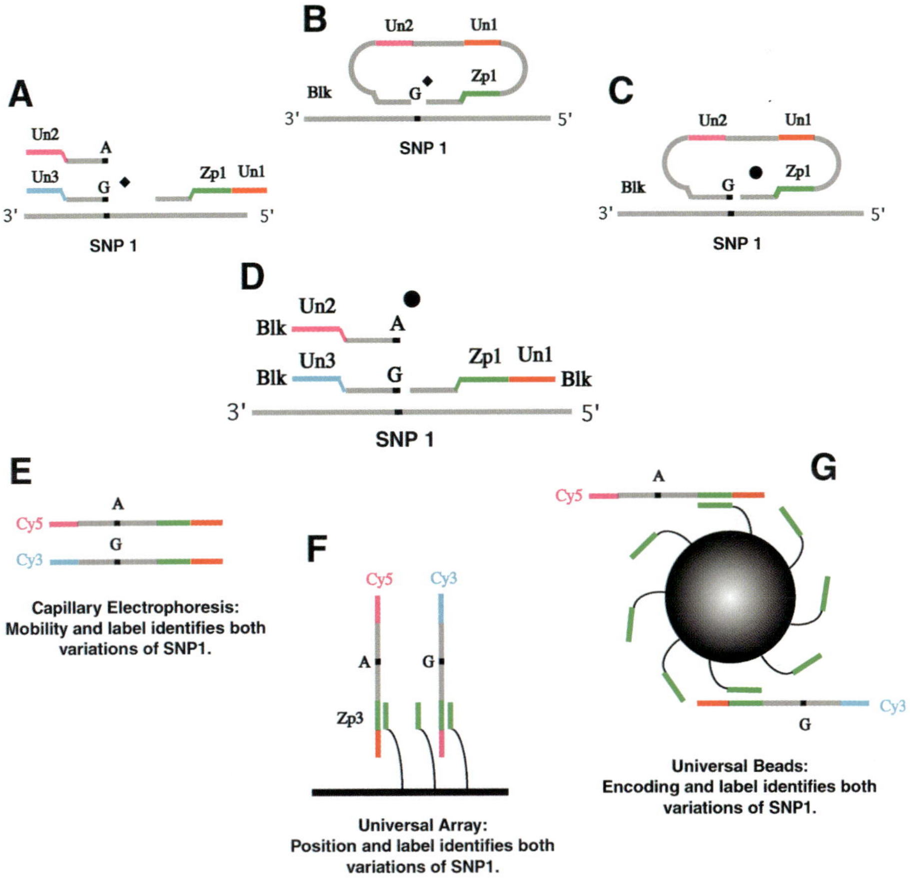

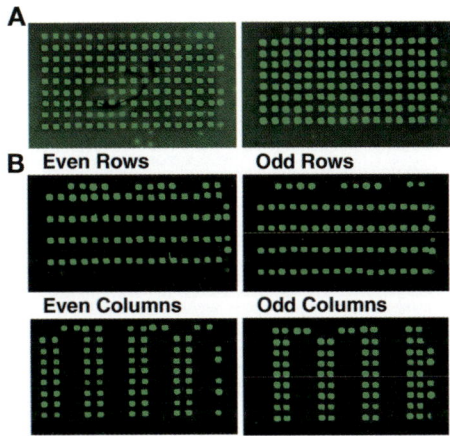

Color Plate 12. Ligase-based multiplexed single-nucleotide polymorphism assays combined with zip-code universal display (Chapter 2, Fig. 5, *see* pp. 36–38).

Color Plate 13. Validation of arrays following spotting (Chapter 2, Fig. 6, *see* pp. 41, 42).

Bisulfite/PCR-PCR/LDR/Universal Array

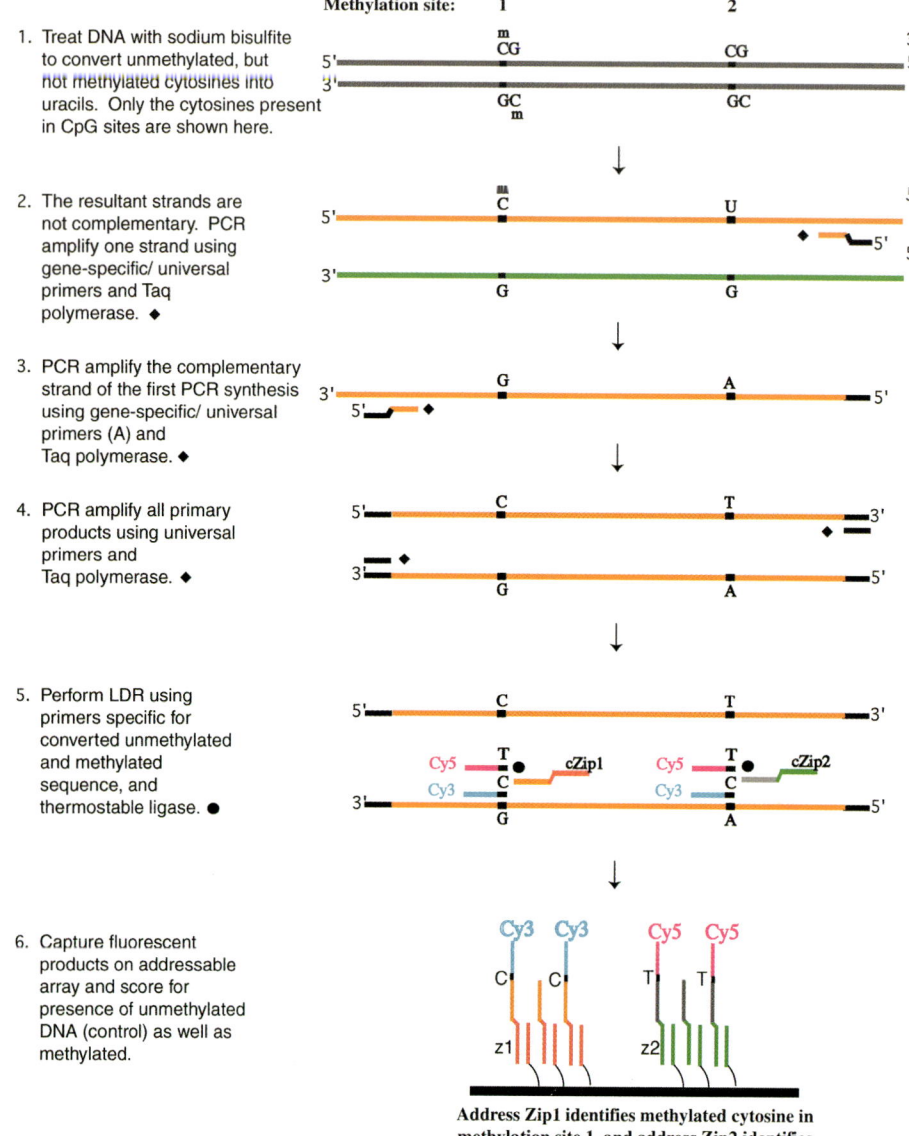

Methylation site: 1   2

1. Treat DNA with sodium bisulfite to convert unmethylated, but not methylated cytosines into uracils. Only the cytosines present in CpG sites are shown here.

2. The resultant strands are not complementary. PCR amplify one strand using gene-specific/ universal primers and Taq polymerase. ◆

3. PCR amplify the complementary strand of the first PCR synthesis using gene-specific/ universal primers (A) and Taq polymerase. ◆

4. PCR amplify all primary products using universal primers and Taq polymerase. ◆

5. Perform LDR using primers specific for converted unmethylated and methylated sequence, and thermostable ligase. ●

6. Capture fluorescent products on addressable array and score for presence of unmethylated DNA (control) as well as methylated.

Address Zip1 identifies methylated cytosine in methylation site 1, and address Zip2 identifies unmethylated cytosines in methylation site 2.

Color Plate 14. Schematic of PCR-PCR/LDR assay (Chapter 2, Fig. 7, *see* pp. 45, 46).

2. Vortex gently, and digest upside down for 2–18 h at 55°C in an incubator.
3. Centrifuge the sample at full speed for 5 min, and then pipet the recommended volume of lysis buffer (provided by the respective RNA extraction kit) into the tube.
4. Mix and lyse the cells by inversion. To improve lysis, incubate at 42°C for 30 min.
5. Centrifuge the sample at full speed for 5 min.
6. Proceed with the RNA extraction procedure as described by the respective manufacturer. If you are not proceeding immediately, store the samples at –80°C in lysis buffer.

3.4.2.3. CELL LYSIS

1. For lysis, mix the cells by inversion, and centrifuge the sample at full speed for 5 min.
2. Proceed with the RNA extraction procedure as described by the respective manufacturer (*See* **Note 6**).

3.4.2.4. EXTRACTION OF RNA

1. It is difficult to estimate the amount of RNA to be expected after extraction, since many factors (like species, cell/tissue type, degradation, fixation, staining, fragmentation, extraction procedure of different manufacturers and others) will influence the outcome.
2. From mouse liver, frozen sections with roughly 5–20 pg of RNA per cell (calculated from extractions of 1000 cells and analysis on an Agilent Bioanalyzer) can be retrieved *(50)*.

3.4.3. Isolation of PolyA(+) mRNA From a Single Captured Cell

Measuring the expression profiles of individual cells is useful in a wide range of research and clinical applications. LMPC combined with real-time PCR for expression profiling allows several investigations of individual cells, as differences in cell state or type are correlated with changes in mRNA expression levels of genes *(51)*.

3.4.3.1. CAPTURE OF THE SAMPLES

1. Cut snap-frozen murine liver tissue in 7-μm serial sections on a cryotome at –25°C.
2. Transfer sections to PALM MembraneSlides, and let air-dry for 10 s.
3. After a 5-min fixation step in 70% ethanol at –20°C, stain the sections with cresyl violet acetate (1%), wash the sections again in increasing ethanol series, and use immediately.
4. Excise selected single cells, and catapult them into collecting caps filled with 10 μL of mRNA lysis/binding buffer.

3.4.3.2. EXTRACTION OF POLYA(+) MRNA

1. Fill sample volume to 100 μL with lysis/binding buffer and isolated mRNA according to the QuickPick™ mRNA nano kit instructions of the manufacturer.
2. Resuspend particle-bound mRNA in 5 μL elution buffer, and use directly as a solid-phase template in downstream RT-PCR.

3.4.3.3. QUALITY CONTROL OF POLYA(+) MRNA BY REAL-TIME PCR

1. For subsequent PCR analyses, use 5 µL of each cDNA solution as templates.
2. PCR amplification of the cDNA may be performed in a LightCycler instrument in 20 µL reaction volumes using protocols and components of the LightCycler-Fast-Start DNA MasterPLUS SYBR Green Kit
3. cDNA-specific primers for the respective gene of interest may be used as a model system.

4. NOTES

1. When you are working with low magnifying objectives, including 40 and 63× long-distance objectives, regular 1-mm-thick glass slides can be used. Because of the short working distance of the 100× magnifying objectives, 0.17-mm thin cover glass slides have to be used.
2. To facilitate easy catapulting, additional adhesive substances or Superfrost + charged slides should only be applied if absolutely necessary for the adhesion of special material (e.g., thick brain sections; up to 50 µm).
3. Working with RNA is more demanding than working with DNA, because of the chemical instability of the RNA and the ubiquitous presence of RNases. Therefore:
   a. Use filtered pipetor tips.
   b. Bake glassware at 180°C for 4 h. (*RNases can maintain activity even after prolonged boiling or autoclaving.*)
   c. Prepare all solutions with DEPC-treated H_2O.
   d. For best results, use either fresh samples or samples that have been quickly frozen in liquid nitrogen or at −80°C. (This procedure minimizes degradation of RNA by limiting the activity of endogeneous RNases.) All required reagents should be kept on ice.
   e. Store RNA, aliquoted in ethanol or RNA buffer, at −80°C. Most RNA is relatively stable at this temperature. Store prepared slides also at −80°C.
   f. RNA is not stable at elevated temperatures; therefore avoid high temperatures (>65°C), since these affect the integrity of RNA.
4. All chemical substances containing amino groups (e.g., TRIS, MOPS, EDTA, HEPES, and so on) cannot be treated directly with DEPC. Prepare these solutions in DEPC-treated H_2O.
5. For best RNA quality, we use frozen sections on PALM MembraneSlides and perform catapulting of the cells directly into RNase-free water containing 0.5% Igepal or into Catapult Buffer without Proteinase K. After catapulting, the samples are mixed with lysis buffer as soon as possible to protect the RNA. If you are using paraffin sections for catapulting, use a buffer containing Proteinase K. It is also possible to work with PALM MembraneSlides NF, which are pretreated and free of contaminations. To reduce the risk of contamination with exogenous RNases, only use special reagents and solutions for RNA isolation, reverse transcription, and RT-PCR. All solutions and tubes used should be prepared with DEPC-treated water or purchased as guaranteed RNase free.

6. Good RNA results are achieved with RNeasyMini and RNeasyMicro kits (Qiagen), Absolutely RNA Nanoprep kit (Stratagene), High Pure RNA Tissue kit (Roche), and Purescript RNA isolation kit (Gentra, distributor: Biozym).

Acknowledgments

The authors thank R. Baumeister (Freiburg, Germany) for the donation of *C. elegans*. We also thank S. Ehnle and B. Haar for excellent technical assistance.

References

1. Westphal, G., Burgemeister, R., Friedemann, G., et al. (2002) Noncontact laser catapulting: a basic procedure for functional genomics and proteomics *Methods Enzymol.* **356,** 80–99.
2. Schütze, K., Becker, B., Bernsen, M., et al. (2002) Tissue microdissection, in *DNA Microarrays (*Bowtell, D. and Sambrook, J., eds.), CSHL Press, New York, pp. 307–313 and 331–356.
3. Burgemeister, R., and Schütze, K. (2003) Tissue microdissection techniques, in *Analysing Gene Expression* (Lorkowski, S. and Cullen, P., eds.), Wiley-VCH, Weinheim, Germany.
4. Roesch, A., Vogt, T., Stolz, W., Dugas, M., Landthaler, M., and Becker, B. (2003) Discrimination between gene expression patterns in the invasive margin and the tumour core of malignant melanomas. *Melanoma Res.* **13,** 503–509.
5. von Eggeling, F., Melle, C., and Ernst, G. (2003) Biomarker discovery by tissue microdissection and ProteinChip® array analysis. *J. Lab. Med.* **27,** 79–84.
6. Cantz, T., Zuckerman, D. M., Burda, M. R., (2003) Quantitative gene expression analysis reveals transition of fetal liver progenitor cells to mature hepatocytes after transplantation in uPA/RAG-2 mice. *Am. J. Pathol.* **162,** 37–45.
7. Tannapfel, A., Anhalt, K., Häusermann, P., et al. (2003) Identification of novel proteins associated with hepatocellular carcinomas using protein microarrays. *J. Pathol.* **201,** 238–249.
8. Thalhammer, S., Kölzer, A., Frösner, G., and Heckl, W. (2000) Laser-based isolation of cells and cell clusters for virus specific PCR analysis. *Eur. Biophys. J.* **29,** 12D-5.
9. Thalhammer, S., Stark, R., Schütze, K., Wienberg, J., and Heckl, W. (1997). Laser microdissection of metaphase chromosomes and characterization by atomic force microscopy. *J. Biomed. Optics* **2,** 115–119.
10. Srinivasan, R. (1986) Ablation of polymers and biological tissue by ultraviolet lasers. *Science* **234,** 559–565.
11. Vogel, A., and Venugopalan, V. (2003) Mechanisms of pulsed laser ablation of biological tissues. *Chem. Rev.* **103,** 577–644.
12. deWitt, A., and Greulich, K. (1995) Wavelength dependence of laser-induced DNA damage in lymphocytes observed by single-cell gel electrophoresis. *J. Photochem. Photobiol.* **30,** 71–76.
13. Bernsen, M., Dijkman, H., de Vries, E., et al. (1998) Identification of multiple mRNA and DNA sequences from small tissue samples isolated by laser-assisted microdissection. *Lab. Invest.* **78,** 1267–1273.

14. Schütze, K., and Lahr, G. (1999) Use of laser technology for microdissection and isolation. *Am. Biotechn. Lab.* **17**(4), 24–30.
15. Mayer, A., Stich, M., Brocksch, D., Schütze, K., and Lahr, G. (2002). Going in vivo with laser microdissection. *Methods Enzymol.* **356**, 25–33.
16. Stich, M., Thalhammer, S., Burgemeister, R., et al. (2003) Live cell catapulting and recultivation. *Pathol. Res. Pract.* **199**, 405–409.
17. Schütze, K., Burgemeister, R., Clement-Sengewald, A., et al. (2003) Non-contact live cell laser micromanipulation using PALM MicroLaser systems P.A.L.M. Scientific Edition No. 11, Life Cell Manual. P.A.L.M., Bernried, Germany.
18. Berns, M., Rounds, D., and Olson, R. (1969). Effects of laser micro-irradiation on chromosomes. *Exp. Cell Res.* **56**, 292–298.
19. Bereiter-Hahn, J. (1971) Melaninbewegung mit Laser untersucht. *Umschau* **16**, 601–602.
20. Meier-Ruge, W., Bielser, W., Remy, E., Hillenkamp, F., Nitsche, R., and Unsold, R. (1976) The laser in the Lowry technique for microdissection of freeze-dried tissue slices. *Histochem. J.* **8**, 387–401.
21. Berns, M., Aist, J., Edwards, J., et al. (1981) Laser microsurgery in cell and developmental biology. *Science* **213**, 505–513.
22. Schütze, K., and Clement-Sengewald, A. (1994) Catch and move—cut or fuse. *Nature* **368**, 667–669.
23. Kubo, Y., Klimek, F., Kikuchi, Y., Bannasch, P., and Hino, O. (1995) Early detection of Knudson's two-hits in preneoplastic renal cells of the Eker rat model by the laser microdissection procedure. *Cancer Res.* **55**, 989–990.
24. Antinori, S., Selman, H., Caffa, B., Panci, C., Dani, G., and Versaci, C. (1996) Zona opening of human embryos using a non-contact UV laser for assisted hatching in patients with poor prognosis of pregnancy. *Hum. Reprod.* **11**, 2488–2492.
25. Clement-Sengewald, A., Schütze, K., Ashkin, A., Palma, G., Kerlen, G., and Brehm, G. (1996) Fertilization of bovine oocytes induced solely with combined laser microbeam and optical tweezers. *J. Assist. Reprod. Gen.* **13**, 259–265.
26. Clement-Sengewald, A., Bucholz, T., and Schütze, K. (2000) Laser microdissection as a new approach to prefertilization genetic diagnosis. *Pathobiology* **68**, 232–236.
27. Clement-Sengewald, A., Schütze, K., Sandow, S., Nevinny, C., and Pösl, H. (2000) PALM® Robot-MicroBeam for laser-assisted fertilization, embryo hatching and single-cell prenatal diagnosis, in *Photomedicine in Gynecology and Reproduction* (Wyss, P., Tadir, Y., Tromberg, B., and Haller, U., eds.), Karger, Basel, pp. 340–351.
28. Becker, I., Becker, K. F., Röhrl, M., Minkus, G., Schütze, K., and Höfler, H. (1996) Single-cell mutation analysis of tumors from stained histologic slides. *Lab. Invest.* **75**, 801–807.
29. Schütze, K., Becker, I., Becker, K. F., et al. (1997). Cut out or poke in—the key to the world of single genes: laser micromanipulation as a valuable tool on the lookout for the origin of disease. *Genet. Anal.* **14**, 1–8.
30. Schütze, K., and Lahr, G. (1998). Identification of expressed genes by laser-mediated manipulation of single cells. *Nat. Biotech.* **16**, 737–742.
31. Meimberg, H., Thalhammer, S., Brachmann, A., et al. (2003) Selection of chloroplasts by laser microbeam microdissection for single-chloroplast PCR. *Biotechniques* **34**, 1238–1243.

32. Fink, L., Seeger, W., Ermert, L., et al. (1998) Real-time quantitative RT-PCR after laser-assisted cell picking. *Nat. Med.* **4,** 1329–1333.
33. Lahr, G. (2000). RT-PCR from archival single cells is a suitable method to analyze specific gene expression. *Lab. Invest.* **80,** 1477–1479.
34. Lehmann, U., Glöckner, S., Kleeberger, W., von Wasielewski, R., and Kreipe, H. (2000) Detection of gene amplification in archival breast cancer specimens by laser-assisted microdissection and quantitative real-time polymerase chain reaction. *Am. J. Pathol.* **156,** 1855–1864.
35. Sirivatanauksorn, Y., Drury, R., Crnogorac-Jurcevic, T., Sirivatanauksorn, V., and Lemoine, N. R. (1999) Laser-assisted microdissection: applications in molecular pathology. *J. Pathol.* **189,** 150–4.
36. Scheidl, S., Nilsson, S., Kalén, M., et al. (2002) mRNA expression profiling of laser microbeam microdissected cells from slender embryonic structures. *Am. J. Pathol.* **160,** 801–813.
37. Sirivatanauksorn, Y., Sirivatanauksorn, V., and Lemoine, N. R. (2002) DNA fingerprinting from cells captured by laser microdissection. *Methods Enzymol.* **356,** 289–294.
38. Xu, B. J., Caprioli, R. M., Sanders, M. E., and Jensen, R. A. (2002) Direct analysis of laser capture microdissected cells by MALDI mass spectrometry. *J. Am. Soc. Mass Spectrom.* **13,** 1292–1297.
39. Kim, J. O., Kim, H. N., Hwang, M. H., et al. (2003) Differential gene expression analysis using paraffin-embedded tissues after laser microdissection. *J. Cell Biochem.* **90,** 998–1006.
40. Fellenberg, J., Krauthoff, A., Pollandt, K., Delling, G., and Parsch, D. (2004) Evaluation of the predictive value of Her-2/neu gene expression on osteosarcoma therapy in laser-microdissected paraffin-embedded tissue. *Lab. Invest.* **84,** 113–121.
41. Micke, P., Bjφrnsen, T., Scheidl, S., et al. (2003). A fluid cover medium provides superior morphology and preserves RNA integrity in tissue sections for laser microbeam microdissection. *J. Pathol.* **202,** 130–138.
42. Bernsen, M. R., Diepstra, J. H., van Mil, P., et al. (2004) Presence and localization of T-cell subsets in relation to melanocyte differentiation antigen expression and tumour regression as assessed by immunohistochemistry and molecular analysis of microdissected T cells. *J Pathol.* **202,** 70–79.
43. Kehr, J. (2003) Single cell technology. *Curr. Opin. Plant Biol.* **6,** 617–621.
44. Rook, M. S., Delach, S. M., Deyneko, G., Worlock, A., and Wolfe, J. L. (2004) Whole genome amplification of DNA from laser capture-microdissected tissue for high-throughput single nucleotide polymorphism and short tandem repeat genotyping. *Am. J. Pathol.* **164,** 23–33.
45. Tao, W., Wilkinson, J., Stanbridge, E., and Berns, M. (1987) Direct gene transfer into human cultured cells facilitated by laser micropuncture of the cell membrane. *Proc. Natl. Acad. Sci. USA* **84,** 4180–4184.
46. Tirlapur, U. K., and König, K. (2002) Targeted transfection by femtosecond laser. *Nature* **418,** 290–291.
47. Tsukakoshi, M., Kurata, S., Nomiya, Y., Ikawa, Y., and Kasuya, T. (1984) A novel method of DNA transfection by laser microbeam cell surgery. *Appl. Physics B.* **35,** 135–140.

48. Weber, G., and Greulich, K. (1992) Manipulation of cells, organelles, and genomes by laser microbeam and optical trap. *Int. Rev. Cytol.* **133,** 1–41.

49. Wiegand, R., Weber, G., Zimmermann, K., Monajembashi, S., Wolfrum, J., and Greulich, K. (1987) Laser-induced fusion of mammalian cells and plant protoplasts. *J. Cell Sci.* **88,** 145–149.

50. Agilent Application Note 5980-EN on www.palm-microlaser.com or at www. chem.agilent.com.

51. BioNobile Application Note TN4100-003 on www.palm-microlaser.com.

2

Applications of the Universal DNA Microarray in Molecular Medicine

Reyna Favis, Norman P. Gerry, Yu-Wei Cheng, and Francis Barany

Summary

Integration of molecular medicine into standard clinical practice will require the availability of diagnostics that are sensitive, rapid, and robust. The backbone technology underlying the diagnostic will likely serve double duty during clinical trials in order to first validate the biomarkers that contribute to both drug response and disease stratification. PCR/LDR/Universal DNA microarray is a promising technology to help drive the transition from the current paradigms of clinical decision making to the new era of personalized medicine. By uncoupling the mutation detection step from array hybridization, this technology becomes fully programmable. It exploits full use of the sensitivity that the ligase detection reaction can provide, while maintaining a rapid read out on a universal microarray. Thus, PCR/LDR/Universal DNA microarray is 50-fold more sensitive and 10-fold more rapid than conventional hybridization-only arrays. The intent of this article is to provide investigators with a perspective on current uses of this approach, as well as to serve as a practical guide to implementation.

Key Words: Microarray; ligase detection reaction; high throughput; multiplexing; cancer genomics; SNP; DNA methylation; mutation detection; thermostable ligase.

1. Introduction

1.1. The Future of Molecular Medicine

Molecular medicine holds the promise of maximizing patient benefit while minimizing risk. In a vision of the future, standard care for a patient would involve the selection of the drug most likely to be efficacious, given the particulars of an individual's indication, and an adjustment of the drug dose to accommodate the patient's drug metabolism (absorption, distribution, metabolism, and excretion [ADME]) profile. Guiding the physician in making these decisions will be information contained on the drug label, as well as results obtained from point-of-care diagnostic kits and/or the diagnostic laboratory.

From: *Methods in Molecular Medicine, Vol. 144, Microarrays in Clinical Diagnostics*
Edited by: T. Joos and P. Fortina © Humana Press Inc., Totowa, NJ

Although the present state of medical care bears little resemblance to this scenario, appropriate technologies to support this standard of care are actively being developed.

To bring molecular medicine a step closer to becoming a reality, the technology platforms of today must have the flexibility to make the transition from clinical trial to diagnostic laboratory. In the recent Food and Drug Administration draft guidance for pharmacogenomic data submission (http://www.fda.gov/cder/guidance/5900dft.pdf), it was recommended that the test and the drug be codeveloped and complete information on the test be submitted to the agency. Thus, a cost-effective strategy is to apply the test used during the clinical trial, very likely in fundamentally the same form, to execute the same function in a diagnostic capability after the drug is launched. The standard requirements for a marketable test include high sensitivity and specificity, reproducibility, rapid throughput, and cost effectiveness; however, clinical trials require extra considerations. Of note, tests must assess only those factors for which subjects have provided consent. A suitable technology platform must provide for the aforementioned requirements, as well as be able to accommodate molecular versatility, since there are numerous indications that would benefit from the use of biomarkers.

Molecular approaches for identifying single-base variants fall into three general strategies:

1. Target amplification, e.g., (real-time polymerase chain reaction [PCR], usually with TaqMan® or molecular beacon probes), whole-genome amplification (WGA), and strand displacement amplification (SDA).
2. Probe amplification, e.g., ligase chain reaction (LCR), ligase detection reaction (LDR), and rolling circle amplification (RCA).
3. Signal amplification, e.g., 3D dendrimer labeling systems, enzymatic cascade reporters, and invader assay (*see* **ref.** *1* for review) (**Fig. 1;** *see* Color Plate 8 following p. 18).

These amplification strategies are often combined with identification of the product sequence by hybridization to its complement on an array. Although many of these approaches have great utility in certain contexts, various drawbacks exist that limit their use in other contexts. For example, direct hybridization schemes cannot always distinguish between closely related sequences and cannot find mutant sequence in an excess of normal sequence (**Fig. 1**, hybridization array on bottom left). Another fundamental stumbling block is that all amplification techniques have the risk of false-positive signal. Whereas PCR amplification suffers from carryover contamination and amplification of false products, LCR and RCA may amplify probe in the absence of the correct target, and signal amplification schemes may create signal arising from nonspecific binding. By combining two such techniques, i.e., PCR and LDR, we have been able to integrate the best features of each approach, achieving high sensitivity while avoiding false positives *(2–14)*. Furthermore, we have been able to overcome the limitations of hybridization arrays by introducing the concept of divergent zip-code sequences with similar T_m values to guide products to their correct addresses on universal arrays (**Fig. 1**, universal arrays on bottom right).

1.2. Development and Proliferation of Ligase-Based Detection Strategies

The ability to cycle ligation reactions efficiently was first made possible in the early 1990s, following the cloning of the *Thermus thermophilus* (*Tth*) DNA ligase-encoding gene by our laboratory *(15,16)*. That same year, we demonstrated that LCR (the exponential amplification of DNA by ligating four adjacent primers on both DNA strands) could be used for the detection of genetic disease. LDR (linear amplification by ligating two adjacent primers on one DNA strand) was subsequently developed as a versatile method for discriminating single-base mutations or polymorphisms in a multiplexed fashion *(17,18)*. LDR is ideal for multiplexing when combined with PCR, since several primer sets can ligate along a gene without the interference encountered in polymerase-based systems. This advance paved the way for LDR/PCR *(2)*, which permitted amplification of the signal from the ligation detection event. The development of the Universal DNA microarray allowed different LDR products tagged with unique "zip-code" sequences to be guided to complementary addresses on a universal DNA chip *(3)*.

The techniques and approaches described above have been shown to be versatile, robust, and accessible, having been successfully validated and extended in other laboratories (*see* **Subheading 1.4.** below). Indeed, another measure of utility and accessibility has been the conversion of ligase-based strategies into commercial products. Our industrial collaborators (Applied Biosystems [ABI], Foster City, CA) and others have subsequently adopted our universal zip-code design and/or ligation assays for distinguishing multiple signals in capillary, liquid, bead, or microchip arrays *(11,19–35)*. For example, ABI has extended the multiplexing utility of LDR (*see* **Subheading 1.4.3.** below).

1.3. Principles of the Universal DNA Microarray

The Universal DNA microarray is a technology platform that provides an alternate strategy in microarray design. It differs from the conventional approaches to microarray technology in that mutation detection and hybridization to the array surface are completely separate events. Since the specificity for the Universal DNA microarray is determined by LDR, it avoids the false negatives and false positives associated with direct DNA hybridization arrays. For high-throughput detection of specific multiplexed LDR products, unique zip-code sequences are

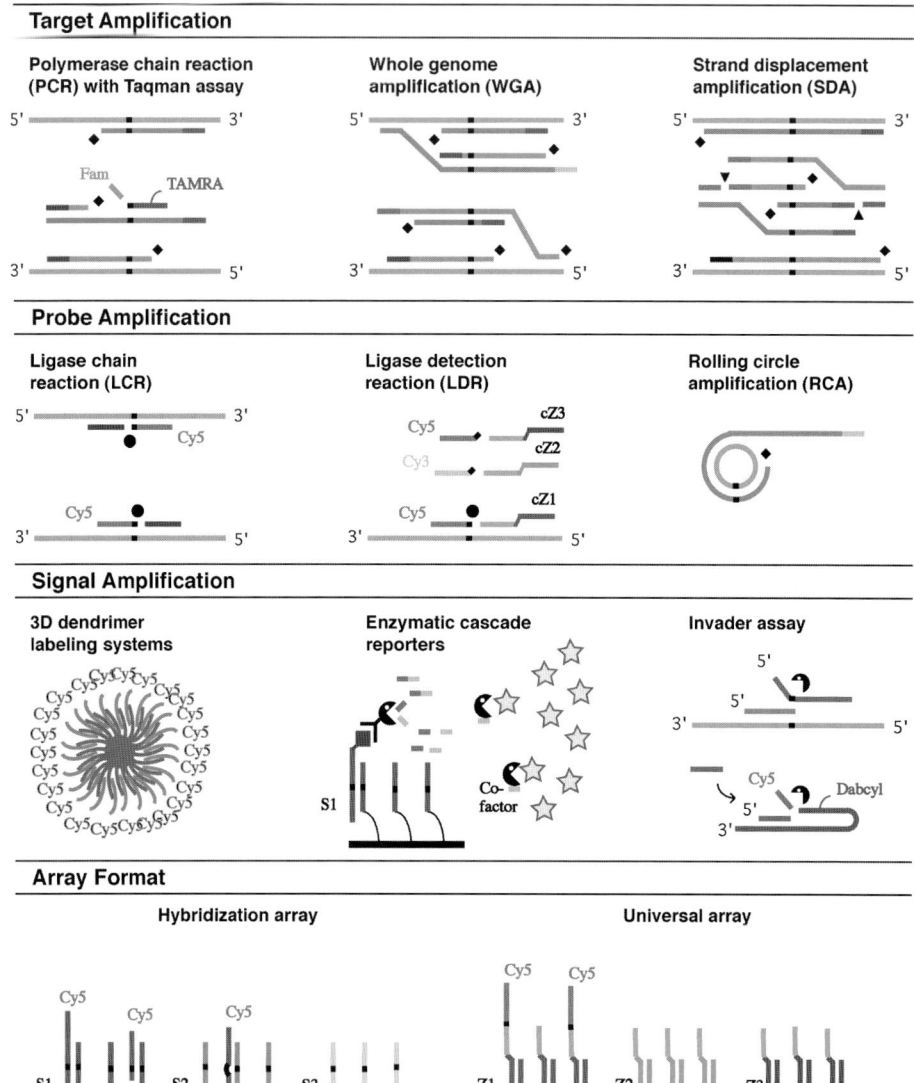

Fig. 1. (Color Plate 8 following p. 18). Overview of target amplification, probe amplification, signal amplification, and arrays used in molecular diagnostics (*see* **ref.** *1* for original references). **Target amplification**: PCR/TaqMan—The 5′–3′ exonuclease activity of Taq polymerase releases a fluorescent group from its quencher during primer extension, allowing real-time monitoring of the PCR reaction. WGA—random oligonucleotide primers are used in conjunction with a processive polymerase with strand displacement

attached to each LDR product, allowing for specific capture at complementary addresses on a DNA microarray *(3)*. The Universal DNA microarray is thus programmable and can accommodate any gene without redesigning the array. The details of the methodology provide explanations as to why this technology is sensitive, rapid, and robust.

The major components of the process are simple to perform and involve PCR followed by LDR and then hybridization of the product to the Universal DNA microarray (**Fig. 2**; *see* Color Plate 9 following p. 18). To amplify multiple genomic regions while minimizing primer-specific differences in efficiency,

---

Fig. 1. *(Continued from previous page)* activity, allowing isothermal, nonspecific amplification total genomic DNA with little amplification bias. SDA—two primers hybridize to each strand of the target DNA and are extended so that the upstream primer displaces the downstream primer. Two more primers anneal to the displaced product and extend in the same way, followed by restriction endonuclease nicking and extension of the double-stranded product. **Probe amplification**: LCR—two primers anneal to each strand of the target DNA region and are ligated if there is perfect complementarity at the junctions. The product molecules can act as templates for the next round of LCR, resulting in exponential amplification. LDR—a fluorescently labeled allele-specific primer and a downstream primer anneal adjacently on one strand of the target DNA. If there is perfect complementarity at the junction, a thermostable ligase joins the primers, and the resulting product is detected on an array or by electrophoresis. RCA—both sides of a long primer hybridize to the target sequence and, when ligated, produce a circular product molecule, which is then amplified by a strand-displacing DNA polymerase. **Signal amplification**: 3D dendrimer labeling systems—a dendrimer has a core that consists of a matrix of double-stranded DNA, and an outer surface with hundreds of single-stranded "arms," which are available for hybridization to a specific sequence or to oligonucleotides that carry signal molecules. Enzymatic cascade reporters—an event, such as hybridization of the correct DNA molecule to a position on an array, stimulates an enzyme to produce a cofactor that is specific for a second enzyme. This second enzyme can, in turn, separate a fluorescent group from its quencher. Invader assay—an allele-specific oligonucleotide and a second oligonucleotide hybridize to the target so there is a single nucleotide overlap at the base in question, and cleavase exonuclease releases the "flap" at the 5′ end of the oligonucleotide only if there is perfect complementarity at the junction. This "flap" is then able to bind to a specific fluorescence resonance energy transfer (FRET) moiety, allowing cleavase to release a fluorescent signal. **Array format**: Hybridization array—biological oligonucleotides that are specific for a particular organism (e.g., cDNAs) are covalently attached to an array surface at specific locations, allowing detection of their labeled complements (e.g., RNAs). This format has difficulty distinguishing all single-base mutations. Universal array—unique oligonucleotides (zip-codes) are located at specific positions on an array surface, allowing fluorescently labeled LDR products, with a tails that are complementary to the zip-codes, to hybridize. Appropriate primer design allows a single Universal array to be used for many different assays.

PCR / PCR / LDR / Universal Array

1. PCR amplify all p53 exons using gene-specific/universal primers and Taq polymerase. ◆

2. PCR amplify all primary products using universal primers and Taq polymerase. ◆

3. Perform LDR using mutation-specific LDR primers, common primers containing complementary zip code sequences, and thermostable ligase. ●

4. Capture fluorescent products on addressable array and score for presence of mutation.

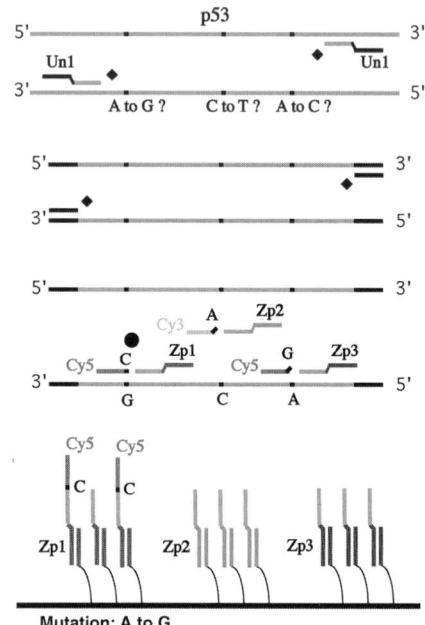

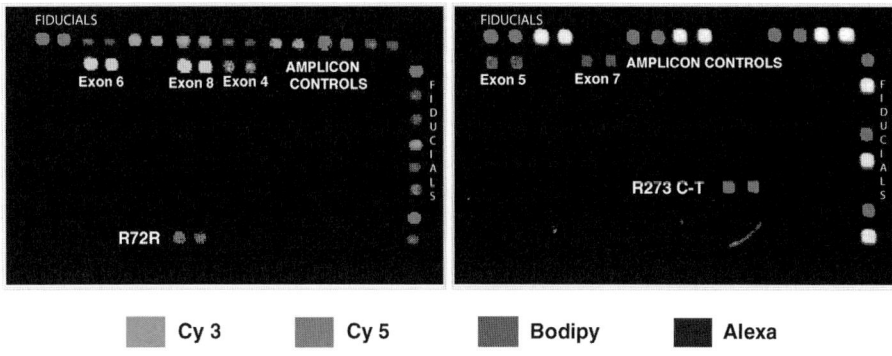

Fig. 2. (Color Plate 9 following p.18) Schematic diagram and results for the detection of multiple mutations using PCR/PCR/LDR/Universal Array. Upper panel shows schematic. p53 exons 4–8 are PCR-amplified in a multiplex format with gene-specific primers bearing 5′ universal sequences, and in a second PCR, amplified simultaneously with universal PCR primers. Multiplexed LDR is performed on all PCR products with fluorescently labeled allele-specific LDR primers and common primers containing complementary zip-code sequences. LDR products are hybridized to universal array containing zipcodes. Address and color of the microarray spot scores are given for the mutation. Lower panel shows chip results. The p53 chip can detect 110 different mutations in exons

two rounds of PCR are used. PCR primers are constructed to have gene-specific portions connected to universal sequences on the 5′ ends. The first round of PCR relies on a limited number of cycles using the gene-specific portions of the primers. The reaction is supplemented with a PCR cocktail containing primers complementary to the universal sequences and the majority of the amplification cycles are completed. Following this multiplex PCR/PCR amplification of the genomic regions of interest, each mutation is simultaneously detected using a thermostable ligase that joins adjacent pairs of oligonucleotides complementary to the sequences of interest. Attached to one of the paired oligonucleotides (referred to as the common oligo) are nongenomic 24-base sequences that are complementary to 24-base zip-code sequences present at known locations on the microarray surface. The remaining oligonucleotide of the pair (referred to as the discriminating oligo) is fluorescently labeled. Ligation occurs only when the sequence at the junction between the paired oligos is exactly complementary to the template sequence. Thus, when the variant of interest is present, ligation joins the oligonucleotide bearing the fluorescent label to the oligonucleotide bearing the zip-code complement. Hybridizing the LDR product to the Universal DNA microarray reveals the presence of a variant. If the variant of interest is present, a fluorescent signal will be visible on the address bearing the zip-code sequence that captures the complementary zip-code on the LDR oligo. If the variant is not present, the LDR oligo with the complementary zip-code will still hybridize to the appropriate address, but no fluorescent signal will be joined to it.

The sensitivity of this system is augmented by the use of a 3D polyacrylamide surface. This surface permits hybridization times of 30–60 min and signal intensities 100-fold better than conventional microarrays (e.g., poly-L-lysine or amino/silane-coated slides) *(3)*.

The most significant advantage of our technique is the ability to separate and therefore optimize mutation identification independently of array hybridization.

Fig. 2. *(Continued from previous page)* 5, 6, 7, and 8. A total of 216 LDR primers were required for detection. The mutation status of each sample and the zip-codes expected to capture signal are indicated at the bottom of each array; fiducials are along the top and right side of all arrays. Two reactions were performed for each sample containing LDR primers that were designed to hybridize to the upper strand or lower strand of *p53* sequence. The array was imaged on a Lumonics ScanArray 5000 to visualize the Cy3, Cy5, and FAM signals. The 16-bit grayscale images for each dye were captured using the MetaMorph Imaging System (Universal Imaging), rendered in color, overlaid, and merged. R72R is a polymorphism in exon 4, and R273 C→T is the mutant signal. PCR, polymerase chain reaction; LDR, ligase detection reaction.

The background signal from each step can be minimized, and consequently, the overall sensitivity and accuracy of our method can be significantly enhanced over other strategies. Direct hybridization DNA microarrays suffer from differential hybridization efficiencies owing either to sequence variation or to the amount of target present in the sample. Consequently, hybridizations are performed at low temperatures, often for several hours to overnight, and this results in increased background noise and false signals caused by mismatch hybridization and nonspecific binding, for example, on small insertions and deletions in repeat sequences *(36–39)*. In contrast, our approach of designing divergent zip-code sequences with similar T_ms, allows for a more stringent and rapid hybridization at 65°C.

1.4. Applications of PCR/LDR and the Universal DNA Microarray

When our approaches are combined with PCR, they have been successfully applied to the simultaneous multiplex detection of numerous genetic diseases (*see* **Subheading 1.4.1.** below). In our own laboratory, the approach has been validated on hundreds of clinical tumor samples during detection of 19 K-*ras* and 110 *p53* gene mutations in non-microdissected tumors, as well as stool, demonstrating the ability to find mutations despite a large quantity of background normal sequence *(5,10,12–14)*. Our approach has the sensitivity to detect 1 in 100 for a *p53* mutation in a wild-type sequence, which is impossible to achieve using standard commercial hybridization chips *(3,12)*.

The Universal DNA microarray allows for the detection of (1) dozens to hundreds of polymorphisms in a single-tube multiplex format, (2) small insertions and deletions in repeat sequences, (3) low-level mutations in a background of normal DNA *(3,5,6,17,18)*, and (4) methylation status of gene promoters. In addition, it requires less manipulation of the DNA. Direct hybridization methods require (1) multiple rounds of PCR or PCR/T7 transcription and (2) processing of PCR-amplified products into fragments or rendering them single stranded. In contrast, our approach allows multiplexed PCR in a single reaction *(18)* but does not require an additional step to convert product into a single-stranded form.

PCR/LDR and the Universal DNA microarray have been successfully employed in studies that required the following capabilities.

1.4.1. Multiplexed Detection of Single-Nucleotide Polymorphisms and Point Mutations

PCR/LDR has been successfully applied to simultaneous multiplex detection of 61 cystic fibrosis alleles *(40,41)*, 6 hyperkalemic periodic paralysis alleles *(42)*, and 20 21-hydroxylase deficiency alleles *(17,43)*. In addition to point mutations,

we demonstrated that PCR/LDR could detect instability within the transform-
ing growth factor-β type II receptor gene and the APCI1307K mononucleotide
repeat allele in DNA derived from both blood and paraffin-embedded tumor
samples *(4,6)*. The cystic fibrosis test is commercialized by our corporate
collaborators, ABI and Celera Diagnostics, and is used throughout the world for
prenatal testing of this inherited disease.

1.4.2. Multiplexed PCR for Amplifying Many Regions of Chromosomal DNA Simultaneously

We have developed a coupled multiplex PCR/PCR/LDR assay for use in
armed forces personnel. This technique was developed to mitigate the problems
of false amplicons, allele dropout, and uneven amplifications, which often mar
attempts to perform highly multiplexed PCR *(18)*. A comparison of LDR profiles of several individuals demonstrated the ability of PCR/LDR to distinguish
both homozygous and heterozygous genotypes at each locus *(18)*. Others have
independently validated the use of PCR/PCR in human identification to amplify 26 loci simultaneously *(44)* or ligase-based detection to distinguish 32 alleles, although the latter was in individual reactions *(45)*. We have also developed
a PCR/PCR/LDR assay to detect the founder Jewish BRCA1 and BRCA2
insertion and deletion mutations associated with breast cancer *(8)*.

1.4.3. Multiplexed LDR/PCR to Determine DNA Copy Number or Score SNPs

We initially developed multiplexed LDR followed by PCR to score chromosomal instability in tumors *(2)*. Others have extended our approach to
detection of deletions in the *DMD* gene, deletions in the *hMLH1* and *hMSH2*
genes, and chromosomal trisomy *(46–48)*. Our corporate collaborators at
ABI have extended our LDR/PCR protocol for typing single-nucleotide
polymorphisms (SNPs) directly on genomic DNA. In their protocol, one of
each LDR primer pair contains a unique zip-code sequence, and locus-specific sequences are flanked by universal primer sequences. Consequently, all
LDR products may be amplified in a single PCR step, and each product may
be identified by its unique zip-code sequence. The products may be rendered
single stranded and hybridized on a universal array (**Fig. 3**; *see* Color Plate
10 following p. 18), or alternatively used to capture premade fluorescently
labeled "zip-chutes" (developed by ABI) (**Fig. 4**; *see* Color Plate 11 following p. 18), each with a unique size, for scoring by electrophoretic separation.
The technique has been validated on 3000 SNPs using 96 genomic DNA
samples *(11)*.

LDR / PCR / Universal Array

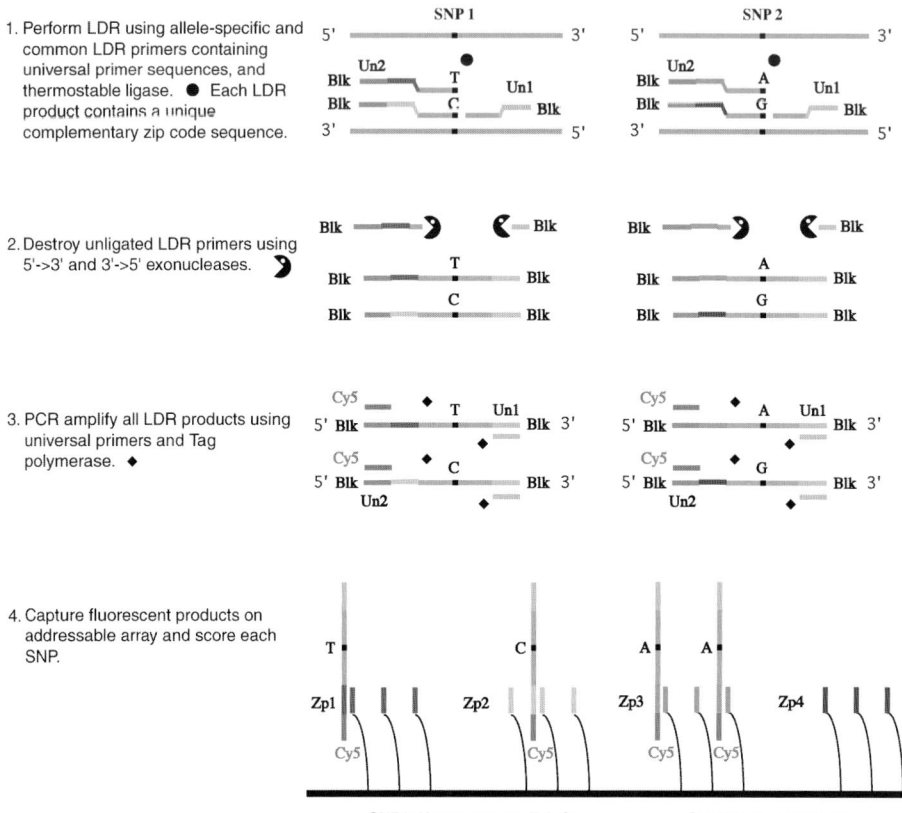

1. Perform LDR using allele-specific and common LDR primers containing universal primer sequences, and thermostable ligase. ● Each LDR product contains a unique complementary zip code sequence.

2. Destroy unligated LDR primers using 5'->3' and 3'->5' exonucleases. ❱

3. PCR amplify all LDR products using universal primers and Taq polymerase. ◆

4. Capture fluorescent products on addressable array and score each SNP.

SNP1: Heterozygous, T & C.   SNP2: Homozygous A.

Fig. 3. (Color Plate 9 following p. 18) Schematic diagram for the detection of multiple single-nucleotide polymorphisms (SNPs) using LDR/PCR/Universal Array. Multiplexed LDR is performed on multiple SNPs using allele-specific primers containing unique zip-code sequences and a 5' universal sequence (Un2) as well as locus-specific primers containing a different universal sequence (Un1) on their 3' ends. Only if there is perfect complementarity at the junction will the ligation product form, thus distinguishing different SNPs. Unligated products are destroyed with λ exonuclease (5'→3') and exonuclease 1 (3'→5'). All remaining LDR products are coamplified simultaneously with universal PCR primers Un2 and Un1. In this illustration, Un2 is labeled with Cy5 on the 5' end, and Un1 may be phosphorylated on the 5' end, allowing for the option to convert PCR products to a single-stranded form with λ exonuclease prior to hybridization. PCR products are hybridized to universal array containing zip-codes. Fluorescent signal at a given address scores for the presence or absence of each SNP. LDR, ligase detection reaction; PCR, polymerase chain reaction.

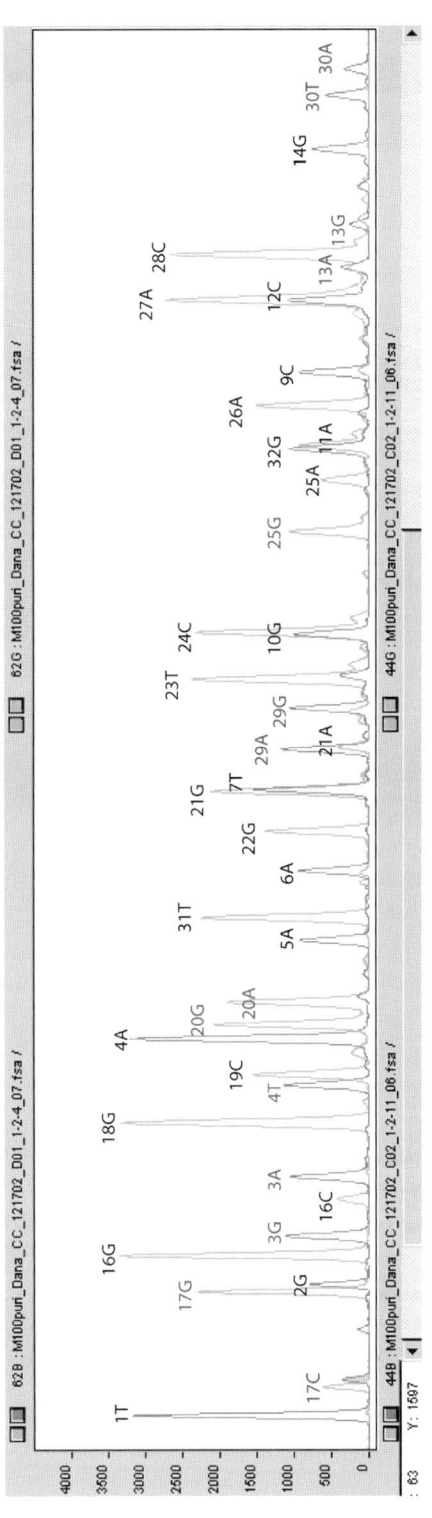

Fig. 4. (Color Plate 11 following p. 18) Electrophoretogram of 30-plex detection in genomic DNA. The data demonstrate the ability of LDR/PCR to characterize 60 alleles simultaneously in an SNP genotyping assay. The blue and green peaks are either FAM or Vic labeled mobility modified zip-chutes. The *x*-axis is the size of zip-chute and the *y*-axis is fluorescent intensity of capillary electrophoresis. The zip-chutes have the same color but different motilities for both of allele-1 and allele-2. For example, 1T (first blue peak) is homozygous, T; 17C (green) and 17G (green) are heterozygous C/G, etc.

1.4.4. Detection of K-ras, BRCA1, BRCA2, and p53 Mutations Using Multiplex PCR/LDR and Zip-Code Capture

Since the zip-code sequences remain constant and their complements can be appended to any set of LDR primers, our zip-code arrays are universal, and we and our cancer collaborators at The Rockefeller University and the Institut Curie have applied this array-based mutation detection to mutations in the K-*ras, BRCA1, BRCA2*, and *p53* genes *(3,8,9,12–14)*. **Figure 2** shows the schematic and results for *p53* mutation analysis, where by 110 mutations could be queried simultaneously. Mutations present at 1% of the wild-type DNA level, or in pooled samples could be distinguished *(8,12)*.

1.5. Back to the Future

The practice of molecular medicine will require technology platforms that can span the progression from clinical trial to diagnostic laboratory and rapidly deliver an answer. En route, the incipient diagnostic will be challenged with a variety of genes containing diverse assortments of genetic variation.

The Universal DNA microarray is a strong candidate to serve this need. Because the platform is programmable, it is robust enough to accommodate changes in the genes, SNPs, or mutations of interest without reengineering the array. In addition, our zip-code concept can be used for displaying mutations on a variety of platforms: universal array surfaces, gel or capillary electrophoresis, and universal encoded beads (*see* **Fig. 5** [Color Plate 12 following p. 18] and **Subheading 3.3.**). This programmability and display versatility are added boons in the context of clinical trials, in which the list of genes for which subjects provide consent for genotyping can vary from trial to trial, the SNP content must reflect the targeted patient/volunteer population, and the number of recruited subjects can vary by orders of magnitude.

We and others have also shown that our assay can detect a greater range of human genetic variation compared with other systems. In addition to SNPs and low-level mutations *(10)*, insertion/deletion mutations *(8)* and length polymorphisms in mononucleotide *(6)* and dinucleotide repeats *(7)* can also be reliably detected—two blind spots for direct hybridization arrays *(49–51)*. Even though this latter type of variation is invisible to most mutation detection technologies, such variations are known to have pharmacological relevance owing to their prevalence in ADME genes. For example, CYP2D6 has numerous insertion/deletions that impact enzyme activity *(52)*; UGT1A1 contains a promoter polymorphism consisting of variable numbers of TA repeats that influences enzyme concentration *(53)*.

This technology is also amenable to the rapid provision of results in a clinical setting. Given that LDR can be performed in 5–10 min *(54)*, a microfabricated

LDR / PCR / Universal Display

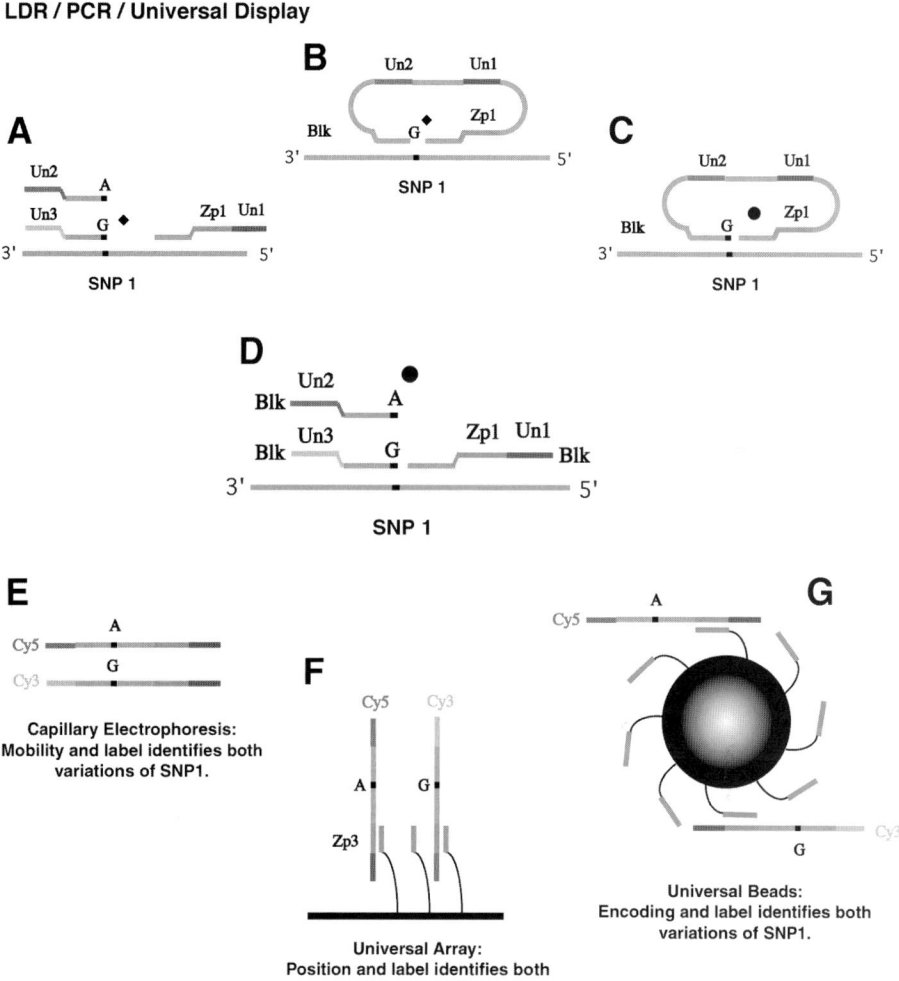

Fig. 5. (Color Plate 12 following p. 18) Ligase-based multiplexed single-nucleotide polymorphism (SNP) assays combined with zip-code universal display. The original LDR/PCR procedure combines ligation of multiple oligonucleotide probes on multiple target sequences followed by PCR coamplification of ligation products using universal primer sequences *(2)*. The generic version for SNP detection is illustrated in the middle of the figure **(D)**. Allele-specific oligonucleotides may hybridize on target DNA adjacent to a downstream locus-specific sequence containing a specific zip-code sequence and will ligate if there is perfect complementarity at the junction. Use of blocked ends allows for exonuclease to degrade unligated probes but not the ligation products. Universal primersUn1, Un2, and Un3 are used to coamplify all the products. A variation of this scheme uses only two universal primers but distinguishes alleles by appending different zip-code sequences to the allele-specific oligonucleotides (as shown in **Fig. 3**). In

device that integrates sample prep with multiplexed genetic variation identification (PCR/LDR) and molecular profiling (LDR/PCR) would be anticipated to provide results in less than 30 min. A reason for optimism lies in the significant progress recently made in developing microfluidic and microchip-based platforms for point-of-care analysis. Although they are beyond the scope of this chapter, platforms that integrate sample preparation *(55–57)*, purification and concentration of the DNA/RNA component *(58,59)*, component delivery *(60–63)*, thermocycling *(58,64,65)*, and capillary or channel-based separation and signal detection *(54,66–80)* will be considered for enriching our assay development.

2. Materials

1. Oligonucleotides (*see* **Note 1**).
2. dNTPs (PE Biosystems, cat. no. 4303441).
3. AmpliTaq Gold (PE Biosystems, cat. no. 4311820).
4. Taq DNA ligase (NEB, cat. no. M0208L).
5. Proteinase K (Qiagen, cat. no. 19131).
6. Three 1-in microscope slides with etched circles (VWR, cat. no. 48349-057 or Erie, cat. no. 2960; *see* **Note 2**).
7. 24 × 50-mm Cover slips (VWR, cat. no. 48393-081; *see* **Note 2**).
8. Corning crystallizing dishes, 170-mm diameter × 90-mm height (VWR, cat. no. 25411-140 or Corning, cat. no. 3140-170; *see* **Note 3**).
9. 20-Slide glass slide racks (VWR, cat. no. 25463-009 or Wheaton, cat. no. 900204).
10. Glass slide rack handle (VWR, cat. no. 25464-001 or Wheaton, cat. no. 900205).
11. 50-Slide rack and staining dish (VWR, cat. no. 25461-024 or Wheaton, cat. no. 900400).
12. Acrylamide (Boehringer Mannheim, cat. no. 1871757).
13. Bis-acrylamide (Boehringer Mannheim, cat. no. 1685830).
14. Acrylic acid (Aldrich, cat. no. 14,723-0).

Fig. 5. *(Continued from previous page)* another variation of this scheme, allele-specific oligonucleotide primers are extended with a polymerase prior to ligation to the common oligonucleotide (**A** *[23]*). Alternatively, the nonligating ends of the upstream and downstream probes may also be blocked by synthesizing a single long probe, which is subsequently linearized after exonuclease digestion in preparation for the PCR step (**B** *[21]*) In one version of this variation, an extra polymerase step is used to add a single base prior to ligation (**B**), although the approach works as well by using straight ligation (**C** *[19]*). In the examples illustrated here, two different fluorescent labels are used for the UN2 and Un3 PCR primers, although schemes using a single label or more may be used. The PCR products may be displayed using gel or capillary electrophoresis (**E** *[2,11]*), universal array surfaces (**F** *[3,8–10,12–14,19,21]*), or universal encoded beads (**G** *[22–24]*).

15. Ammonium persulfate (APS; VWR, cat. no. JT0762-11).
16. 1-[3-(Dimethylamino)propyl]-3-ethylcarbodimide hydrochloride (Aldrich, cat. no. 16,146-2).
17. *N*-hydroxysuccinimide (NHS; Aldrich, cat. no. 13,067-2).
18. 3-(Trimethoxysilyl)propyl methacrylate (Aldrich, cat. no. 44,015-9).
19. Concentrated NH_4OH (VWR, cat. no. JT9721-4).
20. 30% Hydrogen peroxide (Aldrich, cat. no. 21,676-3).
21. Concentrated HCl (VWR, cat. no. JT9535-4).
22. High-performance liquid chromatography (HPLC) grade methanol (VWR, cat. no. JT9093-3).
23. HPLC grade acetone (VWR, cat. no. JT9002-3).
24. HPLC grade chloroform (Fisher, cat. no. C606-4).
25. Triethylamine (Aldrich, cat. no. 47,128-3).
26. $0.4\ M\ K_2HPO_4 / KH_2PO_4$, pH 5.5.
27. $1\ M$ 2-Morpholinoethanesulfonic acid (MES), pH 6.0 (light sensitive; filter-sterilize and store at room temperature; stable for 3–4 mo; do not use if the solution appears yellow).
28. 2X Hybridization buffer: $0.6\ M$ MES, pH 6.0, 20 mM $MgCl_2$, 0.2% sodium dodecyl sulfate (SDS; light sensitive; store at room temperature; stable for 3–4 mo).
29. 1X Wash buffer: $0.3\ M$ bicine, pH 8.0, filter-sterilized, 0.1% SDS (store at room temperature).
30. Drierite (VWR, cat. no. WLC3712T).
31. Cover wells, 9-mm diameter × 1-mm height (Grace, Sunriver, OR).
32. Razor blade.
33. Plastic slide box for hybridization (SPI Supplies, cat. no. 01253A-CF).
34. Dessicator.
35. Shaker.
36. Heat blocks.
37. Hot plates.
38. Thermocycler.
39. Rotating hybridization oven.
40. Cartesian Pixsys 5500 array spotting robot with a quill-type spotter in a controlled atmosphere chamber or similar instrument.
41. Perkin-Elmer ScanArray 5000 or similar instrument.

3. Methods

3.1. Array Fabrication

Below is a description of the process used to produce universal arrays. Commercially available arrays with 3D surfaces are an alternative to in-house production.

3.1.1. Clean Slides to Remove Oxidized Surfaces

1. Assemble slides in the glass slide racks. Put slides in back to back so there are two slides per slot (*see* **Note 4**).

2. In a hood using hot plates, boil 600 mL of ultrapure water in two glass crystallizing dishes partially covered with a glass plate to prevent evaporation. Fill a third
   dish with water, and heat to 60–70°C.
3. When the water boils, add 120 mL concentrated NH_4OH and 120 mL 30% peroxide to the first dish (5:1:1 water/NH_4OH/peroxide; the solution will bubble vigorously). Place racked slides in solution, and boil for 10 min (*see* **Note 5**).
4. Rinse slides in 60–70°C dish of water for 2–3 min. Drain excess water from slide
   racks.
5. Add 120 mL concentrated HCl and 120 mL 30% peroxide to second dish of boiling
   water (5:1:1 water/HCl/peroxide; the solution will bubble vigorously). Add slide
   racks, and boil for 10 min.
6. Rinse slides in water followed by HPLC-grade methanol and finally HPLC-grade
   acetone. Air-dry (*see* **Note 6**).

3.1.2. Silanize Slides

1. Place cleaned slides in 50-slide rack and immerse in 400 mL $CHCl_3$ solution containing 2% γ-(trimethoxysilyl)propyl methacrylate and 0.2% triethylamine. Agitate
   gently at room temperature on a shaker table for 30 min (*see* **Note 7**).
2. Wash slides in 400 mL $CHCl_3$ for 15 min on a shaker table. Repeat wash with fresh
   $CHCl_3$ (*see* **Note 8**).
3. Drain excess $CHCl_3$, blot racked slides on paper towel, and let air-dry in a hood.
   Store slides in slide box until ready to prepare polymer surface. Do not touch surface of slides; handle slides by frosted end only (*see* **Note 9**).

3.1.3. Preparation of Polymer Surface

1. Heat heating blocks to 70°C, and then invert blocks so the solid bottom surface is up.
2. Make up monomer solution. The individual monomer solutions should be made
   fresh (no more than 1or 2 d in advance). The solution is as follows: 200 µL 40%
   w/v acrylamide, 50 µL 40% w/v acrylic acid, 20 µL 1% w/v bis-acrylamide, 80 µL
   10% APS, and 650 µL water.
3. Vortex to mix, and allow mixture to sit at room temperature for 1 h before making
   slides (*see* **Note 10**).
4. Spot 20 µL of monomer solution onto the center of a silanized slide from above.
   a. Cover with a cover slip by lowering the cover slip parallel to the surface of the
      slide until it makes contact with the top of the monomer solution droplet.
   b. Release the cover slip, and allow the monomer solution to spread.
   c. The solution should spread to the edges of the cover slip on its own.
   d. If numerous or large bubbles are caught under the cover slip, it can be slid off,
      the surface gently wiped with a Kimwipe, and the procedure repeated using a
      fresh cover slip (*see* **Note 11**).
5. Place slide on heated blocks for 4.5 min with cover slip side up (*see* **Note 12**).
6. Following polymerization, place the slide into a slide rack immersed in water for
   5 min.

7. Use a razor blade to wedge up one of the short sides of the cover slip, and then slowly peel the cover slip from the polymer.
8. Place slides that have new polymer surfaces in racks. (We use test tube racks.)
9. Rinse the slides under running ultrapure water, use a forced air line to gently blow off excess liquid, and then allow slides to air-dry. Store at room temperature in slide boxes until ready to activate surfaces for spotting (*see* **Note 13**).

3.1.4. Activation of Polymer Surfaces

1. Place polymer-coated slides in a 50-slide rack, and immerse in 400 mL 0.1 *M* potassium phosphate, pH 6.0, containing 0.1 *M* (EDC) 1-[3-(Dimethylamino)propyl]-3-ethylcarbodiimide hydrochloride and 20 m*M* NHS. Agitate gently at room temperature on a shaker table for 30 min.
2. Dunk racked slides in receptacle of ultrapure water approx 20 times to rinse. Replace water, and immerse approx 10 times.
3. Tap racked slides on paper towel to blot excess water, and blow off water using forced air.
4. Heat slides at 65°C in an oven (cracked open slightly for good circulation) until completely dry (30 min to 1 h).
5. Store slides desiccated at room temperature in slide boxes (*see* **Note 14**).

3.1.5. Array Spotting

1. Prepare spotting plates by mixing 5 µL 1000 µ*M* zip-code oligonucleotide solutions with 5 µL 0.4 *M* K_2HPO_4/KH_2PO_4, pH 8.5, in 384 conical well spotting plates (*see* **Note 15**).
2. Place slides in spotter, and set relative humidity to 60–70%. Allow the slides to incubate for 15–20 min (*see* **Note 16**).
3. Spot slides in desired layout (*see* **Note 17**).
4. Following spotting, removed uncoupled oligonucleotides from the polymer surfaces by soaking the slides in 300 m*M* bicine, pH 8.0/300 m*M* NaCl/0.1% SDS for 30 min at 65°C (*see* **Note 18**).
5. Rinse the slides with ultrapure water, and dry as described above in **Subheading 3.1.4.**
6. Store slides desiccated at room temperature in slide boxes (*see* **Note 19**).

3.1.6. Array Quality Control

Spotting failures are detected by staining two newly minted arrays from the beginning and end of a spotting run with SYBR Green II via the method of Battaglia et al. *(81)* (**Fig. 6A**; *see* Color Plate 13 following p. 18) This dye clearly shows whether addresses failed to spot during the array printing process; if critical addresses used by all assays are affected, defective arrays can be discarded. If addresses deeper into the array are affected, these arrays can be set aside for use in small assays that will be unaffected by the missing addresses (e.g., K-*ras* mutation detection only requires the first four addresses).

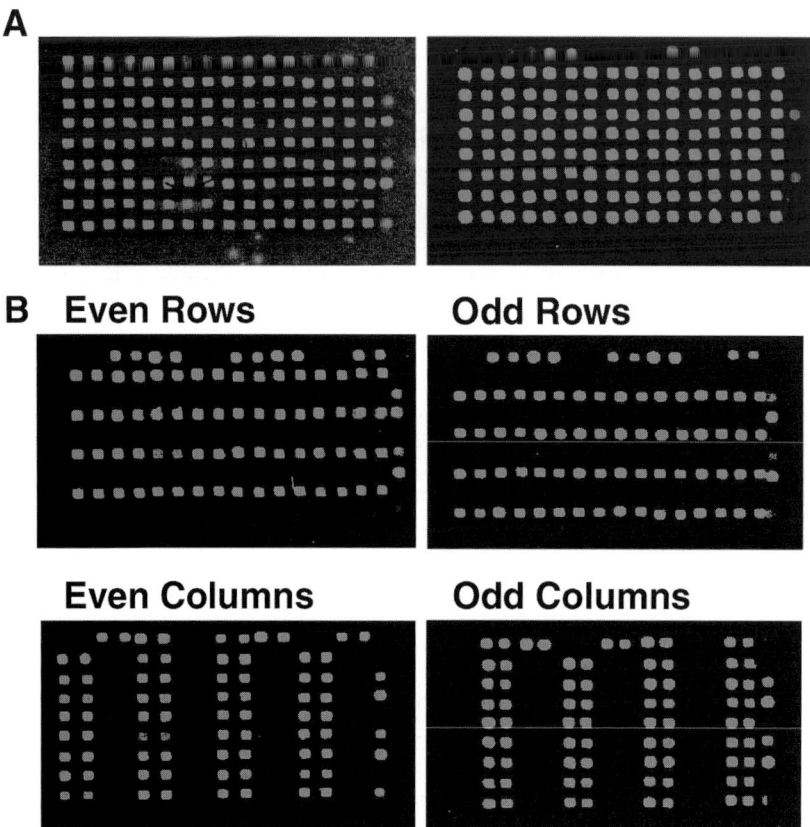

Fig. 6. (Color Plate 13 following p. 18) Validation of arrays following spotting. (**A**) Arrays are stained with a solution of SYBR Green II to determine whether all the zip-codes have been spotted successfully. The left panel shows an array, which has had a spotting failure, and the right panel shows the complete array. (**B**) Arrays are hybridized with mixtures of fluorescein-labeled zip-code complements to look for cross-talk resulting from either well-plate cross-contamination or poor washing of the pins between spotting cycles.

Four random arrays are next chosen from array sets that pass this first round of inspection and are subjected to hybridization with mixtures of fluorescein-labeled zip-code complements (*see* **Subheading 3.2.** for conditions). Mixtures are prepared for the even and odd rows, and the even and odd columns. This test confirms that there is no crosstalk between addresses caused by either well-plate crosscontamination or poor washing of the pins between spotting cycles. **Figure 6B** shows an array that has passed this level of inspection: specific hybridization to odd or even rows and odd or even columns are visible with no extraneous signals.

3.2. PCR/LDR/Array Hybridization

3.2.1. PCR

1. Multiplex PCR is performed with 50–100 ng of genomic DNA in a 25 µL reaction using AmpliTaq Gold and 2 pmol of each gene-specific primer bearing the universal primer on the 5′ ends.
   a. Overlay the reaction with mineral oil and princubate for 10 min at 95°C.
   b. Amplifiy for 15 cycles using conditions optimum for the genes of interest.
   c. Add a second 25 µl aliquot of the reaction mixture through the mineral oil containing 25 pmol universal primer.
   d. Continue cycling for 25 cycles at a higher annealing temperature than that used for the gene-specific primers.
2. Digest the reaction with the addition of 1 µL proteinase K (18 mg/mL) and incubation at 70°C for 10 min. Inactivate the proteinase K by incubating at 95°C for 15 min.
3. Analyze a 1–2 µL aliquot by agarose gel electrophoresis to verify the presence of amplification product of the expected size.

3.2.2. LDR

1. LDR reactions are carried out under oil. The following reactants are combined in a PCR tube for each sample: 1 µL 10X ligase buffer, 1 µL 100 mM dithiothreitol (DTT), 1 µL 10 mM NAD$^+$, 500 fmol of each primer, and 2 µL of PCR product from **Subheading 3.2.1., step 3** in a total volume of 10 µL.
2. Dilute the 40,000 U/mL (0.3 pmol/µL) Tth DNA ligase to 2.5 fmol/µL in 1X ligase buffer. The volume of the diluted ligase should be sufficient to add 10 µL to each reaction.
3. Heat-denature the 10 µL reaction at 94°C for 2 min, and add 10 µL (25 fmol) of diluted Tth DNA ligase.
4. Cycle the reaction for 20 rounds of 94°C for 30 s and 65°C for 4 min.

3.2.3. Array Hybridization

1. Dilute LDR reactions with an equal volume of 2X hybridization buffer to produce a final buffer concentration of 300 mM MES, pH 6.0, 10 mM MgCl$_2$, 0.1% SDS (*see* **Note 20**).
2. Denature the mixture at 94°C for 3 min and chill on ice.
3. Preincubate arrays for 15 min at 25°C in 1X hybridization buffer. Remove the arrays from the buffer and blow the surface dry of excess liquid.
4. Attach cover wells to the arrays and fill with 35 µL of the diluted LDR products (*see* **Note 21**). Use adhesive plastic to seal the cover well openings to prevent drying.
5. Place the arrays in plastic slide holders and humidify using a moistened sponge. Secure the arrays to the rotisseary of the rotating hybridization oven, and incubate for 1 h at 65°C and 20 rpm (*see* **Note 22**).
6. After hybridization, wash the arrays in 300 mM bicine, pH 8.0/0.1% SDS for 10 min at 25°C. Rinse the arrays briefly in water, dry, and scan.

3.3. Variations on a Theme

In addition to the standard PCR/LDR technique outlined in **Subheading 3.2.** above, there are several variations of LDR as an assay tool for mutation identification, SNP detection and DNA methylation analysis.

3.3.1. LDR/PCR

To genotype hundreds of thousands SNPs accurately in multiple samples in a high-throughput format, one variation is to perform LDR on the DNA samples followed by PCR amplification of the ligation products. By performing LDR directly on the DNA samples using primers bearing universal sequences on the 5′ end (*see* **Fig. 3**), ligation products can be subsequently simultaneously amplified using universal primers. There are several advantages to performing LDR prior to PCR amplification: (1) it eliminates the time-consuming steps of design and optimization of multiplex gene-specific PCR primers; (2) it reduces the cost of synthesizing hundreds of fluorescently labeled allele-specific LDR primers, since product labeling can be accomplished using fluorescently labeled universal PCR primers; (3) it reduces the complexity of the assay system to ensure an accurate and efficient DNA analysis by avoiding the common pitfalls associated with multiplex PCR amplification (e.g., formation of primer dimers and other nonspecific amplicons that may interfere with downstream applications); (4) it reduces the time required for LDR primer design and reaction optimization; and (5) it provides an initial linear amplification of the targeted genomic information that is nonbiased, it promotes allelic balance, and it may minimize the need for later PCR cycles that may detract from this balance.

3.3.1.1. LDR/PCR/Universal Microarray

The use of multiplex LDR followed by PCR was initially developed to score chromosomal instability in tumors (*see* **Subheading 1.4.** above). A schematic diagram of multiplex LDR/PCR/Universal Array to determine DNA copy number or score SNPs is shown in **Fig. 3.** In this approach, the universal primer sequences are added to the 5′ end of discriminating LDR primers and to the 3′ end of common LDR primers. After ligase detection reaction, the excess unligated LDR primers and DNA templates can be digested using 5′→3′ and 3′→5′ exonucleases. The ligation products are protected from digestion, since blocking groups are added at both their 5′ and 3′ ends. This exonuclease digestion step reduces the potential of nonspecific hybridization and false-positive results on the universal array readout. The ligation products are simultaneously amplified with universal PCR primers. Only one of the universal PCR primers is fluorescently labeled to serve as the detection signal when these amplicons are captured on a universal array.

3.3.1.2. LDR/PCR/UNIVERSAL DISPLAY

ABI recently extended the LDR/PCR concept with the development of an ultra-high-throughput genotyping method, SNPlex. This technology utilizes multiplexed oligonucleotide ligation assay on genomic DNA. Each LDR primer pair was synthesized with universal primer sequences flanking the locus-specific sequences. A unique zip-code sequence is designed within the LDR primers to uniquely identity each LDR product. The excess LDR primers and genomic DNA are eliminated through enzymatic digestion. Consequently, all LDR products may be amplified in a single PCR step with two universal PCR primers, one of which is biotinylated. Biotinylated amplicons may be rendered single stranded and captured on streptavidin-coated plates. Each single-stranded PCR product may be identified by its unique zip-code sequence through inter-rogation with a set of universal ZipChute probes. These probes have fluores-cence labels, unique complementary zip-code sequences, and ABI mobility modifiers. ZipChute probes can be eluted and electrophoretically separated on an ABI 3730xl DNA analyzer. **Figure 4** shows an example of a 60-plex reac-tion using this approach. This technique has been validated on 3,000 SNPs using 96 genomic DNA samples. Compared with other genotyping platforms, the SNPlex system demonstrates 98.7 and 99.2% concordance with dideoxy sequencing and TaqMan assays, respectively. This variation is an alternative approach to existing genotyping methodologies and has the advantage of a robust detection strategy and low DNA consumption.

3.3.2. Bisulfite/PCR-PCR/LDR/Universal Microarray

One application of LDR/PCR/Universal Array is to study DNA methylation. In particular, this variation focuses on the detection of aberrant promoter methylation occurring at the 5-position of cytosine within the CpG dinu-cleotide. Sodium bisulfite conversion of cytosines to uracils is one of the most commonly used methods to study DNA methylation. 5-Methylcytosines are resistant to conversion, and deamination only occurs on unmethylated cytosines. The modified DNA sequences can then either be amplified and sequenced, or one can perform methylation-specific PCR (MSP) to determine cytosine methylation status. Our bisulfite/PCR-PCR/LDR/Universal Array approach provides a sen-sitive and accurate high-throughput format that can detect methylation status in virtually any gene sequence of interest.

A multiplex PCR-PCR/LDR assay is shown in **Fig. 7**; (*see* Color Plate 14 fol-lowing p. 18) to illustrate this approach. When possible, the gene-specific PCR primers used for multiplexing are designed to avoid CpG sites present in the pro-moter sequences. As a further improvement to accommodate situations in which bisulfite-modified bases cannot be avoided, pyrimidine and purine nucleotide

Bisulfite/PCR-PCR/LDR/Universal Array

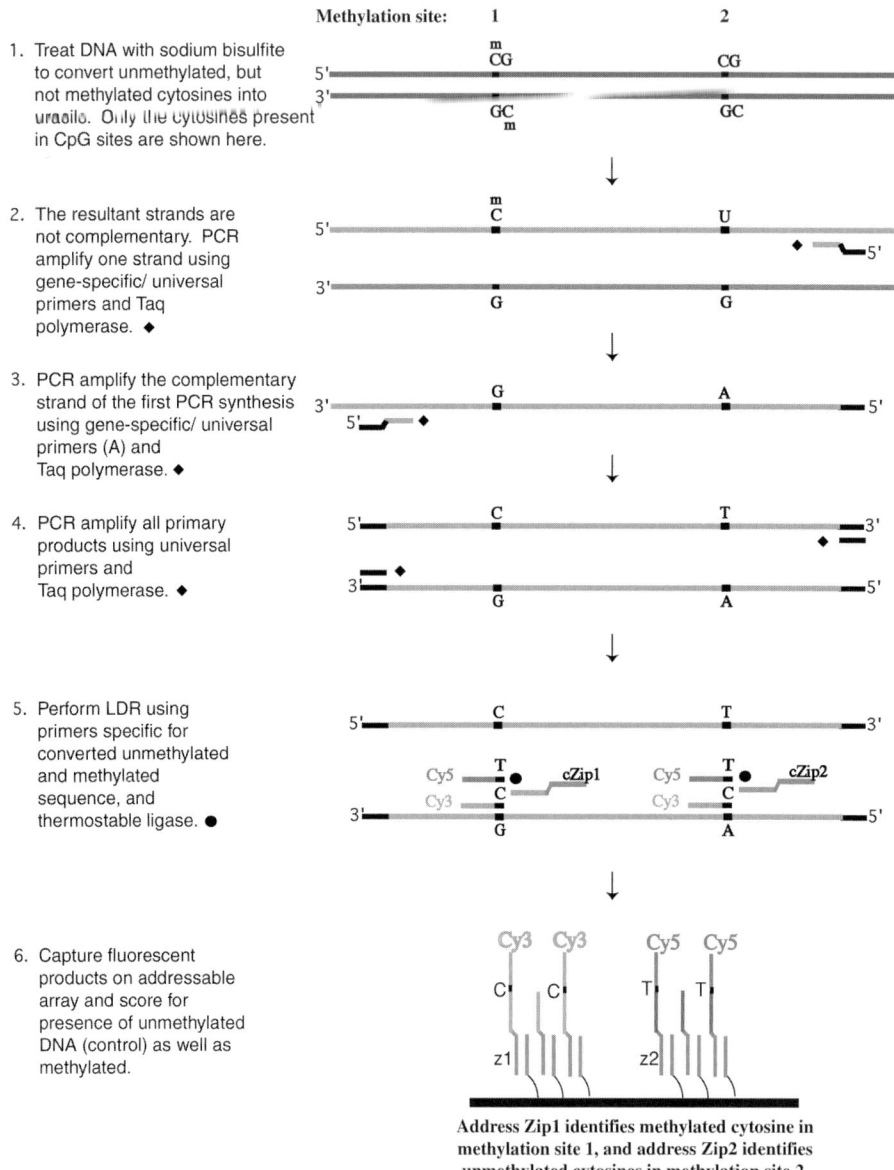

1. Treat DNA with sodium bisulfite to convert unmethylated, but not methylated cytosines into uracils. Only the cytosines present in CpG sites are shown here.

2. The resultant strands are not complementary. PCR amplify one strand using gene-specific/ universal primers and Taq polymerase. ◆

3. PCR amplify the complementary strand of the first PCR synthesis using gene-specific/ universal primers (A) and Taq polymerase. ◆

4. PCR amplify all primary products using universal primers and Taq polymerase. ◆

5. Perform LDR using primers specific for converted unmethylated and methylated sequence, and thermostable ligase. ●

6. Capture fluorescent products on addressable array and score for presence of unmethylated DNA (control) as well as methylated.

Address Zip1 identifies methylated cytosine in methylation site 1, and address Zip2 identifies unmethylated cytosines in methylation site 2.

Fig. 7. (Color Plate 14 following p. 18) Schematic diagram, illustrating the procedure for high-throughput detection of promoter methylation status with the combination of bisulfite treatment, multiplex PCR, multiplex LDR, and universal array approaches. The different fluorescently labeled (Cy3 and Cy5) LDR products are captured on the same addressable array. PCR, polymerase chain reaction; LDR, ligase detection reaction.

analogs are incorporated within the PCR primers. These modified bases, designated P and K, show considerable promise as degenerate bases. The pyrimidine derivative P base pairs with either A or G, whereas the purine derivative K base pairs with either C or T; thus the target DNA can be amplified regardless of its methylation status. Multiple promoter regions are amplified in a two-stage nested PCR reaction simultaneously. The first-stage multiplex amplification uses pairs of gene-specific PCR primers with universal sequences attached to their 5′ ends. The second-stage amplification uses universal PCR primers to amplify the first-stage PCR products. The final PCR products are usually verified on an agarose gel prior to LDR analysis.

The details of LDR primer design have been described in **Subheading 3.** and in **Note 1**. Briefly, three LDR primers are designed for each CpG dinucleotide site. Two discriminating primers are labeled at the 5′ end with either Cy3 or Cy5 and at the 3′ end with a G or A, respectively. The single common primer for the reaction consists of a 5′ phosphate and terminates at the 3′ end with a zip-code complement sequence. Degeneracy is also accommodated in the LDR primers by using pyrimidine and purine nucleotide analogs. After the LDR products are captured on a universal array, the methylated cytosine residues are detected by the presence of Cy3 signals; the presence of unmethylated cytosines is revealed by the presence of Cy5 signals.

Typically, 1–2 μg of genomic DNA in a volume of 40 μL is incubated with 0.2 N NaOH at 37°C for 10 min. Then 30 μL freshly made 10 mM hydroquinone and 520 μL of freshly made 3 M sodium bisulfite, pH 5.0 (Sigma, ACS grade) is added. This mixture is next incubated for 16 h in a DNA thermocycler using alternating cycles of 50°C for 20 min followed by a denaturing step of 85°C for 15 s. The bisulfite-treated DNAs can be desalted using MICROCON centrifugal filter devices (Millipore, Bedford, MA) or, alternatively, cleaned with a Wizard DNA clean-up kit (Promega, Madison, WI). The eluted DNAs are incubated with 1/10 volume of 3 N NaOH at room temperature for 5 min prior to ethanol precipitation. The DNA pellet is then resuspended in 20 μL deionized H$_2$O and stored at 4°C. Bisulfite-modified DNA is stable at 4°C for at least 1 mo.

The current assay is designed to detect the extent of DNA methylation within the promoters of the tumor suppressor genes *p15^INK4b*, *p16^INK4a*, *p19^ARF*, *p21^CIP*, *p27^KIP*, *p53*, and *BRCA1*, as well as the imprinted gene small nuclear ribonucleoprotein N (*SNRPN*). Using the same design parameters, the promoter regions of seven additional genes were chosen to investigate their promoter methylation status in human tumors. These include O6 methyl guanine DNA methyl transferase (*MGMT*), adenomatous polyposis coli (*APC*), retinoic acid receptor (*RARb*), tissue inhibitor metalloproteinase (*TIMP-3*), death-associated protein kinase (*DAPK*), E-cadherin (*ECAD*), glutathione S-transferase (*GSTP1*), and Ras association domain family 1 (*RASSF1*). The hemimethylated SNRPN is used as a positive internal control.

As seen in **Fig. 8** (*see* Color Plate 15 following p. 178), to demonstrate that LDR primers are working properly, genomic DNAs of normal lymphocytes with and without in vitro methylation are included in experiments as controls. DNA extracted from colorectal cancer cell lines SW1116 and DLD1 is also used to validate this strategy. All experiments were performed minimally in duplicate to avoid ambiguity. For each promoter region, three CpG sites were chosen to analyze their methylation status. The presence of Cy5 signals indicates efficient amplification during multiplexing PCR steps. The promoter regions will be considered to be hypermethylated only when at least two CpG sites can be detected by Cy3 labeling. In most cases, the universal arrays provide very high capture specificity. The methylated promoters identified in this method may be reconfirmed by either bisulfite sequencing or uniplex PCR/PCR/LDR under more stringent hybridization conditions on a fresh array in a separate experiment.

In contrast to MSP-based methods, the bisulfite/PCR/LDR approach circumvents the issues of incomplete bisulfite conversion (C to U modification is not 100% efficient) and the potential primer extension of unmethylated DNA by extension of a G:U mismatch. The requirement of scoring methylation at three CpG sites per promoter using LDR should help the assay retain its exquisite specificity.

4. Notes

1. We recommend the software program Oligo for LDR primer design. This program is also useful in designing PCR and multiplex PCR primers. This program calculates T_m using the nearest neighbor method. Gene-specific PCR and LDR primers are generally designed with T_ms around 70°C. To perform multiplex PCR/PCR, gene-specific oligonucleotide primers with universal primer sequence attached to the 5' ends are required. The sequence of the universal primer is 5'-ggagcacgctatc-ccgttagac-3'. LDR discriminating primers are labeled at the 5' end with fluorophores that can be detected by the array-scanning instrument and that have sufficient spectral separation to avoid confounding owing to overlapping signals. The final base on the discriminating primer is the query base. The LDR common primer is modified with a 5' phosphate and the 24-base zip-code complement is appended to the 3' end (*see* **Table 1** for sequences). When synthesized on a 1-μmol scale or larger, zip-code oligonucleotides used for spotting arrays should be gel purified. When synthesized on a smaller scale, a reversed-phase, solid-phase extraction column produces satisfactory results. The zip-code oligonucleotides are synthesized on a 3' amino modifier C3 column (Glen Research) with a spacer C18 (Glen Research) inserted before the first base.

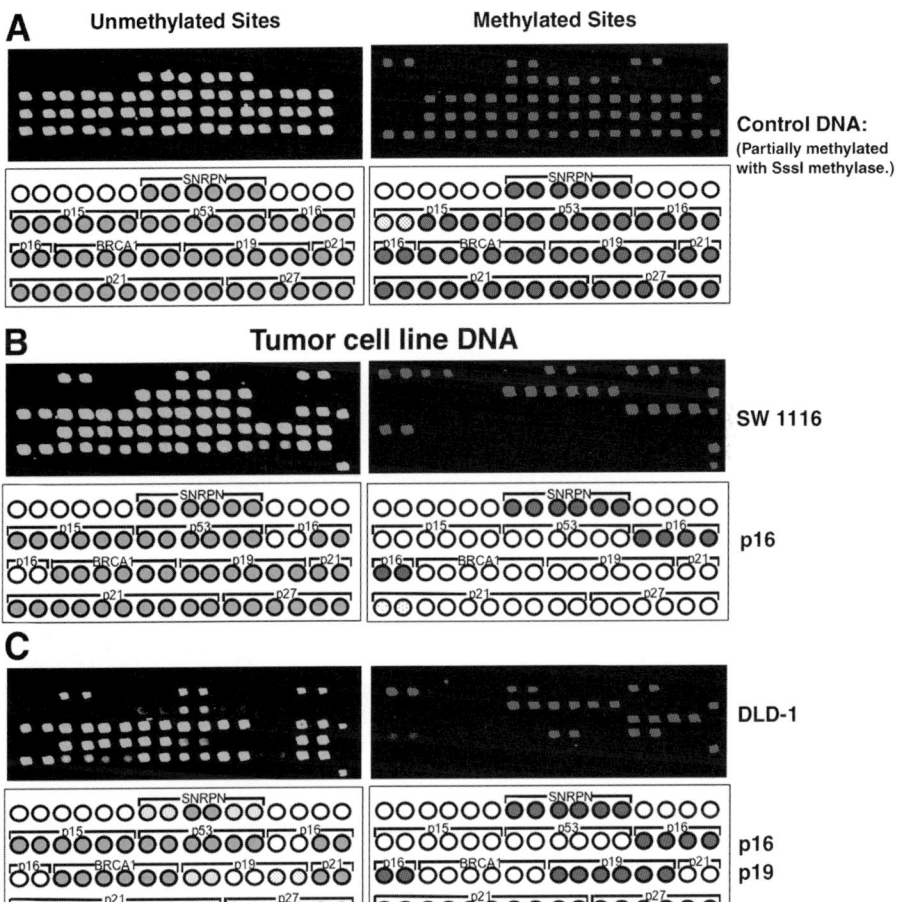

Fig. 8. (Color Plate 15 following p. 178) Universal array images of methylation profiles of selected promoter regions (SNRPN, p15, p16, p19, p21, p27, p53, and BRCA1) in normal and colorectal tumor cell line genomic DNAs. False color green represents the status of unmethylated promoter regions detected by Cy5-labeled LDR primers. False color red represents the status of methylated promoter regions detected by Cy3-labeled LDR primers. **(A)** LDR results of normal human lymphocyte genomic DNAs in the presence (right panel) and absence (left panel) of in vitro methylation using SssI methylase. **(B,C)** The methylation profiles of two colorectal cancer cell line genomic DNAs were analyzed. Among the eight genes that were analyzed in cell line SW1116, Cy3 labeled LDR products only present on the p16 promoter region. This indicates that only the p16 promoter was hypermethylated. The presence of Cy3 signal on both p16 and p19 promoters in cell line DLD-1 indicates that both of these promoters are hypermethylated.

Table 1
Sequences of Zip-Code-Related Oligonucleotides

Zip-code	Zip-code complement
TTGAAATCCAGCGCAAAATCTGCG	CGCAGATTTTGCGCTGGATTTCAA
TTGAAAAGCCTACACGACGGCGAA	TTCGCCGTCGTGTAGGCTTTTCAA
TTGATCTGCCATACGGGCTTACGG	CCGTAAGCCCGTATGGCAGATCAA
TTGACTTGTCCCCAGCACGGCCAT	ATGGCCGTGCTGGGGACAAGTCAA
TTGACGTTGACCAGCCCGTTGCAA	TTGCAACGGGCTGGTCAACGTCAA
TTGACGAAGCTTTCCCCCATGATG	CATCATGGGGGAAAGCTTCGTCAA
TTGAGCAAGGACGACCGCAAACGG	CCGTTTGCGGTCGTCCTTGCTCAA
TTGAGATGACGGACGGTGCGGCAA	TTGCCGCACCGTCCGTCATCTCAA
TTGATCCCATCGAAAGGGACGATG	CATCGTCCCTTTCGATGGGATCAA
TTGATGCGTCTGGGACGTGCCTTG	CAAGGCACGTCCCAGACGCATCAA
TTGACACGTCGTCAGCTCCCGTGC	GCACGGGAGCTGACGACGTGTCAA
TTGACAGCCTGTTGCGGTGCGTCT	AGACGCACCGCAACAGGCTGTCAA
TTGAGTGCGGTACTTGCAGCGATG	CATCGCTGCAAGTACCGCACTCAA
TTGAACGGTCTGCACGTCCCAGCC	GGCTGGGACGTGCAGACCGTTCAA
TGATTCTGGTGCGTGCCAGCCAGC	GCTGGCTGGCACGCACCAGAATCA
TGATTGTCGCTTTCTGACGGAGCC	GGCTCCGTCAGAAAGCGACAATCA
TGATCGTTTGCGGGTATCCCTCGT	ACGAGGGATACCCGCAAACGATCA
TGATCGAAAGGACAGCAGCCTCCC	GGGAGGCTGCTGTCCTTTCGATCA
TGATGCAAGCAACGAACACGCTGT	ACAGCGTGTTCGTTGCTTGCATCA
TGATTGCGAGTGGACCATCGCCAT	ATGGCGATGGTCCACTCGCAATCA
TGATCACGCTTGCCATGGACGGAC	GTCCGTCCATGGCAAGCGTGATCA
TGATGTGCCTCAACGGGTGCAGCC	GGCTGCACCCGTTGAGGCACATCA
TGATGGACCGTTAGCCGATGTTGA	TCAACATCGGCTAACGGTCCATCA
TGATACGGAGGAGGACTGCGTGCG	CGCACGCAGTCCTCCTCCGTATCA
TTAGGATGAGCCAGCCTGCGAGCC	GGCTCGCAGGCTGGCTCATCCTAA
AATCTCGTCGTTTCCCCTCATGCG	CGCATGAGGGGAAACGACGAGATT
AATCGCAACTGTCGTTCACGGTGC	GCACCGTGAACGACAGTTGCGATT
AATCAGGACACGCAGCGACCTGCG	CGCAGGTCGCTGCGTGTCCTGATT
AATCGACCCTGTGTCTGCTTTGCG	CGCAAAGCAGACACAGGGTCGATT
AATCAGCCAAAGCGAAGTGCGATG	CATCGCACTTCGCTTTGGCTGATT
ATACGACCTCGTGAGTTCCCGCAA	TTGCGGGAACTCACGAGGTCGTAT
AAAGCTTGACCTATCGAGCCGTGC	GCACGGCTCGATAGGTCAAGCTTT
AAAGAGCCGCTTGAGTCGAAATCG	CGATTTCGACTCAAGCGGCTCTTT
TCTGCTTGCTCACCTACCATTGCG	CGCAATGGTAGGTGAGCAAGCAGA
TCTGATCGCCTAGGTAACGGGGAC	GTCCCCGTTACCTAGGCGATCAGA
TCTGCAGCGGTACTGTGGACCCAT	ATGGGTCCACAGTACCGCTGCAGA
TCTGAGCCACCTAATCTCCCACGG	CCGTGGGAGATTAGGTGGCTCAGA
TGTCTCGTTCCCACCTCCATTCCC	GGGAATGGAGGTGGGAACGAGACA
TGTCATCGCAGCGAGTCAGCCACG	CGTGGCTGACTCGCTGCGATGACA

(Continued)

Table 1
(Continued)

Zip-code	Zip-code complement
TGTCCCTAACCTGATGGTGCGCAA	TTGCGCACCATCAGGTTAGGGACA
TGTCTGCGGTCTCCATATCGGTGC	GCACCGATATGGAGACCGCAGACA
TGTCCACGCGTTACCTTGTCGATG	CATCGACAAGGTAACGCGTGGACA
TGTCGTGCTCTGACCTTGCGCTCA	TGAGCGCAAGGTCAGAGCACGACA
TGTCACGGAATCGTGCTGCGGCTT	AAGCCGCAGCACGATTCCGTGACA
TCGTTCTGGCTTGGACGCTTCTCA	TGAGAAGCGTCCAAGCCAGAACGA
TCGTTGCGTGTCGGACCTTGGATG	CATCCAAGGTCCGACACGCAACGA
CTTGCGTTGATGCGAATCGTCGAA	TTCGACGATTCGCATCAACGCAAG
CTGTCACGCTCAACCTTCCCCGTT	AACGGGGAAGGTTGAGCGTGACAG
CTGTGTGCCGTTTCGTGTGCAGTG	CACTGCACACGAAACGGCACACAG
CCATGCAATCCCAGGATGTCGGTA	TACCGACATCCTGGGATTGCATGG
CCATCAGCTCTGGCAATGCGGAGT	ACTCCGCATTGCCAGAGCTGATGG
CGAATCTGGGTAAGGAAGCCATCG	CGATGGCTTCCTTACCCAGATTCG
CGAATGTCCTGTCCATCGAATGCG	CGCATTCGATGGACAGGACATTCG
CGAACGTTTACATGCGTCGTAGCC	GGCTACGACGCATGTAAACGTTCG
CGAAAGTGAGCCGCAACTTGGGAC	GTCCCAAGTTGCGGCTCACTTTCG
CGAAGGACAGTGAGTGTGCGCACG	CGTGCGCACACTCACTGTCCTTCG
GCAAAGCCATACCTTGGCTTGCTT	AAGCAAGCCAAGGTATGGCTTTGC
ACCTCTTGCCTACGAACAGCCGAA	TTCGGCTGTTCGTAGGCAAGAGGT
ACCTGCAAGTGCCCATGTGCCCTA	TAGGGCACATGGGCACTTGCAGGT
AGGATCTGGACCGGACTCCCCGAA	TTCGGGGAGTCCGGTCCAGATCCT
AGGAACGGCAGCTACACACGAGCC	GGCTCGTGTGTAGCTGCCGTTCCT
GATGGACCACCTCAGCGCTTGACC	GGTCAAGCGCTGAGGTGGTCCATC
TCCCTTAGGACCCAGCGTCTGTGC	GCACAGACGCTGGGTCCTAAGGGA
TGCGCCATAAAGGACCTTAGCCAT	ATGGCTAAGGTCCTTTATGGCGCA
TGCGAGGACCATGGTAGGTAAGCC	GGCTTACCTACCATGGTCCTCGCA

2. Use only sealed boxes of slides. Opened boxes do not clean well, probably because of to either heavy oxidation or reaction with chemicals present in the lab.
3. Each dish holds two glass slide racks.
4. We use slides with two etched circles when making arrays. A single array is spotted within each circle.
5. This process can be repeated for a second set of slides with the existing solutions before the reagents boil out and too much liquid evaporates. Use fresh solutions for the next rounds.
6. Silanize as soon as possible or leave in water overnight (rinse #1, **step 7**) if you cannot silanize immediately.
7. Do not touch the surface of slides after cleaning; handle slides by frosted end only.

8. If you are preparing a large number of slides, the second wash of one batch can be used as the first wash for the next batch of slides.

9. The silanized slides are stable for several weeks prior to coating with the polymer. For long-term storage, place in a desiccator.

10. Allowing the solution to sit produces more uniform surfaces.

11. If slides were cleaned and silanized properly, the solution should form a relatively small, well-defined droplet on the surface. Do not worry about small voids or bubbles at the edges of the cover slip; you will not be using that area of the polymer.

12. Using two to four blocks, you can heat 8–16 slides simultaneously if you stagger them by 20 s. The two outer slides on the blocks will hang off the edges slightly, but this does not effect the polymerization.

13. The slides are stable for at least 6 mo prior to activation.

14. The activated slides are stable for a minimum of 6 mo if stored properly desiccated.

15. Plates can be sealed and stored refrigerated for spotting, on consecutive days. Prior to spotting, spin the plates to collect all the liquid in the bottom of the wells. For long-term storage, dry the spotting plate, seal, and store at −20°C. The day before spotting, redissolve the wells in the appropriate amount of water (assume ~0.5 μL loss in volume/spotting run), vortex the plate several times, and store refrigerated. The day of spotting, vortex several times, and then spin the plate to collect all the liquid in the bottom of the wells.

16. The slides are very hydroscopic when dry. If not allowed to partially rehydrate prior to spotting, the first few slides will suck all the spotting solution out of the pin.

17. We normally spot each zip-code in duplicate. Additionally, we spot a set of fiducials along the top and one side of the array (*see* bottom panel of **Fig. 2**) for alignment purposes. The fiducials are made on a DNA synthesizer and have the following structure: amino-group-T-T-fluorescent dye. The fiducials should be printed after the zip-code oligonucleotides because the dyes are sticky, and carryover contamination can be a problem. The pins should be sonicated thoroughly after spotting of the fiducials to prevent contaminating the next spotting run with fluorescent dye.

18. This can be done immediately following spotting or any time just prior to hybridization.

19. The spotted slides are stable for a minimum of 6 mo.

20. The LDR can remain in the original reaction tube/plate, and the 2X buffer can be added through the oil. Following denaturation, the tubes/plate should not need to be spun down, and the hybridization mix can be drawn out from under the oil.

21. There are cover wells available in a variety of sizes for different size arrays. The important thing is not to fill the chamber completely so there is an air bubble present to facilitate mixing.

22. Avoid excessive liquid in the sponge, since this may create unnecessary moisture and result in a capillary stream between the adjacent hybridization chambers under cover well gaskets that have not been securely sealed against the array surface. This cross-contamination issue can easily be solved by just slightly dampening the sponge, that is, make the sponge damp but without the ability to squeeze out any liquid.

Acknowledgments

Support for this work was provided by the National Cancer Institute (grouts P01-CA65930 and RO1-CA81467).

References

1. Kirk, B. W., Feinsod, M., Favis, R., Kliman, R. M., and Barany, F. (2002) Single nucleotide polymorphism seeking long term association with complex disease. *Nucleic Acids Res.* **30**, 3295–3311.

2. Barany, F. and Lubin, M. (1997) Detection of nucleic acid sequence differences using coupled ligase detection and polymerase chain reactions. International Patent Application (WO9745559A1).

3. Gerry, N. P., Witowski, N. E., Day, J., Hammer, R. P., Barany, G., and Barany, F. (1999) Universal DNA microarray method for multiplex detection of low abundance point mutations. *J. Mol. Biol.* **292**, 251–262.

4. Khanna, M., Park, P., Zirvi, M., et al. (1999) Multiplex PCR/LDR for detection of K-ras mutations in primary colon tumors. *Oncogene* **18**, 27–38.

5. Khanna, M., Cao, W., Zirvi, M., Paty, P., and Barany, F. (1999) Ligase detection reaction for identification of low abundance mutations. *Clin. Biochem.* **32**, 287–290.

6. Zirvi, M., Nakayama, T., Newman, G., McCaffrey, T., Paty, P., and Barany, F. (1999) Ligase-based detection of mononucleotide repeat sequences. *Nucleic Acids Res.* **27**, e40.

7. Zirvi, M., Bergstrom, D. E., Saurage, A. S., Hammer, R. P. and Barany, F. (1999) Improved fidelity of thermostable ligases for detection of microsatellite repeat sequences using nucleoside analogs. *Nucleic Acids Res.* **27**, e41.

8. Favis, R., Day, J. P., Gerry, N. P., Phelan, C., Narod, S., and Barany, F. (2000) Universal DNA array detection of small insertions and deletions in BRCA1 and BRCA2. *Nat. Biotechnol.* **18**, 561–564.

9. Favis, R. and Barany, F. (2000) Mutation detection in K-ras, BRCA1, BRCA2, and p53 using PCR/LDR and a universal DNA microarray. *Ann. N. Y. Acad. Sci.* **906**, 39–43.

10. Dong, S. M., Traverso, G., and Johnson, C. (2001) Detecting colorectal cancer in stool with the use of multiple genetic targets. *J. Natl. Cancer Inst.* **93**, 858–865.

11. Day, J. P. (2003) The 53rd American Society of Human Genetics Annual Meeting, Poster #1589.

12. Favis, R., Huang, J., Gerry, N., et al. (2003) Harmonized microarray/mutation scanning analysis of TP53 mutations in undissected colorectal tumors. *Hum. Mutat.* **24**, 63–75.

13. Fouquet, C., Antoine, M., Tisserand, P., et al. (2003) Rapid and sensitive p53 alteration analysis in biopsies from lung cancer patients using a functional assay and a universal oligonucleotide array: a prospective study. *Clin. Cancer Res.* **10**, 3479–3489.

14. Overholtzer, M., Rao, P. H., Favis, R., et al. (2003) The presence of p53 mutations in human osteosarcomas correlates with high levels of genomic instability. *Proc. Natl. Acad. Sci. USA* **100**, 11547–11552.

15. Barany, F. and Gelfand, D. H. (1991) Cloning, overexpression and nucleotide sequence of a thermostable DNA ligase-encoding gene. *Gene* **109,** 1–11.

16. Barany, F. (1991) The ligase chain reaction in a PCR world. *PCR Methods Appl.* **1,** 5–16.

17. Day, D. J., Speiser, P. W., White, P. C., and Barany, F. (1995) Detection of steroid 21-hydroxylase alleles using gene-specific PCR and a multiplexed ligation detection reaction. *Genomics* **29,** 152–162.

18. Belgrader, P., Devaney, J. M., Del Rio, S. A., Turner, K. A., Weaver, K. R., and Marino, M. A. (1996) Automated polymerase chain reaction product sample preparation for capillary electrophoresis analysis. *J. Chromatogr. B. Biomed. Appl.* **683,** 109–114.

19. Baner, J., Isaksson, A., Waldenstrom, E., Jarvius, J., Landegren, U., and Nilsson, M. (2003) Parallel gene analysis with allele-specific padlock probes and tag microarrays. *Nucleic Acids Res.* **31,** e103.

20. Consolandi, C., Busti, E., Pera, C., et al. (2003) Detection of HLA polymorphisms by ligase detection reaction and a universal array format: a pilot study for low resolution genotyping. *Hum. Immunol.* **64,** 168–178.

21. Hardenbol, P., Baner, J., Jain, M., et al. (2003) Multiplexed genotyping with sequence-tagged molecular inversion probes. *Nat. Biotechnol.* **21,** 673–678.

22. Iannone, M. A., Taylor, J. D., Chen, J., et al. (2000) Multiplexed single nucleotide polymorphism genotyping by oligonucleotide ligation and flow cytometry. *Cytometry.* **39,** 131–140.

23. Oliphant, A., Barker, D. L., Stuelpnagel, J. R., and Chee, M. S. (2002) BeadArray technology: enabling an accurate, cost-effective approach to high-throughput genotyping. *Biotechniques* **Suppl.,** 56–80, 60–61.

24. Han, M., Gao, X., Su, J. Z., and Nie, S. (2001) Quantum-dot-tagged microbeads for multiplexed optical coding of biomolecules. *Nat. Biotechnol.* **19,** 631–635.

25. Busti, E., Bordoni, R., Castiglioni, B., et al. (2002) Bacterial discrimination by means of a universal array approach mediated by LDR (ligase detection reaction). *BMC Microbiol.* **2,** 27.

26. Chen, J., Iannone, M. A., Li, M. S. T., et al. (2000) A microsphere-based assay for multiplexed single nucleotide polymorphism analysis using single base chain extension. *Genome Res.* **10,** 549–557.

27. Epstein, J. R., Ferguson, J. A., Lee, K. H., and Walt, D. R. (2003) Combinatorial decoding: an approach for universal DNA array fabrication. *J. Am. Chem. Soc.* **125,** 13753–13759.

28. Hirschhorn, J. N., Sklar, P., Lindblad-Toh, K., et al. (2000) SBE-TAGS: an array-based method for efficient single-nucleotide polymorphism genotyping. *Proc. Natl. Acad. Sci. USA* **97,** 12164–12169.

29. Jarvius, J., Nilsson, M., and Landegren, U. (2003) Oligonucleotide ligation assay. *Methods Mol. Biol.* **212,** 215–228.

30. Ladner, D. P., Leamon, J. H., Hamann, S., et al. (2001) Multiplex detection of hotspot mutations by rolling circle-enabled universal microarrays. *Lab. Invest.* **81,** 1079–1086.

31. Mikhailovich, V., Lapa, S., Gryadunov, D., et al. (2001) Identification of rifampin-resistant *Mycobacterium tuberculosis* strains by hybridization, PCR, and ligase detection reaction on oligonucleotide microchips. *J. Clin. Microbiol.* **39,** 2531–2540.

32. Taylor, J. D., Briley, D., Nguyen, Q., et al. (2001) Flow cytometric platform for high-throughput single nucleotide polymorphism analysis. *Biotechniques* **30,** 661–666, 68–169.

33. Zhong, X. B., Lizardi, P. M., Huang, X. H., Bray-Ward, P. L., and Ward, D. C. (2001) Visualization of oligonucleotide probes and point mutations in interphase nuclei and DNA fibers using rolling circle DNA amplification. *Proc. Natl. Acad. Sci. USA* **98,** 3940–3945.

34. Zhong, X. B., Reynolds, R., Kidd, J. R., et. al. (2003) Single-nucleotide polymorphism genotyping on optical thin-film biosensor chips. *Proc. Natl. Acad. Sci. USA* **100,** 11559–11564.

35. Pastinen, T., Raitio, M., Lindroos, K., Tainola, P., Peltonen, L., and Syvanen, A. C. (2000) A system for specific, high-throughput genotyping by allele-specific primer extension on microarrays. *Genome Res.* **10,** 1031–1042.

36. Cronin, M. T., Fucini, R. V., Kim, S. M., Masino, R. S., Wespi, R. M., and Miyada, C. G. (1996) Cystic fibrosis mutation detection by hybridization to light-generated DNA probe arrays. *Hum. Mutat.* **7,** 244–255.

37. Hacia, J. G., Brody, L. C., Chee, M. S., Fodor, S. P., and Collins, F. S. (1996) Detection of heterozygous mutations in BRCA1 using high density oligonucleotide arrays and two-colour fluorescence analysis [see comments]. *Nat. Genet.* **14,** 441–447.

38. Southern, E. M. (1996) DNA chips: analysing sequence by hybridization to oligonucleotides on a large scale. *Trends Genet.* **12,** 110–115.

39. Wang, Y., Blandino, G., Oren, M., and Givol, D. (1998) Induced p53 expression in lung cancer cell line promotes cell senescence and differentially modifies the cytotoxicity of anti-cancer drugs. *Oncogene* **17,** 1923–1930.

40. Grossman, P. D., Bloch, W., Brinson, E., et al. (1994) High-density multiplex detection of nucleic acid sequences: oligonucleotide ligation assay and sequence-coded separation. *Nucleic Acids Res.* **22,** 4527–14534.

41. Eggerding, F. A., Iovannisci, D. M., Brinson, E., Grossman, P., and Winn-Deen, E. S. (1995) Fluorescence-based oligonucleotide ligation assay for analysis of cystic fibrosis transmembrane conductance regulator gene mutations. *Hum. Mutat.* **5,** 153–165.

42. Feero, W. T., Wang, J., Barany, F., et al. (1993) Hyperkalemic periodic paralysis: rapid molecular diagnosis and relationship of genotype to phenotype in 12 families. *Neurology* **43,** 668–673.

43. Day, D. J., Speiser, P. W., Schulze, E., et al. (1996) Identification of non-amplifying CYP21 genes when using PCR-based diagnosis of 21-hydroxylase deficiency in congenital adrenal hyperplasia (CAH) affected pedigrees. *Hum. Mol. Genet.* **5,** 2039–2048.

44. Lin, Z., Cui, X. and Li, H. (1996) Multiplex genotype determination at a large number of gene loci. Proc. *Natl. Acad. Sci. USA* **93,** 2582–2587.

45. Nickerson, D. A., Whitehurst, C., Boysen, C., Charmley, P., Kaiser, R., and Hood, L. (1992) Identification of clusters of biallelic polymorphic sequence-tagged sites (pSTSs) that generate highly informative and automatable markers for genetic linkage mapping. *Genomics* **12**, 377–387.
46. Schouten, J. P., McElgunn, C. J., Waaijer, R., Zwijnenburg, D., Diepvens, F., and Pals, G. (2002) Relative quantification of 40 nucleic acid sequences by multiplex ligation-dependent probe amplification. *Nucleic Acids Res.* **30**, e57.
47. Gille, J. J., Hogervorst, F. B., Pals, G., et al. (2002) Genomic deletions of MSH2 and MLH1 in colorectal cancer families detected by a novel mutation detection approach. *Br. J. Cancer* **87**, 892–897.
48. White, S., Kalf, M., Liu, Q., et al. (2002) Comprehensive detection of genomic duplications and deletions in the DMD gene, by use of multiplex amplifiable probe hybridization. *Am. J. Hum. Genet.* **71**, 365–374.
49. Ahrendt, S., Halachmi, S., Chow, J., et al. (1999) Rapid p53 sequence analysis in primary lung cancer using an oligonucleotide probe array. *Proc. Natl. Acad. Sci. USA* **96**, 7382–7387.
50. Wikman, F. P., Lu, M. L., Thykjaer, T., et al. (2000) Evaluation of the performance of a p53 sequencing microarray chip using 140 previously sequenced bladder tumor samples. *Clin. Chem.* **46**, 1555–1561.
51. Wen, W. H., Bernstein, L., Lescallett, J., et al. (2000) Comparison of TP53 mutations identified by oligonucleotide microarray and conventional DNA sequence analysis. *Cancer Res.* **60**, 2716–2722.
52. Ingelman-Sundberg, M., Daly, A. K., and Nebert, D. W. (2003). Feb. 18, 2003 Ed. Karolinska Institute, Vol. 2003.
53. Lampe, J. W., Bigler, J., Horner, N. K., and Potter, J. D. (1999) UDP-glucuronosyltransferase (UGT1A1*28 and UGT1A6*2) polymorphisms in Caucasians and Asians: relationships to serum bilirubin concentrations. *Pharmacogenetics* **9**, 341–349.
54. Wabuyele, M. B., Farquar, H., Stryjewski, W., et al. (2003) Approaching real-time molecular diagnostics: single-pair fluorescence resonance energy transfer (spFRET) detection for the analysis of low abundant point mutations in K-ras oncogenes. *J. Am. Chem. Soc.* **125**, 6937–6945.
55. Belgrader, P., Okuzumi, M., Pourahmadi, F., Borkholder, D. A., and Northrup, M. A. (2000) A microfluidic cartridge to prepare spores for PCR analysis. *Biosens. Bioelectron* **14**, 849–852.
56. Soper, S. A., Ford, S. M., Xu, Y., et al. (1999) Nanoliter-scale sample preparation methods directly coupled to polymethylmethacrylate-based microchips and gel-filled capillaries for the analysis of oligonucleotides. *J. Chromatogr. A* **853**, 107–120.
57. Taylor, M. T., Belgrader, P., Furman, B. J., Pourahmadi, F., Kovacs, G. T., and Northrup, M. A. (2001) Lysing bacterial spores by sonication through a flexible interface in a microfluidic system. *Anal. Chem.* **73**, 492–496.
58. Belgrader, P., Benett, W., Hadley, D., et al. (1998) Rapid pathogen detection using a microchip PCR array instrument. *Clin. Chem.* **44**, 2191–2194.

59. Broyles, B. S., Jacobson, S. C., and Ramsey, J. M. (2003) Sample filtration, concentration, and separation integrated on microfluidic devices. *Anal. Chem.* **75,** 2761–2777.
60. Belgrader, P., Elkin, C. J., Brown, S. B., et al. (2003) A reusable flow-through polymerase chain reaction instrument for the continuous monitoring of infectious biological agents. *Anal. Chem.* **75,** 3114–3118.
61. Groisman, A., Enzelberger, M., and Quake, S. R. (2003) Microfluidic memory and control devices. *Science* **300,** 955–958.
62. Handique, K., Burke, D. T., Mastrangelo, C. H., and Burns, M. A. (2000) Nanoliter liquid metering in microchannels using hydrophobic patterns. *Anal. Chem.* **72,** 4100–4109.
63. Liu, J., Hansen, C., and Quake, S. R. (2003) Solving the "world-to-chip" interface problem with a microfluidic matrix. *Anal. Chem.* **75,** 4718–4723.
64. Belgrader, P., Benett, W., Hadley, D., Richards, J., Stratton, P., Mariella, R. Jr., and Milanovich, F. (1999) PCR detection of bacteria in seven minutes. *Science* **284,** 449–450.
65. Lagally, E. T., Medintz, I., and Mathies, R. A. (2001) Single-molecule DNA amplification and analysis in an integrated microfluidic device. *Anal. Chem.* **73,** 565–570.
66. Khandurina, J., McKnight, T. E., Jacobson, S. C., Waters, L. C., Foote, R. S., and Ramsey, J. M. (2000) Integrated system for rapid PCR-based DNA analysis in microfluidic devices. *Anal. Chem.* **72,** 2995–3000.
67. Wang, Y., Vaidya, B., Farquar, H. D., et al. (2003) Microarrays assembled in microfluidic chips fabricated from poly(methyl methacrylate) for the detection of low-abundant DNA mutations. *Anal. Chem.* **75,** 1130–1140.
68. Burns, M. A., Johnson, B. N., Brahmasandra, S. N., et al. (1998) An integrated nanoliter DNA analysis device. *Science* **282,** 484–487.
69. Culbertson, C. T., Jacobson, S. C. and Ramsey, J. M. (2000) Microchip devices for high-efficiency separations. *Anal. Chem.* **72,** 5814–5819.
70. Emrich, C. A., Tian, H., Medintz, I. L., and Mathies, R. A. (2002) Microfabricated 384-lane capillary array electrophoresis bioanalyzer for ultrahigh-throughput genetic analysis. *Anal. Chem.* **74,** 5076–5083.
71. Haab, B. B. and Mathies, R. A. (1999) Single-molecule detection of DNA separations in microfabricated capillary electrophoresis chips employing focused molecular streams. *Anal. Chem.* **71,** 5137–5145.
72. Hansen, C. and Quake, S. R. (2003) Microfluidics in structural biology: smaller, faster em leader better. *Curr. Opin. Struct. Biol.* **13,** 538–544.
73. Henry, A. C., Tutt, T. J., Galloway, M., et al. (2000) Surface modification of poly(methyl methacrylate) used in the fabrication of microanalytical devices. *Anal. Chem.* **72,** 5331–5337.
74. Hearps, A., Zhang, Z., and Alexandersen, S. (2002) Evaluation of the portable Cepheid SmartCycler real-time PCR machine for the rapid diagnosis of foot-and-mouth disease. *Vet. Rec.* **150,** 625–628.
75. Hong, J. W. and Quake, S. R. (2003) Integrated nanoliter systems. *Nat. Biotechnol.* **21,** 1179–1183.

76. McWhorter, S. and Soper, S. A. (2000) Near-infrared laser-induced fluorescence detection in capillary electrophoresis. *Electrophoresis* **21,** 1267–1280.

77. McWhorter, S. and Soper, S. A. (2000) Conductivity detection of polymerase chain reaction products separated by micro-reversed phase liquid chromatography. *J. Chromatogr. A* **883,** 1–9.

78. Scherer, J. R., Kheterpal, I., Radhakrishnan, A., Ja, W. W., and Mathies, R. A. (1999) Ultra-high throughput rotary capillary array electrophoresis scanner for fluorescent DNA sequencing and analysis. *Electrophoresis* **20,** 1508–1517.

79. Thomas, G. A., Williams, D. L., and Soper, S. A. (2001) Capillary electrophoresis-based heteroduplex analysis with a universal heteroduplex generator for detection of point mutations associated with rifampin resistance in tuberculosis. *Clin. Chem.* **47,** 1195–203.

80. Wabuyele, M. B., Ford, S. M., Stryjewski, W., Barrow, J., and Soper, S. A. (2001) Single molecule detection of double-stranded DNA in poly(methylmethacrylate) and polycarbonate microfluidic devices. *Electrophoresis* **22,** 3939–3948.

81. Battaglia, C., Salani, G., Consolandi, C., Rossi Bernardi, L., and De Bellis, G. (2000) Analysis of DNA microarrays by non-destructive fluorescent staining using SYBR® Green II. *Biotechniques* **29,** 78–81.

3

Sensitive Detection of SARS Coronavirus RNA by a Novel Asymmetric Multiplex Nested RT-PCR Amplification Coupled With Oligonucleotide Microarray Hybridization

Zhi-wei Zhang, Yi-ming Zhou, Yan Zhang, Yong Guo, Sheng-ce Tao, Ze Li, Qiong Zhang, and Jing Cheng

Summary

We have developed a sensitive method for the detection of specific genes simultaneously. First, DNA was amplified by a novel asymmetric multiplex PCR with universal primer(s). Second, the 6-carboxytetramethylrhodamine (TAMRA)-labeled PCR products were hybridized specifically with oligonucleotide microarrays. Finally, matched duplexes were detected by using a laser-induced fluorescence scanner. The usefulness of this method was illustrated by analyzing severe acute respiratory syndrome (SARS) coronavirus RNA. The detection limit was 10^0 copies/μL. The results of the asymmetric multiplex nested reverse transcription-PCR were in agreement with the results of the microarray hybridization; no hybridization signal was lost as happened with applicons from symmetric amplifications. This reliable method can be used to the identification of other microorganisms, screening of genetic diseases, and other applications.

Key Words: Polymerase chain reaction (PCR); multiplex PCR; asymmetric PCR; universal primer; severe acute respiratory syndrome (SARS) coronavirus; microarray.

1. Introduction

Multiplex Polymerase chain reaction (PCR) was designed to amplify multiple target sequences using more than one pair of primers in the reaction. It has the potential to save a considerable amount of time and effort without compromising test utility and additional instruments.

Since its first report in 1988 *(1)*, multiplex PCR has become a rapid and convenient screening procedure in both clinical and research laboratories. It has been successfully applied to gene deletion analysis *(2)*, mutation and polymorphism analysis *(3)*, mRNA quantitative analysis *(4)*, RNA detection *(5,6)* and

From: *Methods in Molecular Medicine, Vol. 144, Microarrays in Clinical Diagnostics*
Edited by: T. Joos and P. Fortina © Humana Press Inc., Totowa, NJ

genome sequencing *(7)*. For infectious disease diagnosis, multiplex PCR has been a valuable tool for the identification of viruses *(8,9)*, bacteria *(10,11)*, parasites *(12)*, and bacterial drug-resistance genes *(13,14)*.

The development of an efficient multiplex PCR protocol usually requires careful design of primers and many rounds of optimization. The common problems encountered in multiplex PCR are spurious amplification products, uneven or no amplification of some target sequences, and difficulties in reproducing the results. A successful multiplex PCR assay needs the following parameters to be set properly: relative concentration of the primers, concentration of the PCR buffer, balance between the magnesium chloride and dNTP concentrations, cycling temperatures, and amount of template and DNA polymerase. An optimal combination of annealing temperature and buffer condition is essential to ensure high specificity of multiplex PCR. Magnesium chloride concentration needs to be proportional to the amount of dNTPs. Adjusting primer concentration for each target sequence is also essential *(15,16)*. Henegariu et al. *(17)* presented a step-by-step protocol for multiplex PCR, after study, of some of these factors.

Preferential amplification of one target sequence over another (bias in template-to-product ratios) is a known phenomenon in multiplex PCR; it mainly occurs because multiplex PCR has a limited supply of enzymes and nucleotides. All primer pairs compete for the same pool of supplies, but their amplification efficiencies are different. Amplification biases that were strongly dependent on the primers and, to a lesser degree, the templates, have been described *(17)*.

We have developed a new strategy for the optimization of multiplex PCR to overcome the problem of preferential amplification of one target sequence over another. Two universal sequences irrelevant to the targets were added to the 5′ termini of the specific primers. The extra universal primers, whose sequences were identical with the ones added into the specific primers, were used in the multiplex PCR reaction together (**Fig. 1A**). Ideally, the universal primers can reduce the amplification biases of multiplex PCR, so the optimization of multiplex PCR becomes much easier than before *(18)*.

Asymmetric PCR is often used to generate single-stranded DNA (ssDNA). The method is especially useful for hybridizing PCR product against probes such as the ones used in microarray hybridizations. Oligonucleotide microarray has provided a powerful platform for nucleic acid analysis *(19–22)*. Hybridization of labeled nucleic acid targets with microarrays of surface-immobilized oligonucleotide probes was the central event in the detection of nucleic acids on microarrays *(23)*. Before hybridization, the ssDNA targets were prepared by using denaturation of the PCR products or by other methods. Only one of the two DNA strands was available for hybridization with the immobilized probes; the other one competed with the probes for the target and therefore was regarded as the interfering strand *(24)*. What was worse was that

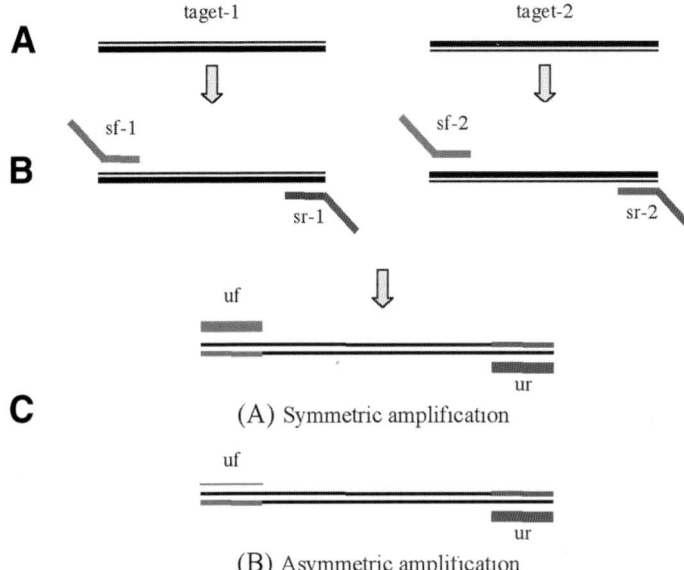

Fig. 1. Schematic representation of the multiplex PCR used in this work. (**a**), Targets for amplification. Only two targets are shown here. (**b**), Gene-specific primer amplifications, which are the same as ordinary PCR except that two different universal sequences irrelevant to the targets were added to the 5′ end of the specific primers. (**c**), Universal primer amplification. In this reaction, the amplicons of the gene-specific primers served as templates. sf, the forward gene-specific primer; sr, the reverse gene-specific primer; uf, the forward universal primer; ur, the reverse universal primer. All primers are added to one tube. (**A**) For symmetric amplification, uf and ur are used with equal molar amounts. (**B**) For asymmetric PCR, the uneven primers uf and ur were used, and in some cases uf was even absent from the reactions.

the annealing effect of the two complementary strands was dominant, because of their faster kinetics and higher thermodynamic stabilities. Kawai et al. *(25)* reported that the sensitivity with ssDNA targets was fivefold higher than that with boiled double-stranded DNA (dsDNA) targets, when they were hybridized with oligonucleotide probes. Thus preparation of ssDNA targets was preferred for high efficient hybridization on oligonucleotide microarrays.

In the conventional asymmetric PCR for ssDNA preparation, the two primers are present in different molar amounts. When the primer in the limited amount is exhausted, an excess of ssDNA will be produced in each cycle *(26)*. Erdogan et al. *(27)* have successfully applied the single-stranded targets produced by asymmetric PCR to a single-nucleotide polymorphism detection system. However, they considered this method to have the disadvantage that the products appeared as a serious smear of bands in agarose gel *(27)*. To improve the ssDNA production efficiency of this method, Kaltenboeck et al. *(28)* designed a two-step asymmetric PCR, in which a symmetric PCR for dsDNA was first preformed under optimal conditions, and then a single primer was used to generate ssDNA targets by using the purified double-stranded PCR products as template. Although it was time consuming, this two-step method was still applied to ssDNA target preparation *(28–30)*.

However, the conditions are extraordinarily involved in ordinary multiplex PCR, if the asymmetric PCR is performed in a parallel fashion in the same reaction. There are few references for asymmetric multiplex amplification. This issue may be simplified by modifying the universal primer-mediated multiplex PCR using disproportional universal primers **(Fig. 1B)**.

Severe acute respiratory syndrome (SARS) is a new infectious disease of humans, first recognized in late February 2003 in Hanoi, Vietnam. The disease spread rapidly, with cases reported from 29 countries on five continents over 4 months *(31–38)*. By July 3, 2003, this epidemic had resulted in 8439 reported cases globally, of which 812 were fatal (http://www.who.int/csr/sars/country/2003 _07_03). Rapid and sensitive laboratory confirmation of SARS coronavirus (CoV) infection was important for managing patient care and for preventing nosocomial transmission. Although serological testing was reliable as a retrospective diagnostic method, diagnosis of the infection in the early phase of the illness was important for patient care.

In this study, SARS CoV RNA was amplified by using multiplex nested reverse transcription-PCR (RT-PCR), followed by microarray hybridization. During the multiplex PCR, several different ratios of the universal primers were used, and the results were compared.

2. Materials

1. Oligonucleotide primers (Sangon, Shanghai, China).
2. Oligonucleotide probes (Sangon).
3. Oligonucleotide microarray (CapitalBio, Beijing, China).
4. SARS CoV RNA (provided by Professor Tao Hung at the Virological Institute, Chinese Center for Disease Control and Prevention).
5. Normal human total RNA (TW-Times, Beijing, China).
6. Roche LightCycler system (Roche, Mannheim, Germany).
7. RealArt HPA-CoV LC RT Reagents (Artus, Hamburg, Germany).
8. One-step RT-PCR kit (TaKaRa, Dalian, Liaoning, China).
9. PTC-225 thermal cycler (MJ Research, Miami, FL).
10. 2X Master mixture (TW-Times).
11. Deoxyuridine triphosphate (dUTP, Sangon).
12. Uracil-DNA glycosylase (Invitrogen, Carlsbad, CA).
13. Mineral oil (Sigma).
14. DL2000 DNA molecular weight marker (TaKaRa).

15. UVP system (Ultraviolet Products, Cambridge, UK).
16. AminoSlide™ slide (CapitalBio).
17. SARSarray™ slide (CapitalBio).
18. HybriCassettes™ (CapitalBio).
19. SmartCover™ (CapitalBio).
20. Dimethyl sulfoxide (DMSO).
21. GenePix 4000B (Axon, Union City, CA).
22. Hybridization solution: 5.6X standard saline citrate (SSC), 9.1X Denhart's solution, 0.36% sodium dodecyl sulfate (SDS), 50 nM of probe 127; make fresh as required.
23. Washing solution A: 2X SSC and 0.2% SDS.
24. Washing solution B: 0.2X SSC.

3. Methods

The methods described below outline (1) the construction of the microarray; (2) the quantification and dilution of SARS CoV RNA; (3) the multiplex nested RT-PCR amplification, symmetric and asymmetric; (4) the agarose gel electrophoresis; (5) the microarray hybridization; and (6) the fluorescence scanning.

3.1. Microarray

An overlapping of 70-nucleotide (nt) segment with the T_m value set at 88°C and with the least free energy in the hairpin structure and dimer was designed as the probe following the rule from http://www.westburg.nl/download/array-poster.pdf to minimize cross-hybridization. To increase the immobilization efficiency of the probes, 10 thymidines were added to the 5′ end of each probe.

After an initial screening test, four oligonucleotides were chosen as the probes for identifying SARS CoV (probes 11, 24, 40, and 44). Additionally, several probes were also included for control purposes. The quality control (QC) probe was used to confirm the efficiency of the attachment chemistry on the surface of the substrate. For all tests, this probe always generates a strong and consistent fluorescence signal. The internal control (IC) probe, which hybridizes to the amplicons of its inner primers, was designed to guarantee that the entire nested RT-PCR process operates as expected. The external control (EC) probe was used to monitor the efficiency of the hybridization process. The negative control (NC) probe was an oligonucleotide whose sequence was irrelevent to any amplicons of the multiplex nested RT-PCR. The blank control (BC) probe was DMSO spotted on the substrate, to ensure that no signal is detected on these spots, indicating no carryover of the previously spotted samples. The sequence information of the probes is listed in **Table 1**.

The probes were suspended in 50% DMSO at a concentration of 10 µM and printed on glass slides modified with amino groups (AminoSlide). Four subarrays

Table 1
List of Probes

Probe	Sequence origin	Sequence (5'-3')	Position [b, c]
Quality control			
QC	Treponema pallidum	T$_{10}$-TGTCTTCCACCAGGAGTCAGCAGAGTGCTTGGTGCCATAAC-HEX [a]	13,254–13,294
EC	Arabidopsis thaliana	T$_{10}$-AAAGTTAAAGCAGACCGAAGTGGATTGCGAGTATTTGAAAAGATGTGTTGAGAAATTAACGGAAGAGAA	561–629
127	Arabidopsis thaliana	TAMRA-TTCTCTTCCGTTAATTTCTCAACACATCTTTCAAATACTCGCAATCCACTTCGGTCGTCTTTAACTTT	561–629
IC	GAPD	T$_{10}$-CGTCAAGGCTGAGAACGGGAAGCTTGTCATCAATGGAAATCCCATCACCATCTTCCAGGAGCGAGATCCC	252–321
NC	GAPD	T$_{10}$-AGTGGTGGACCTGACCTGCCGTCTAGAAAAAACCTGCCAAATATGATGACATCAAGAAGGTGGTGAAGCAG	798–867
SARS detection			
11	ORF1a	T$_{10}$-CTACGTAGTGAAGCTTTCGAGTACTACCATACTCTTGATGAGAGTTTTCTTGGTAGGTACATGTCTGCTT	5074–5143
24	ORF1a	T$_{10}$-TCATAGCTAACATCTTTACTCCTCTTGTGCAACCTGTGGGTGCTTTAGATGTGTCTGCTTCAGTAGTGGC	9272–9341
40	Nucleocapsid protein	T$_{10}$-GAGGTGGTGAAACTGCCCTCGCGCTATTGCTGCTAGCACAGATTGAACCAGCTTGAGAGCAAAGTTTCTGG	28,760–28,829
44	Spike glycoprotein	T$_{10}$-CACCTGGAACAAATGCTTCATCTGAAGTTGCGTGTTCTATATCAAGATGTTAACTGCACTGATGTTTCTAC	23,245–23,314

[a] HEX, hexa-chloro-6-carboxyfluorescein.
[b] The position of the inner primers does not include the polyT in the 5' ends.
[c] The reference sequences for QC, EC,127, IC, and NC probes and SARS detection probes are AE001232, BT006112, NM_002046, NM_002046, and NC_004718, respectively (GenBank accession numbers).

64

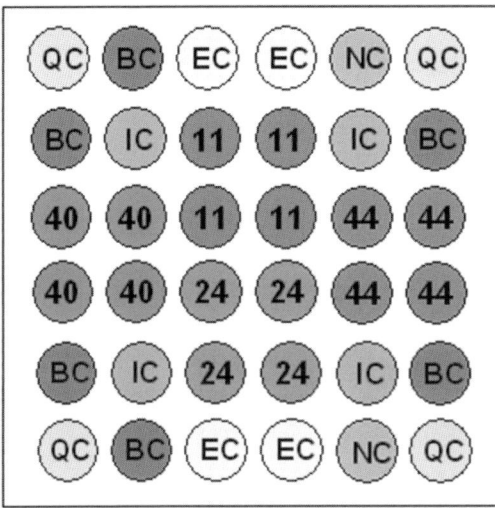

Fig. 2. The illustration of the microarray designed for detection of SARS CoV. BC, DMSO spotted as blank control; QC, Hex-labeled oligonucleotide used for quality control of surface chemistry; IC, internal control probe for nested RT-PCR and hybridization; EC, external control probe for hybridization; NC, negative control probe for hybridization; 11 and 24, probes selected from SARS CoV's ORF 1a; 40, probe selected from the ORF of SARS CoV for nucleocapsid protein; 44, probe selected from the ORF of SARS CoV for spike glycoprotein.

of the 6×6 probes were printed on each slide. The design of the microarray is illustrated in **Fig. 2**.

3.2. SARS CoV RNA

3.2.1. Quantification of the SARS CoV RNA

Real-time RT-PCR was performed to quantify the virus RNA on a Roche LightCycler system using RealArt HPA-CoV LC RT Reagents. The result is shown in **Fig. 3**. The quantification standards (a–d) were 10^1, 10^2, 10^3, and 10^4 copies/μL SARS viral RNA, respectively. Between the log concentration of quantification standards and their crossing points, a regression curve has been drawn but not shown ($r = -1.00$). According to the curve, the concentration of the sample RNA was 10^3 copies/μL. In **Fig. 3**, we can see that the curve of the sample RNA (s) almost overlaps the curve of quantification standard (c). Because the sample RNA has been diluted 10^5 times before quantification, the concentration of the original RNA is approx 10^8 copies/μL.

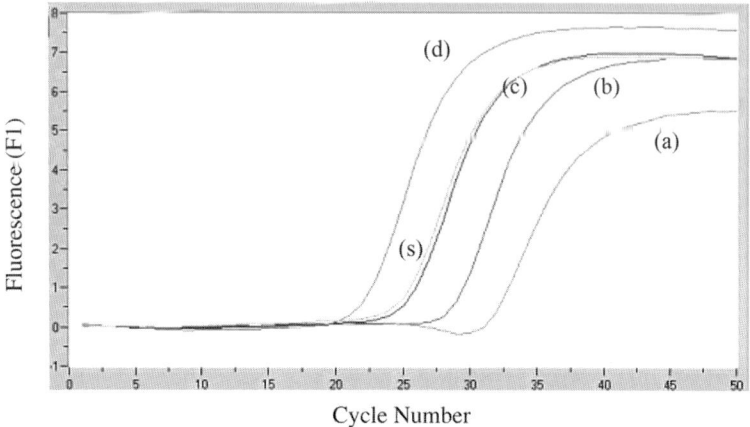

Fig. 3. Quantitative analysis of SARS CoV RNA by means of real-time RT-PCR. (a–d), quantification standards (HPA-CoV LC QS 1–4) supplied within the kit by the manufacturer. Concentrations of SARS virus RNA in (a–d) are 10^1, 10^2, 10^3, and 10^4 copies/μL, respectively. (s), sample for quantification. Before real-time RT-PCR, the SARS viral RNA for quantification was diluted 10^5 times.

3.2.2. Dilution of the SARS CoV RNA

1. The normal human total RNA (TW-Times) was spiked into the SARS CoV RNA. The RNA mixture was prepared by mixing equal amounts of viral RNA (10^5 copies/μL) and human total RNA (16.7 μg/μL).
2. The mixture was then 10-fold diluted continuously with diethyl pyrocarbonate (DEPC)-treated deionized water and served as templates for subsequent amplification.

3.3. Multiplex Nested RT-PCR

3.3.1. Primers (See **Note 7**)

1. Sequence data for SARS CoV were obtained from the curated database in GenBank. The unique and conserved regions of SARS CoV were selected by aligning the released SARS CoV sequences and the latest nonredundant nucleic acid sequences in the NCBI database (ftp://ftp.ncbi.nih.gov).
2. To allow detection of SARS CoV, multiple regions from the open reading frame (ORF) of replicase 1a, spike glycoprotein, and nucleocapsid protein were selected as the targets for hybridization detection.
3. To amplify the four segments from these three ORFs in the genome of SARS CoV, four sets of outer and inner primers were designed, as listed in **Table 1**. The four outer primer sets were designed in the selected genes using Primer3 *(39)* by setting the optimal T_m to 67°C and the size of PCR product from 400 to 1200 bp. Two universal primers were allowed to bind to the 5′ end of the designed inner primers for efficient labeling of the PCR product.

4. To perform nested PCR, the inner primer pair was designed to prime at the region that was approx 200 bp distant from both ends of each selected PCR product.
5. The primers were aligned using BLASTN with the latest nonredundant nucleic acid sequence database downloaded from the ftp site of the NCBI to avoid mispriming. The primers were aligned using BLASTN with all the SARS CoV genomes available to avoid mutations in the primers
6. The primers for amplifying the human glyceraldehyde-3-phosphate dehydrogenase (GAPD) gene from human RNA were designed as the internal control for monitoring the entire amplification process.
7. The universal primers were end-labeled with a fluorescent dye, 6-carboxytetramethylrhodamine (TAMRA; *see* **Note 8**). Primers information is given in **Table 2**.

3.3.2. One-Step RT-PCR

The first-round reaction of the nested RT-PCR is one-step RT-PCR. The second-round reaction is PCR using the products from the first round as the templates.

1. The one-step RT-PCR kit from TaKaRa was used for the first-round reaction, whose conditions were as follows. The total volume for each reaction was 10 μL including 1X One Step RNA PCR buffer, 5 mM MgCl$_2$, 0.8 U/μL RNase inhibitor, 0.1 U/μL AMV RTase XL, 0.1 U/μL AMV-optimized *Taq*, 0.5 μM of all five pairs of outer primers, 1 mM each dNTP, and 4.5 μL RNA mixture (the SARS CoV RNA concentration was 10^3, 10^2, 10^1, and 10^0 copies/μL, respectively), or 4.5 μL DEPC-treated deionized water as the blank control and 4.5 μL of human total RNA as the negative control.
2. Mineral oil (20 μL) was added to cover the reaction fluid.
3. The reactions were performed on a PTC-225 thermal cycler. The thermal conditions were as follows: 1 cycle at 50°C for 30 min; 1 cycle at 94°C for 3 min; 30 cycles at 94°C for 30 s, 55°C for 30 s, and 72°C for 1 min; 1 cycle at 72°C for 10 min.

3.3.3. Nested PCR, Symmetric and Asymmetric

1. During the RT-PCR reaction, the reaction mixture for the second-round nested PCR was prepared. A 2X master mixture was used, and the conditions were as follows. The total volume for each mixture was 40 μL, including 1.25X master mixture, extra 2.5 μM MgCl$_2$, 0.5 mM dUTP, 0.0125 U/μL uracil-DNA glycosylase, 0.25 μM inner primers, and 1.25 μM universal reverse primer (ur). For the universal forward primer (uf), four different concentrations were used, including 1.25, 0.05, 0.0125, and 0 μM, so that the ratios of the primer uf to the ur were 1:1, 1:25, 1:100, and 0:1, respectively (*see* **Notes 1–3, 9**). These reaction mixtures were kept on ice before use.
2. After the RT-PCR reaction, the prepared reaction mixture of the nested PCR was added directly to the fluid of the first round, mixed well, and centrifuged briefly.
3. The PCR reaction was performed on a PTC-225 thermal cycler. Thermal conditions were as follows: 1 cycle at 37°C for 10 min; 1 cycle at 68°C for 10 min; 1 cycle at 94°C for 10 min; 30 cycles at 94°C for 30 s, 60°C for 30 s, and 72°C for 1 min; 1 cycle at 72°C for 10 min.

Table 2
List of Primers

Primers [a]	Sequence (5'-3') [b]	Position [c, d]	Amplicon size (bp)
Outer			
11-of	GCATCGTTGACTATGGTGTCCGATTCT	4433–4459	1071
11-or	ACATCACAGCTTCTACACCCGTTAAGGT	5476–5503	
24-of	GCTGCATTGGTTTGTTATATCGTTATGC	8542–8569	1097
24-or	ATACAGAATACATAGATTGCTGTTATCC	9611–9638	
40-of	CCTCGAGGCCAGGGCGTTCC	28,321–28,340	634
40-or	CACGTCTCCCAAATGCTTGAGTGACG	28,929–28,954	
44-of	TTAAATGCACCGGCCACGGTTTG	23,001–23,023	455
44-or	CCAGCTCCAATAGGAATGTCGCACTC	23,430–23,455	
IC-of	ATGGGGAAGGTGAAGGTCGG	76–95	310
IC-or	TGGTGAAGACGCCAGTGGAC	366–385	
Inner			
11-if	usf-AGCCGCTTGTCACAATGCCAATT[b]	4520–4542	897
11-ir	usr-CATCACCAAGCTCGCCAACAGTT[b]	5355–5377	
24-if	usf-TAGCCAGCGTGGTGGTTCATACAA	8709–8732	836
24-ir	usr-CTCCCGGCAGAAAGCTGTAAGCT	9483–9505	
40-if	usf-TCCTCATCACGTAGTCGCGGTAATTC	28,678–28,703	257
40-ir	usr-GGCTTTTTAGATGCCTCAGCAGCA	28,872–28,895	
44-if	usf-ATGCACCGGCCACGGTTTGTG	23,005–23,025	388
44-ir	usf-ATGCGCCAAGCTGGTGTGAGTTGA	23,330–23,353	
IC-if	usf-CGTATTGGGCGCCTGGTCAC	112–131	275
IC-ir	usr-CCAGCATCGCCCCACTTGAT	328–347	
Universal			
uf	TAMRA-tcacttgcttccgttgagg	/	/
ur	TAMRA-ggtttcggatgttacagcgt	/	/

[a] o, outer primer; i, inner primer; f, forward primer; r, reverse primer; u, universal primer; IC, inner control.

[b] usf, the forward universal sequence 5'-tcacttgcttccgttgagg-3'; usr, the reverse universal sequence 5'-ggtttcggatgttacagcgt-3'.

[c] The position of the inner primers does not include the universal sequences usf and usr.

[d] The reference sequence for SARS gene-specific primers and inner control primers are NC_004718 and NM_002046, respectively (GenBank accession numbers).

3.4. Agarose Gel Electrophoresis

1. A 1.5% agarose gel was prepared by melting 1.5 g of agarose in 100 mL of 1X TBE buffer. Ethidium bromide (30 µg) was added to the agarose solution before it became solidified.

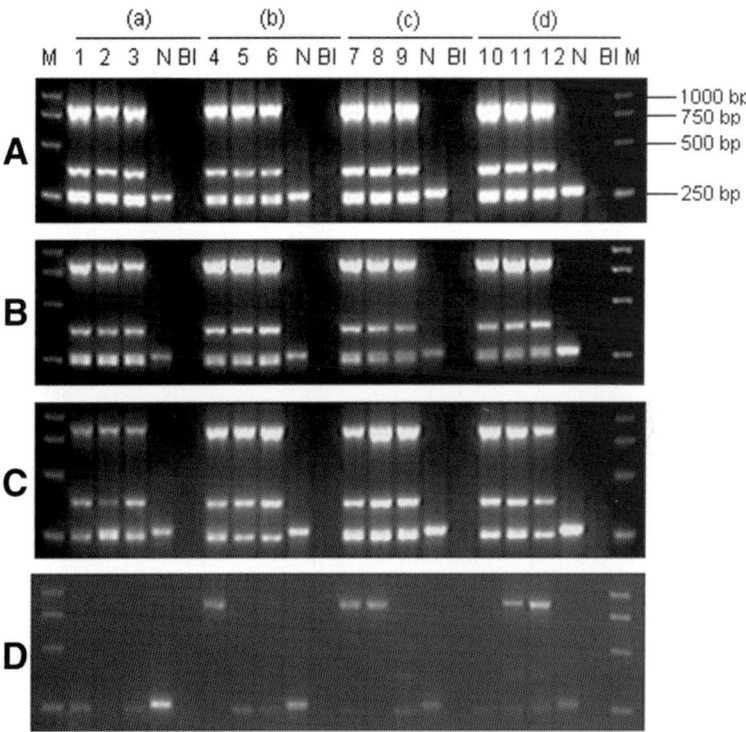

Fig. 4. Electrophoretic results of multiplex nested RT-PCR amplicons. **(A–D)**, SARS CoV RNA and human total RNA at concentrations of 10^3, 10^2, 10^1, and 10^0 copies/μL or 167, 16.7, 1.67, and 0.167 ng/μL as the templates for RT-PCR in lanes 1–12 (three parallel reactions were performed). (a–d), Ratios of universal primers uf to ur were 1:1, 1:25, 1:100, and 0:1, respectively. Lane M, molecular size marker; lane N, negative control, for which the same reaction was performed but without the addition of the virus RNA; lane B1, blank control, in which reaction DEPC-treated H_2O was used as the template. The arrow shows the visible amplicon (257 bp) for the pair of primers 40 in this reaction.

2. PCR products (2 μL) were loaded onto the gel along with the DL2000 DNA molecular weight marker.
3. The gel was run at 8 V/cm for 90 min and then photographed using a UVP system.

The electrophoretic results of the multiplex nested RT-PCR are shown in **Fig. 4** (*see* **Note 5**). No band was seen in all blank controls (lanes B1). A band of the predicted size (275 bp) was observed in all negative controls (lanes N). For other samples, it appeared that the asymmetric amplification was more efficient than the symmetric amplification (D). When the concentration of SARS viral RNA was 10^0 copies/μL, the bands for the inner controls (275 bp) became invisible. The reason was that for the inner control template, the human total RNA spiked into the viral RNA was too little (0.167 ng/μL) to be amplified effectively. Similarly, only part of the bands for SARS CoV amplification was visible when the concentration of SARS viral RNA was 10^0 copies/μL. However, when the higher concentrations of RNA were used as templates, the bands of three parallel multiplex nested RT-PCR reactions were similar to each other, regardless of symmetric or asymmetric amplifications.

3.5. Microarray Hybridization

3.5.1. Hybridization

1. To prepare the hybridization sample, 6.8 μL of the amplicon was added to 8.2 μL of the hybridization solution. The mixture was centrifuged briefly.
2. To denature the dsDNA, the hybridization samples were heated at 95°C for 5 min followed by snap chilling on ice for 3 min.
3. The SARSarray slide with four reaction wells was placed into the HybriCassettes preloaded with 200 μL of double-distilled water to reduce the evaporation.
4. The SmartCover with four molded sample-loading holes was placed on top of the slide.
5. The denatured DNA samples were applied to the individual reaction well on the glass slide through the loading holes on the plastic cover.
6. The sealed cartridge was placed in a 67°C water bath for 2.5 hs.

3.5.2. Washing

1. When the hybridization was completed, the slide was removed from the cartridge, and the cover slip was removed.
2. The slide was washed sequentially in the prewarmed (45°C) washing solution A for 6 min, (twice) followed by washing solution B for 6 min, (twice). Afterward, the slide was rinsed in the double-distilled water and dried by centrifugation at 110g for 2 min.

3.6. Fluorescence Scanning

The slide was scanned using the GenePix 4000B. The scanning conditions were as follows: wavelength, 532 nm; laser power, 33%; pixel size, 10 μm; photomultiplier tube, 550 Vo; brightness, 90%, and contrast, 90%.

Part of the hybridization images is shown in **Fig. 5**. Every QC or EC probe was positive, and every BC or NC probe was negative in all hybridization images, indicating high-quality performance and excellent reproducibility. For the BC PCR, no hybridization signal was visible for the other probes, but the IC probe had a positive signal for the NC PCR. For SARS detection, the hybridization signal of probe 40 was always invisible when symmetric PCR was performed (a), although bands of amplicons with the right size (257 bp) were seen in **Fig. 4**. However, once the concentration of primer uf to primer ur was disproportional, hybridization signals were observed [(b), (c) or (d); *see* **Notes 4–6**].

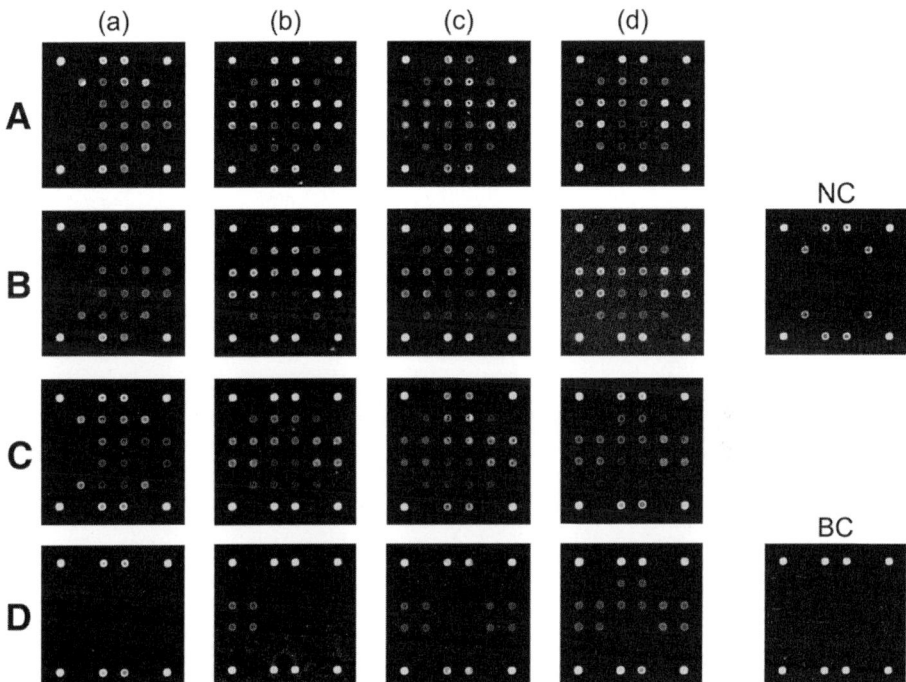

Fig. 5. Part of the hybridization results. (**A–D**), SARS CoV RNA and human total RNA at concentrations of 10^3, 10^2, 10^1, and 10^0 copies/µL or 167, 16.7, 1.67, and 0.167 ng/µL, respectively, were used as templates for RT-PCR. (a–d) In multiplex PCR the ratios of universal primers uf to ur were 1:1, 1:25, 1:100, and 0:1, respectively. In the three parallel multiplex nested RT-PCR reactions, only one hybridization image (lane 3, 6, 9, and 12 in **Fig. 4**) was shown. NC, negative control; BC, blank control. All NC (BC) had the same hybridization images, so only one is shown here. For probe 40, no hybridization signal was observed in (a), but signal was seen in (b–d).

All hybridization results of the SARS detection probes are shown in **Table 3**. From the table we can see that there was at least one SARS CoV-specific probe, whose hybridization result was positive in all cases for the asymmetric amplification, so its limit for SARS viral RNA detection should be 10^0 copies/µL. For the symmetric amplification, because of the hybridization signal loss of probe 40, the detection limit was 10-fold less than that of asymmetric amplification, which was 10^1 copies/µL. (*See* **Notes 9–10**.)

4. Notes

1. Preferential amplification of one target sequence over another is perhaps the biggest problem in multiplex PCR. This is mainly because multiplex PCR has a limited supply, such as the polymerase and the nucleotides, all primer pairs compete for them, and sometimes the amplification efficiency of one primer pair is

Zhang et al.

Table 3
Recordings of Microarray Analyses of SARS Detection Probes[a]

Concentration of virus RNA (copies/mL)	Probe no	1:1			1:25			1:100			0:1		
		1	2	3	1	2	3	1	2	3	1	2	3
10^3	11	+	+	+	+	+	+	+	+	+	+	+	+
	24	+	+	+	+	+	+	+	+	+	+	+	+
	40	−	−	−	+	+	+	+	+	+	+	+	+
	44	+	+	+	+	+	+	+	+	+	+	+	+
10^2	11	+	+	+	+	+	+	+	+	+	+	+	+
	24	+	+	+	+	+	+	+	+	+	+	+	+
	40	−	−	−	+	+	+	+	+	+	+	+	+
	44	+	+	+	+	+	+	+	+	+	+	+	+
10^1	11	+	+	+	+	+	+	+	+	+	+	+	+
	24	+	+	+	+	+	+	+	+	+	+	+	+
	40	−	−	−	+	+	+	+	+	+	+	+	+
	44	+	+	+	+	+	+	+	+	+	+	+	+
10^0	11	−	+	−	+	−	−	+	+	−	−	+	+
	24	−	−	−	−	−	−	−	−	−	−	−	−
	40	−	−	−	−	+	+	+	−	+	+	+	+
	44	−	−	−	−	−	−	−	−	+	−	+	+

[a] +, positive hybridization signal; −, negative hybridization signal.

quite different from that of another. The most curious factor of multiplex PCR is the competition of the primers involved. The development of an efficient multiplex PCR requires strategic planning and multiple attempts to optimize the reaction conditions. Approaches have been reported with varying degree of success. For example, Shuber et al. *(40)* added an unrelated sequence to the 5′ end of the primers to reduce the differences in annealing efficiencies of different loci. Henegariu et al. *(17)* optimized the final concentration of different primers step by step and followed by adjusting the amplification buffer. However, none of the approaches are both time saving and effective enough.

2. Our approach is unique in the utilization of the universal primers in one reaction tube. With a pair of universal primers, the preferential amplification in ordinary multiplex PCR may be avoided. If amplification of one pair of gene-specific primers is inefficient, supplemental amplification would be performed by the universal primers **(Fig. 1)**. The higher the amplification efficiency of gene-specific primers, the less the universal primers participate in the reaction. So the universal primers probably act as a balancer in the novel multiplex PCR, and different targets may be amplified with nearly equal opportunity. In other word, all the primer pairs in a multiplex PCR should allow similar amplification efficiencies for their respective targets.

3. The universal primer-mediated multiplex PCR has a unique advantage. It makes the optimization of multiplex PCR more straightforward than ever. The concentration of the primers does not need to be adjusted repeatedly to overcome the amplification bias. In this work, all inner primers were with the same concentration, 0.2 μM. In our laboratory the concentration for gene-specific primers has been used successfully without optimization for over 2 y already. Attention is only paid to the concentration of the universal primer(s), which should be higher than the concentration of the gene-specific primers. In this study the concentration of the universal primers for symmetric multiplex PCR was up to 1 μM, which confirmed that the primers were competitive with gene-specific primers in the reaction and that enough labeled strand was produced for hybridization.

4. At the initial stage of this research on SARS viral RNA detection, only symmetric multiplex PCR was performed. It was puzzling that no hybridization signal was detectable for probe 40 (*see* [a] in **Fig. 5**). A PCR reaction was performed in which only the pair of inner primers 40 were used. A sharp and bright band with the predicted size was observed on agarose gel, but no hybridization signal was visible (data not shown). So we excluded the possibility of primer interactions in the multiplex PCR reaction. The PCR product was subsequently sequenced, and no mutation was found (data not shown). At that time we were certain that the problem was loss of hybridization signal. To resolve this problem, asymmetric PCR was taken into account automatically. As expected, when asymmetric multiplex PCR was performed, strong fluorescence signal for probe 40 was obtained on the microarrays.

5. Asymmetric PCR is a powerful tool for generating ssDNA. Without self-annealing, ssDNA hybridization with probes should be more efficient than dsDNA, even if dsDNA is denatured by boiling or alkaline treatment before hybridization. Some researchers have reported that hybridization efficiency was much greater with the single-stranded products compared with to the boiled double-strand PCR products *(41,42)*. In traditional asymmetric PCR, the primer length or concentration is usually out of proportion. However, if several targets are amplified simultaneously by use of those tactics, the outcomes are often disappointing owing to ineffectiveness or nonspecific amplification. By comparison, the novel asymmetric multiplex PCR reported here is preeminent, because only the universal primers are involved during asymmetric amplification. As shown in **Fig. 4** [(b), (c), (d) vs (a)], with asymmetric PCR amplification efficiency was no less than with symmetric PCR. Above all, the hybridization signal for probe 40 was on, owing to the asymmetric amplification (**Fig. 5** [b–d]).

6. Why only probe 40 lost the hybridization signal when symmetric PCR was performed is an interesting problem. We have analyzed the GC content of the sequence for all five inner primer amplicons and discovered that the amplicon hybridizing with probe 40 possesses the highest (49.0%) GC content. The GC content for other sequences is from 40.9 to 41.9%. As we know, two strands of DNA rich in G and C will hold to each other more tightly, and they tend to reassociate by themselves after denaturation. This phenomenon may shed partial light on the signal loss for probe 40. Possibly the sequence context of dsDNA also played a role in the signal

loss. The loss of signal may be pernicious for hybridization analysis because false-negative results may be reported. The problem could be resolved by using asymmetric amplification whereby ssDNAs are produced. In this study, the detection limit of the microarray assay for symmetric amplification was 10-fold less than that of asymmetric amplification by virtue of the signal loss of probe 40.

7. Although the asymmetric multiplex PCR reported here works well, a few issues should be kept in mind. The design of the gene-specific primers should comply with the basic principles of conventional multiplex PCR. Special attention to primer design parameters (such as homology of primers with their target nucleic acid sequences, primer length, and GC content) has to be considered *(40,43–47)*, because the overall success of a specific amplification depends on the rate at which primers anneal to their targets, and the rate at which annealed primers are extended along the desired sequence during the early cycles of amplification. The universal primer should be designed to have an appropriate T_m with the gene-specific primers (excluding the universal sequence), to have the least homology with the target nucleic acid sequences, and to avoid the formation of dimmers with the gene-specific primers. Occasionally, even if all these factors are considered, the result may not be the same as predicted owing to the unforeseen interactions of primers. The problem could be solved by changing one of the gene-specific primers.

8. A fluorescence molecule, TAMRA, was labeled to the 5′ end of the universal primers in our study. The hybridization results are in agreement with the asymmetric multiplex PCR results (**Table 3** and **Fig. 4**), so we consider the labeling is adequate for scanning. However, there is a hidden trouble. Some DNA strands produced by the gene-specific primers should also be able to hybridize with the corresponding probes printed on the microarrays, but no fluorescence signal is detectable under these circumstances. To increase the hybridization signal, the gene-specific primers may also be end-labeled.

9. SARS is a serious respiratory illness with significant morbidity and mortality rate *(31–37)*. Its diagnosis depends mainly on the clinical findings of an atypical pneumonia not attributed to any other cause and a history of exposure to a suspect or probable case of SARS, or to the respiratory secretions and other bodily fluids of individuals with SARS. Definitive diagnosis of this novel CoV relies on classic tissue culture isolation, followed by electron microscopy studies to identify the virus in cell culture, which is technically very demanding. Serological testing for increasing titer against SARS-associated CoV was shown to be highly sensitive and specific *(32)* but was not suitable for quick and early laboratory diagnosis. Molecular tests have also been attempted for the detection of this virus or to confirm infection *(48,49)*. However, only one target was detected in one test using the existing methods, so a higher risk of false negatives was inevitably encountered. Our method is quite different from the existing methods: asymmetric multiplex nested RT-PCR amplification followed by microarray hybridization. The assay was sufficiently sensitive that 10^0 copies/µL viral RNA could be detected. We have used this method to realize the early sensitive detection of SARS virus from clinic samples (unpublished

work). In actual application, the universal primer uf was omitted. In other words, the ratio of 0:1 for universal primer uf to ur was adopted for efficiency and convenience.

10. In summary, we have established a sensitive and versatile asymmetric multiplex PCR method. Combining with the microarray assay, this method may be applicable in a number of fields, such as identification of microorganisms, detection of the drug-resistant genes, and diagnosis of genetic diseases.

Acknowledgments

The authors thank Prof. Tao Hung for providing the SARS CoV RNA. We are grateful for the contributions from Hua-fang Gao, Can Wang, Di Jiang, Li Rong, Qing-mei Ma, Hong-li Lu, Tian-tian Cai, Hua-wei Yang, Chuan-zan Zhao, Yan-hua Liu, and Dong Wang.

This work was supported by grants from the National Hi-Tech Program of China, Department of Science and Technology of China (2002AA2Z2011 and 200310230055) and by a grant from the China Postdoctoral Science Foundation (2003034158).

References

1. Chamberlain, J. S., Gibbs, R. A., Ranier, J. E., Nguyen, P. N., and Caskey, C. T. (1988) Deletion screening of the Duchenne muscular dystrophy locus via multiplex DNA amplification. *Nucleic Acids Res.* **16,** 11141–11156.

2. Sieber, O. M., Lamlum, H., Crabtree, M. D., et al. (2002) Whole-gene APC deletions cause classical familial adenomatous polyposis, but not attenuated polyposis or "multiple" colorectal adenomas. *Proc. Natl. Acad. Sci. USA* **99,** 2954–2958.

3. Moutou, C., Gardes, N., and Viville, S. (2002) Multiplex PCR combining deltaF508 mutation and intragenic microsatellites of the CFTR gene for pre-implantation genetic diagnosis (PGD) of cystic fibrosis. *Eur. J. Hum. Genet.* **10,** 231–238.

4. Zimmermann, K., Schogl, D., Plaimauer, B., and Mannhalter, J. W. (1996) Quantitative multiple competitive PCR of HIV-1 DNA in a single reaction tube. *BioTechniques* **21,** 480–484.

5. Jin, L., Richards, A., and Brown, D. W. (1996) Development of a dual target-PCR for detection and characterization of measles virus in clinical specimens. *Mol. Cell. Probes* **10,** 191–200.

6. Zou, S., Stansfield, C., and Bridge, J. (1998) Identification of new influenza B virus variants by multiplex reverse transcription-PCR and the heteroduplex mobility assay. *J. Clin. Microbiol.* **36,** 1544–1548.

7. Tettelin, H., Radune, D., Kasif, S., Khouri, H., and Salzberg, S. L. (1999) Optimized multiplex PCR: efficiently closing a whole-genome shotgun sequencing project. *Genomics* **62,** 500–507.

8. Druce, J., Catton, M., Chibo, D., et al. (2002) Utility of a multiplex PCR assay for detecting herpesvirus DNA in clinical samples. *J. Clin. Microbiol.* **40,** 1728–1732.

9. Robert, P. Y., Traccard, I., Adenis, J. P., Denis, F., and Ranger-Rogez, S. (2002) Multiplex detection of herpesviruses in tear fluid using the "stair primers" PCR method: prospective study of 93 patients. *J. Med. Virol.* **66,** 506–511.

10. Osek, J. (2002) Rapid and specific identification of Shiga toxin-producing *Escherichia coli* in faeces by multiplex PCR *Lett. Appl. Microbiol.* **34,** 304–310.

11. Sloan, L. M., Hopkins, M. K., Mitchell, P. S., et al. (2002) Multiplex LightCycler PCR assay for detection and differentiation of *Bordetella pertussis* and *Bordetella parapertussis* in nasopharyngeal specimens. *J. Clin. Microbiol.* **40,** 96–100.

12. Harris, E., Kropp, G., Belli, A., Rodriguez, B., andAgabian, N. (1998) Single-step multiplex PCR assay for characterization of New World Leishmania complexes. *J. Clin. Microbiol.* **36,** 1989–1995.

13. Strommenger, B., Kettlitz, C., Werner, G., and Witte, W. (2003) Multiplex PCR assay for simultaneous detection of nine clinically relevant antibiotic resistance genes in *Staphylococcus aureus*. *J. Clin. Microbiol.* **41,** 4089–4094.

14. Oliveira, D. C., and Lencastre, H. H. (2002) Multiplex PCR strategy for rapid identification of structural types and variants of the mec element in methicillin-resistant *Staphylococcus aureus*. *Antimicrob. Agents Chemother.* **46,** 2155–2161.

15. Markoulatos, P., Siafakas, N., and Moncany, M. (2002) Multiplex polymerase chain reaction: a practical approach. *J. Clin. Lab. Anal.* **16,** 47–51.

16. Elnifro, E. M., Ashshi, A. M., Cooper, R. J., and Klapper, P. E. (2000) Multiplex PCR: optimization and application in diagnostic virology. *Clin. Microbiol. Rev.* **13,** 559–570.

17. Henegariu, O., Heerema, N. A., Dlouhy, S. R., Vance, G. H., and Vogt, P. H. (1997) Multiplex PCR: critical parameters and step-by-step protocol, *BioTechniques* **23,** 504–511.

18. Tao, S., and Cheng, J. (2003) Methods and compositions for optimizing multiplex PCR primers. China Patent PCT03/00335.

19. Yershov, G., Barsky, V., Belgovskiy, A., et al. (1996) DNA analysis and diagnostics on oligonucleotide microchips. *Proc. Natl. Acad. Sci. USA* **93,** 4913–4918.

20. Drobyshev, A., Mologina, N., Shik, V., Pobedimskaya, D., Yershov, G., and Mirzabekov, A. (1997) Sequence analysis by hybridization with oligonucleotide microchip: identification of beta-thalassemia mutations. *Gene* **188,** 45–52.

21. Haviv, I., and Campbell, I. G. (2002) DNA microarrays for assessing ovarian cancer gene expression. *Mol. Cell. Endocrinol.* **191,** 121–126.

22. Kim, I. J., Kang, H. C., Park, J. H., et al. (2002) RET oligonucleotide microarray for the detection of RET mutations in multiple endocrine neoplasia type 2 syndromes. *Clin. Cancer Res.* **8,** 457–463.

23. Riccelli, P. V., Merante, F., Leung, K. T., et al. (1994) Hybridization of single-stranded DNA targets to immobilized complementary DNA probes: comparison of hairpin versus linear capture probes. *Nucleic. Acids Res.* **29,** 996–1004.

24. Guo, Z., Guilfoyle, R. A., Thiel, A. J., Wang, R., and Smith, L. M. (1994) Direct fluorescence analysis of genetic polymorphisms by hybridization with oligonucleotide arrays on glass supports. *Nucleic. Acids Res.* **22,** 5456–5465.

25. Kawai, S., Maekawajiri, S., and Yamane, A. (1993) A simple method of detecting amplified DNA with immobilized probes on microtiter wells. *Anal. Biochem.* **209,** 63–69.
26. Gyllensten, U. B., and Erlich, H. A. (1988) Generation of single-stranded DNA by the polymerase chain reaction and its application to direct sequencing of the HLA-DQA locus. *Proc. Natl. Acad. Sci. USA* **85,** 7652–7656.
27. Erdogan, F., Kirchner, R., Mann, W., Ropers, H. H., and Nuber, U. A. (2001) Detection of mitochondrial single nucleotide polymorphisms using a primer elongation reaction on oligonucleotide microarrays. *Nucleic. Acids Res.* **29,** E36.
28. Kaltenboeck, B., Spatafora, J. W., Zhang, X., Kousoulas, K. G., Blackwell M., and Storz, J. (1992) Efficient production of single-stranded DNA as long as 2 kb for sequencing of PCR-amplified DNA. *BioTechniques* **12,** 164–168.
29. Gorelov, V. N., Roher, H. D., and Goretzki, P. E. (1994) A method to increase the sensitivity of mutation specific oligonucleotide hybridization using asymmetric polymerase-chain reaction (PCR). *Biochem. Biophys. Res. Commun.* **200,** 365–369.
30. Scott, D. L., Clark, C. W., Fyffe, A. E., Walker, M. D., and Deah, K. L. (1998) The differentiation of *Phytophthora* species that are pathogenic on potatoes by an asymmetric PCR combined with single-strand conformation polymorphism analysis. *Lett. Appl. Microbiol.* **27,** 39–44.
31. World Health Organization. (2003) Severe acute respiratory syndrome (SARS). *Wkly. Epidemiol. Rec.* **78,** 81–83.
32. Peiris, J. S., Lai, S. T., Poon, L. L., et al. (2003) Coronavirus as a possible cause of severe acute respiratory syndrome. *Lancet* **361,** 1319–1325.
33. Ruan, Y. J., Wei, C. L., Ee, A. L., et al. (2003) Comparative full-length genome sequence analysis of 14 SARS coronavirus isolates and common mutations associated with putative origins of infection. *Lancet* **361,** 1779–1785.
34. Ksiazek, T. G., Erdman, D., Goldsmith, C. S., et al. (2003) A novel coronavirus associated with severe acute respiratory syndrome. *N. Engl. J. Med.* **348,** 1953–1966.
35. Marra, M. A., Jones, S. J., Astell, C. R., et al. (2003) The genome sequence of the SARS-associated coronavirus. *Science* **300,** 1399–1404.
36. Drosten, C., Gunther, S., Preiser, W., et al. (2003) Identification of a novel coronavirus in patients with severe acute respiratory syndrome. *N. Engl. J. Med.* **348,** 1967–1976.
37. World Health Organization Multicentre Collaborative Network for Severe Acute Respiratory Syndrome (SARS) Diagnosis (2003) A multicentre collaboration to investigate the cause of severe acute respiratory syndrome. *Lancet* **361,** 1730–173.
38. Rota, P. A., Oberste, M. S., Monroe, S. S., et al. (2003) Characterization of a novel coronavirus associated with severe acute respiratory syndrome. *Science* **300,** 1394–1399.
39. Rozen, S. and Skaletsky, H. (2000) Primer3 on the WWW for general users and for biologist programmers. *Methods Mol. Biol.* **132,** 365–386.
40. Shuber, A. P., Grondin, V. J., and Klinger, K. W. (1995) A simplified procedure for developing multiplex PCRs. *Genome Res.* **5,** 488–493.

41. Kawai, S., Maekawajiri, S., and Yamane, A. (1993) A simple method of detecting amplified DNA with immobilized probes on microtiter wells. *Anal. Biochem.* **209,** 63–69.

42. Guo, Z., Guilfoyle, R. A., Thiel, A .J., Wang, R., and Smith, L. M. (1994) Direct fluorescence analysis of genetic polymorphisms by hybridization with oligonucleotide arrays on glass supports. *Nucleic. Acids Res.* **22,** 5456–5465.

43. Innis, M. A., Gelfand, D. H., Shinsky, J. J., and White, T. J. (eds.) (1989) *PCR Protocols: A Guide to Methods and Applications.* Academic Press, San Diego, CA.

44. Wu, D. Y., Ugozzoli, L., Pal, B. K., Qian, J., and Wallace, R. B. (1991) The effect of temperature and oligonucleotide primer length on the specificity and efficiency of amplification by the polymerase chain reaction. *DNA Cell Biol.* **10,** 233–238.

45. Dieffenbach, C. W., Lowe, T. M. J., and Dveksler, G. S. (1993) General concepts for PCR primer design. *PCR Methods Appl.* **3,** S30–S37.

46. Mitsuhashi, M. (1996) Technical report: part 2. Basic requirements for designing optimal PCR primers. *J. Clin. Lab. Anal.* **10,** 285–293.

47. Robertson, I. M., and Walsh-Weller, J. (1998) An introduction to PCR primer design and optimisation of amplification reactions. *Methods Mol. Biol.* **98,** 121–154.

48. Zhai, J. M., Briese, T., Dai, E., et al. (2004) Real-time polymerase chain reaction for detecting SARS coronavirus, Beijing, 2003. *Emerg. Infect. Dis.* **10,** 300–303.

49. Yam, W. C., Chan, K. H., Poon, L. L. M., et al. (2003) Evaluation of reverse transcription-PCR assays for rapid diagnosis of severe acute respiratory syndrome associated with a novel coronavirus. *J. Clin. Microbiol.* **41,** 4521–4524.

4

Genotyping Single-Nucleotide Polymorphisms by Minisequencing Using Tag Arrays

Lovisa Lovmar and Ann-Christine Syvänen

Summary

The need for large-scale and high-throughput methods for SNP genotyping has rapidly increased during the last decade. Our system, presented here, combines the highly specific genotyping principle of minisequencing with the advantages of a microarray format that allows highly multiplexed and parallel analysis.

Cyclic minisequencing reactions with fluorescently labeled dideoxynucleotides (ddNTPs) are performed in solution using multiplex PCR product as template and detection primers, designed to anneal immediately adjacent and upstream of the SNP site. The detection primers carry unique 5′ tag sequences and oligonucleotides complementary to the tag sequence, cTags, are immobilized on a microarray. After extension, the tagged detection primers are allowed to hybridize to the cTags; then the fluorescent signals from the array are measured, and the genotypes are deduced according to the label incorporated. The "array of arrays" format of the system, accomplished by a silicon rubber grid giving separate reaction chambers, allows either 80 or 14 samples to be analyzed for up to 200 or 600 SNPs, respectively, on a single microscope slide.

Key Words: Single-nucleotide polymorphism (SNP); genotyping; minisequencing; microarray; multiplex; PCR; "array of arrays."

1. Introduction

Genomic nucleotide substitutions that are present in more than 1% of the alleles in a population are denoted single-nucleotide polymorphisms (SNPs) and are the most abundant form of genetic variation (*1*). Following the completion of the nucleotide sequence of the human genome (*2,3*), a large interest in SNPs has arisen owing to their potential use as markers when one is searching for genetic factors underlying complex, multifactorial disorders. Consequently, the need for high-throughput methods for SNP genotyping has increased.

From: *Methods in Molecular Medicine, Vol. 144, Microarrays in Clinical Diagnostics*
Edited by: T. Joos and P. Fortina © Humana Press Inc., Totowa, NJ

Several reaction principles and assay formats have been developed (*see* **ref** *4* for a review of genotyping techniques). One of the reaction principles most often used in high-throughput systems today is minisequencing, in which a DNA polymerase is allowed to extend a detection primer by a single nucleotide at the position of the SNP *(5)*. The system, presented in this chapter, combines the highly specific genotyping principle of minisequencing with the advantages of a microarray format that allows highly multiplex and parallel analysis.

The Tag array minisequencing system utilizes generic capture oligonucleotides (cTags) that are immobilized on a microarray. Multiplex cyclic minisequencing reactions with fluorescently labeled dideoxynucleotides (ddNTPs) are performed in solution using detection (minisequencing) primers, designed to anneal immediately adjacent and upstream of the SNP site. The primers carry 5′ Tag sequences complementary to one of the arrayed cTags, and the SNPs are genotyped by hybridizing the extended detection primers to their corresponding cTags with known locations on the array. The incorporated fluorescently labeled ddNTPs allow deduction of the genotypes of each SNP based on measurement of the signal intensities by fluorescence scanning of the arrays *(6,7)*. The use of generic Tag sequences allows universal, non-SNP-specific array designs. The "array of arrays" format described below is accomplished by a silicon rubber grid forming separate reaction chambers for multiple samples on a single microscope slide. Either 80 or 14 samples can be simultaneously analyzed for up to 200 or 600 SNPs, respectively *(8–10)* (**Fig. 1**).

The procedure described in detail in the Methods section outlines (1) selection of appropriate SNPs; (2) design of oligonucleotides for polymerase chain reaction (PCR), immobilization on the microarray, and SNP genotyping; (3) preparation of microarrays; (4) manufacturing of the silicon rubber grid; (5) the genotyping reaction; and (6) data analysis and genotype interpretation. The main steps of the assay are illustrated in **Fig. 2**.

Several alterations and modifications of the method are possible, and a number of suggestions are given in the Notes section. The protocol is given under the assumption that the reader will use the instrumentation, reagents, and consumables specified in the Materials section, but other equivalent procedures are also feasible.

2. Materials

2.1. Instrumentation

1. Access to arraying facilities or purchased customized arrayed slides. We use a ProSys 5510A instrument (Cartesian Technologies, Huntingdon, UK) with Stealth Micro Spotting Pins (TeleChem, Sunnyvale, CA).
2. Facilities for programmed thermal cycling.
3. Multichannel pipete and a pipeting robot (optional).
4. Centrifuge for microtiter plates (recommended).

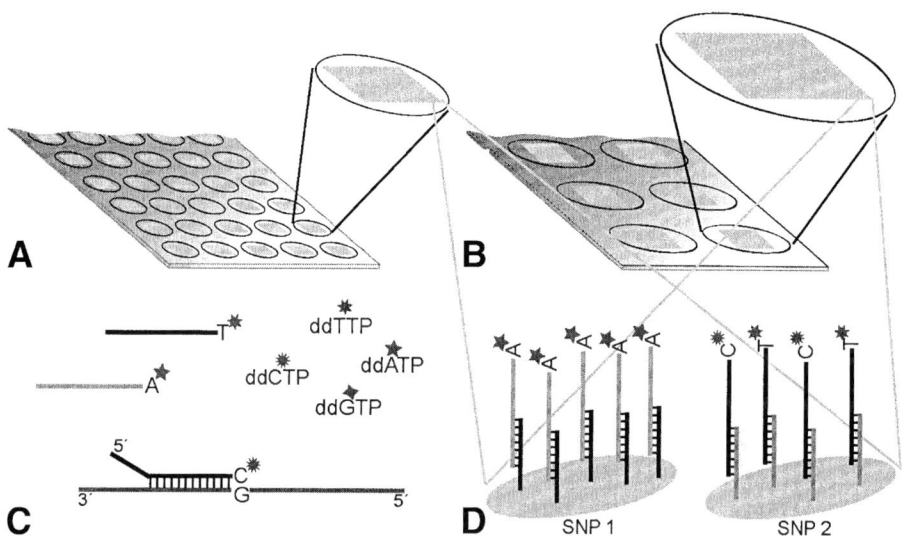

Fig. 1. Principle of Tag array minisequencing. Schematic views of about one-third of two arrayed slides with the subarrays in either a 384-well (A) or 96-well (B) conformation. One of the subarrays for each slide containing up to 200 (**A**) or 600 (**B**) cTags is showed enlarged (**C**). The principle of the minisequencing reaction is illustrated with a minisequencing primer carrying a 5' Tag sequence that has annealed to its target and has been extended with a labeled ddCTP at the position of the single-nucleotide polymorphism (SNP). (**D**) The Tag sequences of the extended minisequencing primers are allowed to hybridize to their complementary cTags arrayed as spots in the subarrays. The genotypes are deduced by measuring the fluorescence of the incorporated nucleotides. Part of one subarray, with the result for two SNPs, is shown. This sample is homozygous (A/A) for SNP 1 and heterozygous (C/T) for SNP 2.

5. Minisequencing reaction rack (**Fig. 3**).
6. Heat block at 42°C.
7. Array scanner and software for signal analysis. We use the ScanArray Express system (PerkinElmer Lifesciences, Boston, MA).

2.2. Reagents and Consumables

All reagents should be of standard molecular biology grade. Use sterile distilled or deionized water.

2.2.1. Oligonucleotide Primers and Nucleotides

1. PCR primers.
2. Minisequencing primers with 5' Tag sequences.
3. cTags with a 3' end 15-T residue spacer and a 3' amino group.
4. Reaction control templates: four oligonucleotides differing in one internal nucleotide position.

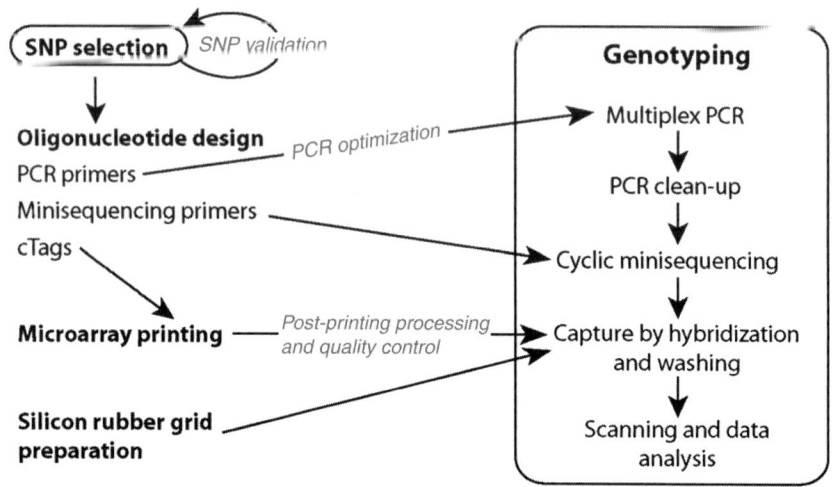

Fig. 2. Flow chart illustrating the main steps of the procedure for genotyping of single-nucleotide polymorphisms (SNPs) by minisequencing using Tag arrays. PCR, polymerase chain reaction.

5. Reaction control minisequencing primer complementary to the reaction control templates and with a 5′ Tag sequence.
6. Spot control oligonucleotide: fluorescently labeled cTag.
7. Hybridization control oligonucleotide: a fluorescently labeled Tag sequence (optimally two controls should be used).
8. Print-control oligonucleotide, fluorescently labeled and designed to hybridize to any cTag.
9. Fluorescent dideoxynucleotides (Texas Red-ddATP, TAMRA-ddCTP, R110-ddGTP, and Cy5-ddUTP; PerkinElmer Life Sciences). Keep light-protected, working aliquots in at 4°C, and store stock solutions at −20°C.

2.2.2. Enzymes

1. Hot-start DNA-polymerase (preferable). We use AmpliTaq Gold (Applied Biosystems, Foster City, CA) or Accuprime Taq (Invitrogen, Carlsbad, CA).
2. Exonuclease I (Amersham Biosciences, Uppsala, Sweden).
3. Shrimp alkaline phosphatase (Fermentas, Vilnius, Lithuania).
4. DNA polymerase compatible with fluorescently labeled ddNTPs. We use ThermoSequenase (Amersham Biosciences) or KlenThermase™ (GeneCraft).

2.2.3. Buffer Solutions

1. Standard PCR reagents or reagents optimized for multiplex PCR.
2. 2X Printing buffer: 300 mM phosphate buffer, pH 8.5. Store at room temperature up to 1 month.

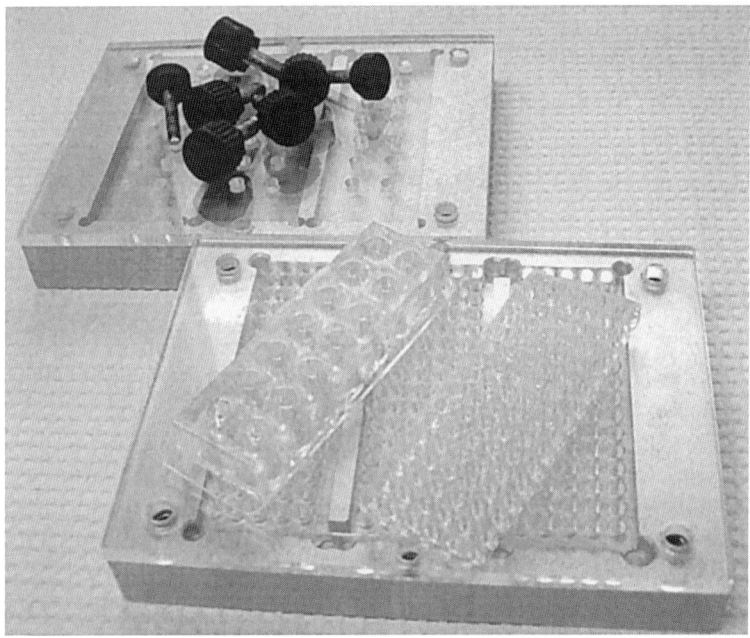

Fig. 3. The arrayed slide is covered with a silicon rubber grid to give separate reaction chambers and placed in a custom-made, heat-conducting aluminum rack. Two silicon rubber grids, one 384-well (right) and the other 96-well (left), are shown on top of a reaction rack. A plexiglas cover with drilled holes through which the reaction chambers are accessible is tightly screwed on top of the assembly, thus ensuring correct positioning of the silicon grid during hybridization.

3. Blocking solution: 50 mM ethanolamine, 100 mM Tris-HCl, pH 9.0, and 0.1% sodium dodecyl sulfate (SDS). Prepare directly before use. Ethanolamine is highly corrosive and should be handled according to safety instructions.
4. Washing solutions: (I) 4X standard saline citrate (SSC) and 0.1% SDS; (II) 2X SSC and 0.1% SDS; and (III) 0. 2X SSC. (20X SSC: 3 M NaCl, 300 mM sodium citrate, pH 7.0.)
5. 1M Tris-HCl, pH 9.5.
6. 50 mM MgCl$_2$.
7. 1% (v/v) Triton X-100.
8. Hybridization solution: 6.25X SSC.

2.2.4. Consumables

1. 384- or 96-well V-bottomed microtiter plates (ABgene, Epsom, UK).
2. CodeLink™ Activated Slides (Amersham Biosciences).
3. Elastosil RT 625 A and B (polydimethyl siloxan; Wacker-Chemie, München, Germany).

3. Methods

3.1. SNP Selection

SNPs can be identified either experimentally or in databases, for example in dbSNP (http://www.ncbi.nlm.nih,gov/SNP/). Database searches for SNPs may be aimed at genes of interest, candidate chromosomal regions, or randomly distributed SNPs with known allele frequencies, depending on the aim of the project (*see* **Note 1**).

3.2. Oligonucleotide Design

3.2.1. PCR Primers

The PCR fragments should optimally be short, i.e., 100–150 bp. Design PCR primers flanking the SNPs of interest using available software. Primer3 is freely available on the internet (http://frodo.wi.mit.edu/cgi-bin/primer3/primer3_www.cgi; *see* **Note 2**).

The sequence of each PCR fragment should be "blasted" against the genome sequence (http://www.ncbi.nlm.nih.gov/BLAST/) and give a single hit only to the intended region (*see* **Note 3**).

3.2.2. Minisequencing Primers

Minisequencing primers are designed to anneal immediately adjacent to and upstream of the SNP position. The minisequencing primers should be approx 20 bases long and have a melting temperature of 55–60°C to ensure specificity in the cyclic primer extension reaction.

At the 5′ end of each minisequencing primer, add the Tag sequences, which should be complementary to the cTags that will be arrayed onto the microarray. The Tags should be 20 bases long, have similar melting temperature, and not be complementary to either each other, to the gene-specific part of the minisequencing primers, or to the human genome (*6*). The Affymetrix GeneChip® Tag Collection can be used as a source for Tag sequences (Affymetrix, Santa Clara, CA). Minisequencing primers from both forward and reverse strands are often helpful as internal controls for the genotyping results (*see* **Note 4**).

3.2.3. Complementary Tag Sequences

The complementary Tag sequences (cTags) have 15 3′ T-residues as a spacer and a 3′ amino group to allow covalent attachment of the cTags to the slides.

3.2.4. Control Oligonucleotides

We recommend the use of a number of oligonucleotides as controls for the different steps of the procedure (*10*).

1. To control for the spotting procedure, a fluorescently labeled cTag may be included in the array. Also a print-control oligonucleotide designed to hybridize to any cTag (5′-AAA AAA AAA ANN NNN NNN NN— Fluorophore -3′) is recommended for use on some subarrays or microarrays from each batch.
2. As a minisequencing reaction control, a minisequencing primer is useful that is complementary to four synthesized single-stranded oligonucleotide templates differing at one nucleotide position mimicking the four possible alleles of a SNP. Add the control templates to the minisequencing reaction up to a final concentration of 1.5 nM. A corresponding cTag should be included in the array.
3. To control for the hybridization reaction, a fluorescently labeled oligonucleotide complementary to an arrayed cTag is used. Optimally use two differently labeled hybridization control oligonucleotides, and add them in an alternating pattern over the microarray to ensure that no leaking between wells has occurred.

3.3. Preparation of Microarrays

3.3.1. Microarray Printing

1. Dissolve the cTags in printing buffer to a final concentration of 25 μM (*see* **Note 5**).
2. If they are not used immediately or if they are to be reused, store the cTags at –20°C, but limit freeze-thawing cycles to 10.
3. Prepare the arrays by contact-printing the cTag oligonucleotides onto CodeLink Activated slides using the ProSys 5510A instrument with SMP3 pins that deliver 1 nL of the cTag solution to the slides as spots with diameters of 125–150 μm and a center-to-center distance of, for example, 200 μM (*see* **Note 6**).
4. To use the "array of arrays" format, print spots in a subarray pattern corresponding to the spacing of wells in a 384-well microtiter plate (*see* **Note 7** and **Fig. 1**).
5. After arraying, mark the position of some of the subarrays on the back side of the slides using a diamond pen.

3.3.2. Postprinting Processing of the Microarray Slides

Process the slides according to the instructions of the manufacturer. The protocol for CodeLink Activated Slides is given below.

1. Prepare an incubation chamber with 75% relative humidity. Add as much solid NaCl to water as needed to form a 1-cm-deep slurry at the bottom of a plastic container with an airtight lid.
2. After printing, keep the arrays in the incubation chamber for 4–72 h.
3. Prepare the blocking solution, and preheat it to 50°C.
4. Deactivate the excess of amine-reactive groups by immersing the arrayed slides for 30 min in the blocking solution at 50°C.
5. Rinse twice with dH$_2$O. Immerse the slides in washing solution I for 30 min at 50°C. (At least 10 mL per slide should be used.) Rinse again with dH$_2$O.
6. Spin-dry the slides for 5 min at about 90 *g*. Store the slides desiccated at 20°C until use.

3.3.3. Quality Control of Printing Procedure

For each batch of printed slides, it is useful to analyze a few subarrays by hybridization as a quality control of the spots. After deactivation of the slides, hybridize the 3′ fluorescently labeled print-control oligonucleotide to some subarrays at 300 nM concentration in 6X SSC for 5 min with subsequent washing and scanning as described below under **Subheadings 3.5.4.** and **3.5.5.**

3.4 Preparation of Reusable Silicon Rubber Grid

A grid of silicon rubber reaction chambers is made using inverted V-bottomed microtiter plates as the mold (**Fig. 3**).

1. Add the two Elastosil RT 625 components in a 50-mL Falcon tube in a mass ratio of 9:1 (i.e., 46.8 g of A and 5.2 g of B); then rotate and turn the tube by hand until the components are fully mixed (approx 30 min; *see* **Note 8**).
2. Pour the mixture onto an inverted V-bottomed 384-well microtiter plate, leaving about 1–2 mm of the tip of the wells uncovered. Allow the silicon rubber to harden at least overnight at room temperature (*see* **Note 9**).
3. Remove the silicon rubber grid from the plate, and use a scalpel to cut the silicon rubber into pieces of the same size as microscope slides, with the wells matching the printed subarrays.

The silicon rubber grid is reusable; wash it with water, and allow it to dry after each use.

3.5. Genotyping

3.5.1. Multiplex PCR and Clean-Up

1. Amplify genomic DNA samples and PCR negatives by multiplex PCR according to an optimized protocol. The success of the amplification may be verified on a 1% agarose gel for a subset of the samples.
2. For each sample, pool the multiplex PCR products (*see* **Note 10**).
3. Prepare a master mix of the exonuclease (ExoI) and alkaline phosphatase (sAP) reagents for clean-up of the PCR products (*see* **Table 1** and **Note 11**).
4. Add 3.4 µL of the clean-up mixture to give a total volume of 10.5 µL.
5. Incubate at 37°C for 30–60 min.
6. Inactivate the enzymes by heating to 85°C for 15 min.

3.5.2. Cyclic Minisequencing

1. Prepare a master mix with minisequencing reagents (*see* **Table 2** and **Note 12**).
2. After the clean-up step, add 4.5 µL of minisequencing reaction mixture to give a total volume of 15 µL.
3. Perform the minisequencing reactions in a thermal cycler using an initial 3-min denaturation step at 96°C followed by, for example, 33 cycles of 20 s at 95°C and 20 s of 55°C in a thermocycler (*see* **Note 13**).

Table 1
PCR Clean-Up Reagents

Reagent	Volume per reaction (µL)	Final concentration
PCR products	7.1	
50 mM MgCl$_2$	1.6	7.62 mM [a]
1 M Tris-HCl, pH 9.5	0.5	0.05 M
20 U/µL Exonuclease I	0.3	0.57 U/µL
1 U/µL Shrimp alkaline phosphatase	1.0	0.10 U/µL
Total volume	10.5	

[a] The true final concentration of MgCl$_2$ is higher depending on the contribution from the PCR products.

Table 2
Minisequencing Reagents

Reagent	Volume per reaction (µL)	Final concentration
100 nM of each pooled minisequencing primer	1.50	10.0 nM
100 µM Fluorescently labeled ddNTPs [a]	4×0.015	0.10 µM
1% Triton X-100	0.30	0.02%
32 U/µL ThermoSequenase	0.03	0.064 U/µL
H$_2$O	2.61	
Total volume	15.00	

[a] Texas Red-ddATP, TAMRA-ddCTP, R110-ddGTP, and Cy5-ddUTP.

3.5.3. Capture by Hybridization

1. Position a silicon rubber grid over the arrayed slide according to the diamond pen markings. Place the arrayed slides into the custom-made aluminum reaction rack, and tighten the plexiglas cover (**Fig. 3**). Preheat the assembly to 42°C on a heat block (*see* **Note 14**).
2. Add 7 µL of the hybridization solution to each minisequencing reaction to a final volume of 22 µL. It is recommended to include hybridization control oligonucleotides at 0.25 nM concentrations in the hybridization mixture.
3. Transfer 20 µL of each sample to a separate reaction chamber on the microscope slide. A multichannel pipete is convenient for this step.
4. Hybridize for 2.5–3 h at 42°C in a humid environment, formed, for example, by placing a wet tissue on the plexiglas lid and covering it with plastic film and aluminum foil.

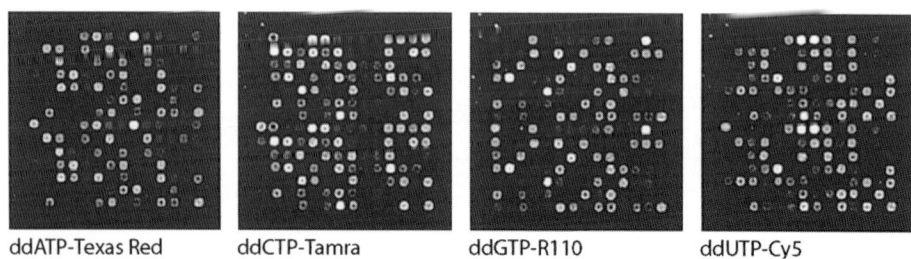

| ddATP-Texas Red | ddCTP-Tamra | ddGTP-R110 | ddUTP-Cy5 |

Fig. 4. Scanning results from genotyping one sample for 45 SNPs in one subarray, after cyclic minisequencing with ddATP, ddCTP, ddGTP, and ddUTP labeled with Texas Red, TAMRA, R110, and Cy5, respectively. Each cTag was spotted as horizontal duplicates, and both polarities of the SNPs were analyzed *(10)*.

3.5.4. Washing

1. Prepare the three washing solutions. Preheat solution II to 42°C.
2. After hybridization, take the slides from the reaction rack, and rinse briefly with solution I, at 20–25°C.
3. Wash the slides twice for 5 min with solution II at 42°C and twice for 1 min with solution III, at 20–25°C in 50-mL Falcon tubes.
4. Spin-dry the slides for 5 min at about 90 g. and store them protected from light.

3.5.5. Fluorescence Scanning

If allowed by the scanner used, balance the signal intensity from each laser channel so that no signals are saturated and the signals from the four fluorophores are as equal as possible. Balancing is feasible if a reaction control with signals from all four fluorophores has been included. **Figure 4** shows an example of a scanned array.

3.6. Data Analysis and Genotype Assignment

A quantification program such as the one supplied with ScanArray Express handles the scanning images and quantitates the signals from each spot. The raw data are collected in an Excel sheet.

1. Subtract the background, measured either around the spots or at negative control spots, i.e., spotted cTags without corresponding tagged primers, from the signals measured in each channel.
2. Assign the genotypes of the SNPs in each sample by calculating the ratios between the signals from one of the alleles and the sum of the signals from both the alleles: Signal Allele 1/(Signal Allele 1 + Signal Allele 2).
3. A scatter plot with this ratio on the horizontal axis and the sum of the signals from both alleles on the vertical axis may be used for assigning the genotypes **(Fig. 5)**. This scatter plot should give three distinct genotype clusters with the

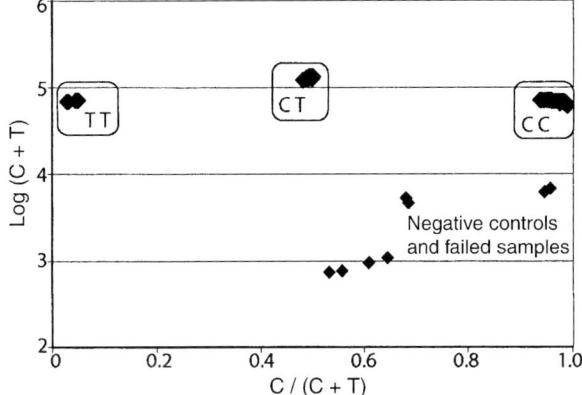

Fig. 5. Scatter plot for one SNP with a C/T variation analyzed in 80 samples, i.e., one slide with a 384-well format and 5 × 16 subarrays. The logarithms of the sum of the fluorescence signals from both alleles in each sample are plotted on the vertical axis. The signal ratios between the signal from one allele divided by the sum of the signals from both alleles are plotted on the horizontal axis. Each sample generates two signals since the cTags are spotted in duplicate in the subarray. The three distinct clusters represent the three genotypes; in this example two negative controls and two failed samples fall outside the clusters.

   homozygote samples clustering at each side and the heterozygotes in the middle.
   The ratios may vary between SNPs, depending on the sequence surrounding it,
   the type of nucleotide incorporated, and the signal intensity of the fluorophores
   (*see* **Note 15**).
4. When using the ScanArray Express or QuantArray program for signal analysis, or
   if the signal quantitation output files have been converted to fit their format, the
   genotyping results can be visualized using the SNPSnapper software customized
   for this method (http://www.bioinfo.helsinki.fi/snpsnapper).

4. Notes

1. Many of the SNPs in databases have not been validated or may not be polymorphic in
   the population from which the study samples originate. The SNP allele frequencies in
   a particular population may be determined by analyzing pooled DNA samples using
   quantitative minisequencing in microtiter plates, or in the microarray format *(7,11)*.
2. A touchdown PCR procedure may be used *(12)*. One strategy when designing
   primers for multiplex PCR is to aim at as similar primer melting temperature and
   G/C content as possible. Complementary 3′ sequences in the primers can be avoided
   by designing primers with the same 3′ terminal nucleotides *(13)*. Other options are
   to introduce common tails on the 5′ ends of all PCR primers and to amplify subsequently with one common primer for all the fragments at an elevated temperature
   (14) or to use universal 5′ sequences, making the PCR primers eligible for the same
   reaction conditions *(15)*.

3. We recommend excluding SNPs located in repetitive elements identified by the RepeatMasker program (http://www.repeatmasker.org).

4. To avoid strong hairpin-loop structures, evaluate the complete minisequencing primer, including the Tag sequence. Secondary structures that involve the 3′ end of a primer may lead to misincorporation of nucleotides. A primer design software that predicts secondary structures (mfold:http://www.bioinfo.rpi.edu/~zukerm/ or NetPrimer: http://www.premierbiosoft.com/netprimer) can be used.

5. The array may also be manufactured with immobilized minisequencing primers. In this assay variant, the genotyping reaction is performed directly on the array surface *(16)*.

6. Microarrays may be purchased from a commercial supplier or manufactured in-house. There are several different slide types and attachment chemistries. Some of them have been tested in our system *(17)*.

7. The number of spots in each subarray can be varied by changing the subarray pattern from a 384- to a 96-well format; thus the number of spots in each subarray is increased, but the maximum number of samples that can be simultaneously analyzed is decreased *(9)*.

8. Elastosil RT601 may be used instead of RT625 to give a slightly harder silicon rubber to decrease deformation of the wells when the rack lid is tightened. If large subarrays, utilizing all available surface, are printed, deformation of the wells may cause the cTags in the corners to be covered by the silicon. The softer, RT625, silicon sticks better to the glass surface and decreases the risk of leakage between wells.

9. Depending on the number of SNPs to be explored, an inverted 96-well microtiter plate may be used as well as a silicon rubber mold to allow larger subarrays *(9)*.

10. Instead of multiplex PCR, single-fragment PCR can be used with subsequent pooling of the amplified fragments, possibly after concentration using ethanol precipitation or spin dialysis. Also, if large numbers of multiplex PCR products are pooled, it may be advantageous to concentrate the pool prior to the subsequent steps.

11. Alkaline phosphatase inactivates the remaining dNTPs, and exonuclease I degrades the single-stranded PCR primers, which would disturb the subsequent minisequencing reactions.

12. Cy5-ddUTP can be used at a 1.5–2-fold higher concentration than the other ddNTPs to compensate for its lower incorporation efficiency. Instead of using four differently labeled nucleotides in the same reaction, depending on the available microarray scanner, a single label or two labels may be used in four or two separate reactions, respectively.

13. If the fluorescent signals obtained are weak, the number of cycles may be increased. We have used up to 99 cycles.

14. Background problems can arise if the hybridization chamber is not kept humid; lack of humidity causes the samples on the slide to dry out.

15. The flanking sequences as well as the fluorophores attached to the dideoxynucleotides affect the efficiency and sequence specificity of nucleotide incorporation by the DNA polymerase. The different properties of the fluorophores, such as

molar extinction coefficients, emission spectra, and quantum yield, as well as nonspecific background, also affect the signal intensities and signal ratios obtained *(7)*.

Acknowledgments

This protocol is the result of the combined work effort of both former and present members of the Molecular Medicine research group at the Department of Medical Sciences at Uppsala University. The group has received financial support from the Swedish Research Council, the Wallenberg Foundation, and the European Commission (FP5 and FP6).

References

1. Sachidanandam, R., Weissman, D., Schmidt, S. C., et al. (2001) A map of human genome sequence variation containing 1.42 million single nucleotide polymorphisms. *Nature* **409,** 928–933.
2. Venter, J. C., Adams, M. D., Myers, E. W., et al. (2001). The sequence of the human genome. *Science* **291,** 1304–1351.
3. Lander, E. S., Linton, L. M., Birren, B., et al. (2001). Initial sequencing and analysis of the human genome. *Nature* **409,** 860–921.
4. Syvanen, A. C. (2001). Accessing genetic variation: genotyping single nucleotide polymorphisms. *Nat. Rev. Genet.* **2,** 930–942.
5. Syvanen, A. C., Aalto-Setala, K., Harju, L., Kontula, K., and Soderlund, H. (1990). A primer-guided nucleotide incorporation assay in the genotyping of apolipoprotein E. *Genomics* **8,** 684–692.
6. Hirschhorn, J. N., Sklar, P., Lindblad-Toh, K., et al. (2000). SBE-TAGS: an array-based method for efficient single-nucleotide polymorphism genotyping. *Proc. Natl. Acad. Sci. USA* **97,** 12164–12169.
7. Lindroos, K., Sigurdsson, S., Johansson, K., Ronnblom, L., and Syvanen, A. C. (2002) Multiplex SNP genotyping in pooled DNA samples by a four-colour microarray system. *Nucleic Acids Res.* **30,** e70.
8. Pastinen, T., Raitio, M., Lindroos, K., Tainola, P., Peltonen, L., and Syvanen, A. C. (2000). A system for specific, high-throughput genotyping by allele-specific primer extension on microarrays. *Genome Res.* **10,** 1031–1042.
9. Fredriksson, M., Barbany, G., Liljedahl, U., Hermanson, M., Kataja, M., and Syvanen, A. C. (2004). Assessing hematopoietic chimerism after allogeneic stem cell transplantation by multiplexed SNP genotyping using microarrays and quantitative analysis of SNP alleles. *Leukemia* **18,** 255–266.
10. Lovmar, L., Fredriksson, M., Liljedahl, U., Sigurdsson, S., and Syvanen, A. C. (2003) Quantitative evaluation by minisequencing and microarrays reveals accurate multiplexed SNP genotyping of whole genome amplified DNA. *Nucleic Acids Res.* **31,** e129.
11. Lagerstrom-Fermer, M., Olsson, C., Forsgren, L., and Syvanen, A. C. (2001). Heteroplasmy of the human mtDNA control region remains constant during life. *Am. J. Hum. Genet.* **68,** 1299–1301.

12. Don, R. H., Cox, P. T., Wainwright, B. J., Baker, K., and Mattick, J. S. (1991). 'Touchdown' PCR to circumvent spurious priming during gene amplification. *Nucleic Acids Res.* **19,** 4008.

13. Zangenberg, A. P., Saiki, R. K., and Reynolds, R. (1999). Multiplex PCR; optimization guidelines, in *PCR Applications* (Innis, M. A., Gelfand, D. H., and Sninsky, J. J., eds.), Academic Press, London, UK, pp. 73–94.

14. Brownie, J., Shawcross, S., Theaker, J., et al. (1997). The elimination of primer-dimer accumulation in PCR. *Nucleic Acids Res.* **25,** 3235–3241.

15. Shuber, A. P., Grondin, V. J., and Klinger, K. W. (1995). A simplified procedure for developing multiplex PCRs. *Genome Res.* **5,** 488–493.

16. Liljedahl, U., Karlsson, J., Melhus, H., et al. (2003). A microarray minisequencing system for pharmacogenetic profiling of antihypertensive drug response. *Pharmacogenetics* **13,** 7–17.

17. Lindroos, K., Liljedahl, U., Raitio, M., and Syvanen, A. C. (2001). Minisequencing on oligonucleotide microarrays: comparison of immobilisation chemistries. *Nucleic Acids Res.* **29,** E69.

5

Single-Nucleotide Polymorphism and Mutation Identification by the Nanogen Microelectronic Chip Technology

Maurizio Ferrari, Laura Cremonesi, Pierangelo Bonini, Barbara Foglieni, and Stefania Stenirri

Summary

The present chapter describes a microarray technology developed by Nanogen Inc., for the identification of DNA variations based on the use of microelectronics.

The NMW 1000 NanoChip™ Molecular Biology Workstation allows the active deposition and concentration of charged biotinylated molecules on designated test sites. The DNA at each pad is then hybridized with specific oligonucleotide probes, complementary to normal or mutant sequences, that labeled with Cy3 or Cy5 dyes, respectively. The array is imaged, and fluorescence signals are scanned, monitored, and quantified by highly developed, digital image-processing procedures. The experimental steps to be performed for the development and execution of a microchip assay are described. Attention is focused on the fundamental aspects of probe design, and guidelines and useful suggestions are given. Protocols for sample preparation, addressing, reporting, and data analysis are also detailed.

Key Words: Microarray; microelectronic; Nanogen; SNPs; mutation.

1. Introduction

Molecular diagnostics is being revolutionized by the development of highly advanced technologies for DNA testing. Through miniaturization of the test platform, DNA microarrays allow high-throughput analysis of genetic information in large sample populations *(1)*.

A microarray technology based on the use of microelectronics for the identification of SNPs and mutations in genes involved in genetic diseases was developed by Nanogen (San Diego, CA). The active electronic approach allows the acceleration of transport of DNA molecules to any selected position on the

From: *Methods in Molecular Medicine, Vol. 144, Microarrays in Clinical Diagnostics*
Edited by: T. Joos and P. Fortina © Humana Press Inc., Totowa, NJ

array surface. Moreover, the concentrating effect produced by active addressing greatly facilitates the hybridization reaction *(2–6)*.

Through the use of electric fields as an independent parameter, charged samples are actively deposited to designated test sites onto a cartridge. The NanoChip® is a multisite, individually controlled, 10×10 array of microelectrodes coated with a thin hydrogel permeation layer containing streptavidin. Biotinylated amplicons are automatically placed on the cartridge and electronically addressed to selected pads by positive bias direct current, where they remain embedded through interaction with streptavidin. The DNA at each pad is then hybridized to a mixture of stabilizers and Cy5- and Cy3-labeled oligonucleotide probes, specific for either the wild-type or the mutant sequence. Single-base pair mismatched probes are then preferentially denatured with the application of an increasing temperature on the cartridge.

After this step of thermal stringency, the array is imaged and the fluorescence quantified. Fluorescence signals emanating from positive test sites are scanned, monitored, and quantified by highly developed, digital image processing procedures. An advanced multitask system controls all specific aspects of machine operation, including assay execution, fluorescent signal detection, signal processing, and data analysis. The fluorescence signal ratios of the reporter probes, discriminating among mutant, heterozygous, and wild-type samples, allow assessment of the patient genotype.

The technology can easily be designed for detection of many DNA variations located within the same DNA amplicon. Moreover, in view of large-scale screening applications, several alternative multiplexing features can be successfully developed, reducing time and costs. In particular, to load more than one sample on the same pad in a single addressing step, conditions for multiplex PCR reactions can be set up. Alternatively, single PCR reactions can be pooled, purified, and addressed to the same pad as a unique sample.

In addition, successful results can also be obtained by readdressing the same pad several times with different amplicons or, alternatively, by multiple simultaneous hybridizations of the same chip with different sets of probes. The combination of all such properties may allow the user to maximize the number of mutations typed per sample and/or to increase the number of addressable test sites, as needed *(7)*.

Many application of Nanogen technology for DNA variation identification have been reported in different fields *(6,8–12)*. The technology has the potential for several additional applications such as unknown mutation detection *(13)*, analysis of simple sequence repeats *(14)*, analysis of anchored *in situ* amplification *(15–17)*, expression profiling *(18,19)*, immunoassays, and preparation of DNA/RNA from bacteria for pathogen screening *(20–22)*.

2. Materials

1. NMW 1000 NanoChip Molecular Biology Workstation, composed of a Loader, in which polymerase chain reaction (PCR) fragments are addressed to the microchip, and a Reader, which processes and scans the cartridge chip.
2. NanoChip Cartridge.
3. High-performance liquid chromatography (HPLC)-purified oligonucleotides: PCR primers (biotinylated and unmodified), Cy3- and Cy5-labeled wild-type and mutant reporters, stabilizer oligonucleotides.
4. PCR reagents.
5. Agarose gel electrophoresis equipment.
6. 96-Well format desalting plates (Multiscreen-PCR Plates MANU 030, Millipore, Bedford, MA).
7. Vacuum manifold (Multiscreen Separation System, Millipore).
8. Conductivity meter.
9. Nunc 96-/384-well plates (Nalge Nunc, Rochester, NY).
10. 50 mM L-histidine: prepare in a sterile or autoclaved recipient, filter using a 0.2-μm porosity filter, check the conductivity (optimal range < 100 μS/cm), and store at 2–8°C in the dark. Prepare fresh every week.
11. 100 mM L-histidine: prepare in a sterile or autoclaved recipient, filter using a 0.2-μm porosity filter, check the conductivity (optimal range < 100 –200 μS/cm), and store approx 5-mL aliquots at −20°C for up to 2 mo.
12. 0.1 N NaOH. Prepare fresh every week.
13. High salt buffer: 500 mM sodium chloride and 50 mM sodium phosphate, pH 7.4 (prepared from a 1 M sodium phosphate dibasic/monobasic 4:1 stock solution). Filter using a 0.2-μm filter, and store at room temperature.
14. Low salt buffer: 50 mM sodium phosphate, pH 7.0 (prepared from a 1 M sodium phosphate dibasic/monobasic 7:3 stock solution). Filter using a 0.2-μm filter, and store at room temperature.

3. Methods

3.1. PCR Primer Set, Stabilizer, and Probe Design

The SNP assay requires four components: the amplicon, the stabilizer oligo, and the two reporters, each specific for either the wild-type or mutant allele (**Fig. 1**). The detection method is based on hybridization of a biotinylated single-stranded amplified DNA molecule to the probe mixture. Reporters are designed to be complementary at their 3′ end to the polymorphic or mutated target DNA nucleotide and are labeled at their 5′ end with a Cy3/Cy5 fluorophore (*see* **Note 1**). The stabilizer oligo is designed contiguous to the 3′ end of the reporter oligo. The format of the reporter-stabilizer juxtaposition allows one to exploit the base-stacking energy, which is generated between neighboring nucleotides along a DNA strand. Hence, when a reporter is specifically hybridized to the target,

**Table 1
Base-Stacking Energy Calculated In the 5′→3′ Orientation[a]**

	A/A	A/C	A/G	A/T	T/A	T/C	T/G	T/T
Energy	−5.37	−10.5	−6.78	−6.57	−3.82	−9.81	−6.57	−5.37
	C/A	C/C	C/G	C/T	G/A	G/C	G/G	G/T
Energy	−6.57	−8.26	−9.69	−6.78	−9.81	−14.6	−8.26	−10.5

[a]In this example the base-stacking energy is between the 3′ end of the reporter and the 5′ end of the stabilizer (G/A = −9.81).

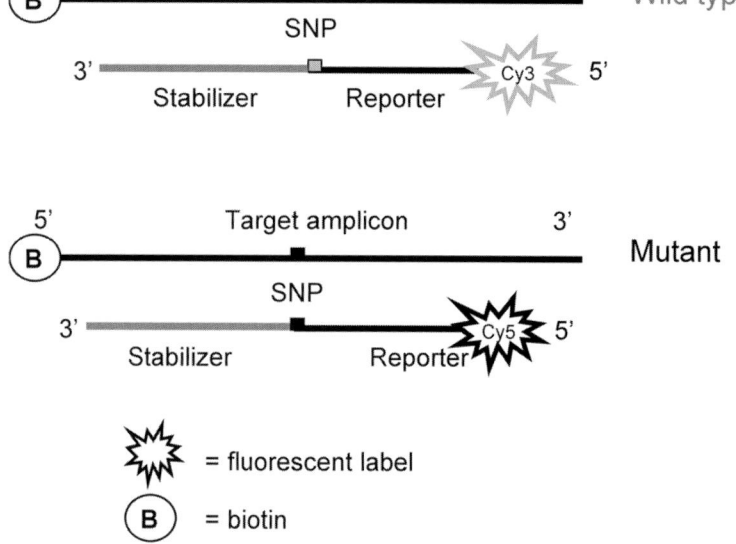

Fig. 1. Schematic representation of the base-stacking single-nucleotide polymorphism (SNP) detection assay format.

the base-stacking interactions across the junction between reporter and stabilizer make this hybridization more stable. The values of base-stacking energies for each base pair combination are listed in **Table 1**.

For optimal assay performance, the base-stacking energies and the melting temperatures between the reporters should be very close, to allow the same hybridization behavior with respect to the specific allele.

In some cases a base-stacking design cannot be exploited, owing to unbalanced base-stacking energies between either wild-type or mutant reporters and stabilizer.

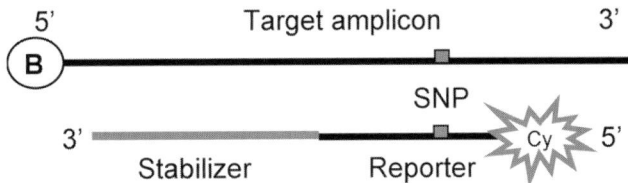

Fig. 2. Schematic representation of the dot-blot single-nucleotide polymorphism (SNP) detection assay format.

In this situation it would be more appropriate to use a dot-blot format with the base variation located internally to the probe sequence **(Fig. 2)**. This approach would also be advisable in case a polymorphism is located near the SNP of interest, which could interfere with the stability of the stabilizer-reporter complex.

Recently a universal probe reporting system has been developed *(23)*, which allows labeling of only two oligonucleotides (universal probes), regardless of the number of DNA variations to be tested (*see* **Note 2**).

PCR primers are designed to amplify the genomic region surrounding the SNP/mutation of interest; one of the two primers (forward or reverse) is 5'-biotinylated. Suggested fragment length ranges between 150 and 500 bp (*see* **Note 3**). Fragments shorter than 137 bp require a desalting procedure different from the recommended Millipore MultiScreen-PCR filtration plate. Within the amplified fragment, the DNA variation should be located 70–80 bp from the biotinylated end.

Stabilizer and reporter oligos should be designed to avoid secondary structures of single-stranded target DNA that could interfere with hybridization. To predict secondary structures, the fragment sequence can be analyzed with the help of a web-free program available at http://www.bioinfo.rpi.edu/Ezukerm/rna. **Figure 3** gives an example of an assay designed with this software.

The classic DNA variation detection format requires the use of two reporter probes, one specific for the wild-type allele and the other for the mutated one.

3.1.1. Oligo Design Guidelines

For the design of both PCR primers and detection oligonucleotides, the program Oligo Analyzer 3.0 (Integrated DNA Technologies; http://www.idtdna.com) can be used. Secondary structures such as stem-loops, hairpins, dimers, and others. should be avoided. All oligonucleotides should be DHPLC purified.

3.1.1.1. REPORTERS

Reporters are designed to have:

1. The 3' end complementary to the base involved in the DNA alteration.

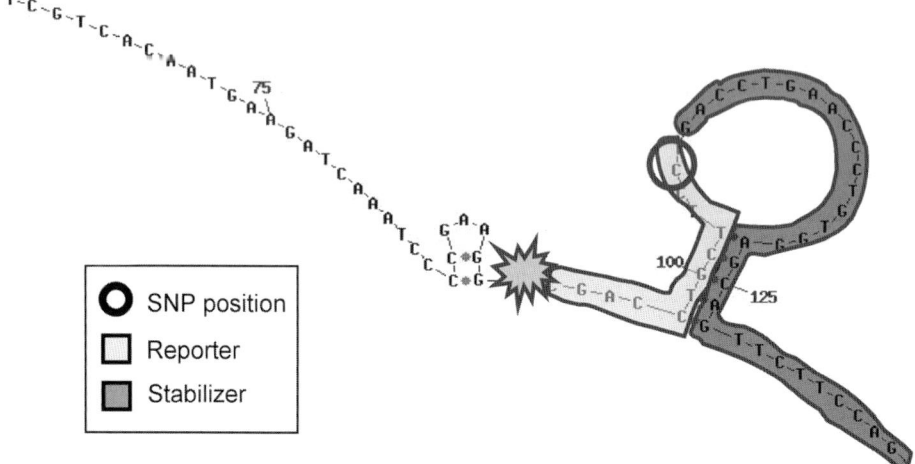

Fig. 3. Example of secondary structure computer prediction. SNP, single-nucleotide polymorphism.

2. 5′ Cy3- and 5′ Cy5 labeling of wild-type and mutant reporters, respectively.
3. Melting temperature (T_m) of 35–40°C and length of 10–14 bp depending on the GC content of the region. The T_ms of the two reporters should be very similar (within 2°C) and can be balanced by using reporter oligos of different lengths.
4. Maximum negative and balanced base-staking energy (between stabilizer/wild-type reporter and stabilizer/mutant reporter).

3.1.1.2. STABILIZERS

Stabilizers are designed to have:

1. The 5′ end contiguous to the 3′ end of the reporter oligo.
2. Melting temperature ≥60°C and length ranging between 30 and 35 bp, depending on the GC content of the region.
3. Maximum negative and balanced base-staking energy.

3.2. Sample Preparation

Genomic DNA can be extracted by using commercial kits or standard procedures; PCR amplification can be performed by a standard protocol, without any particular reagent limitation (dimethyl sulfoxide, DNA polymerase, and so on). Thermal cyclers requiring mineral oil overlay which can interfere with subsequent operative steps should be avoided. The final PCR volume should be 50–100 μL.

After the amplification process, each amplicon is purified and desalted by the use of a 96-well format Multiscreen-PCR Plate coupled with the Multiscreen Separation System according to the following protocol:

1. Add 100 µL of distilled water to each PCR sample (50–100 µL), and mix.
2. Transfer the sample into a desalting plate, without touching the filter membrane.
3. Place the desalting plate on the vacuum manifold, and apply vacuum until all liquid is pulled through the filter (approx 10 min at 10 mmHg).
4. Turn the vacuum pump off, add 100 µL of distilled water to each well, and repeat **step 3**.
5. Remove the desalting plate from the vacuum manifold, and add 70 µL of distilled water. (This final volume is enough for two loading processes.)
6. Place the plate on an orbital shaker for 10–15 min: (If no shaker is available, let the plate stand for 10 min on the bench, and then mix the sample vigorously 10 times.)
7. Mix each sample by pipeting twice, and transfer it into a clean PCR tube. Store samples at 4–8°C (*see* **Note 4**).

Sample conductivity is checked by using the conductivity meter supplied with the Nanogen instrumentation (optimal range < 100 µS/cm). If the conductivity is out of range, an additional washing with 100 µL of distilled water can be performed in new wells of the desalting plate. Gel agarose inspection for amplification and purification checking can be randomly performed on approx 10% of samples.

3.3. Microchip Addressing

Sample are loaded into a Nunc 96-/384-well plate.

The SNP assay includes a negative control (50 m*M* histidine) for background signal subtraction and a heterozygous control specific for each mutation to be analyzed. The heterozygous control is prepared in the same way as the samples, following the protocols described in **Subheading 3.2.** above. For development of the assay, we suggest loading and analyzing each sample in duplicate on two different pads.

1. Loading guidelines:
   a. Mix 30 µL of each sample with 30 µL of 100 m*M* histidine to a final concentration of 50 m*M*.
   b. Pipet 60 µL of 0.1 *N* NaOH into a plate-well.
   c. Create a new Loader Map File.
   d. Load the plate and a new cartridge into the Loader, and start the process.
2. Fill in a new Loader Map File by designating the addressing site on the cartridge for each sample in the loader.
3. The negative control is addressed according to a Capture submap (120 V, 1 min), and all samples are addressed according to a Target/Amplicon submap (120 V, 2 min). The NaOH is addressed to the cartridge as a Passive submap (3 min) after all samples and controls. This final step denatures PCR products and removes the nonbiotinylated single-stranded DNA, making the biotinylated one available for subsequent hybridization steps.

4. Before starting the loading protocol, the Loader performs a conductivity test auto matically to check for activation of addressed pads. Failed pads cannot be included in the loading map.
5. As an additional quality check, the instrument provides the activation values for each addressing step. The expected activation range is 300 600 nA.

The same chip can be readdressed with different samples more than once (*see* **Note 5**).

3.4. Cartridge Reporting

1. After electronic addressing for binding of biotinylated products to streptavidin-containing gel pads, the cartridge is manually incubated for 3 min at room temperature with 100 µL of a mixture specific for each SNP, containing stabilizers and reporters (1 µM each) in high salt buffer (*see* **Note 6**).
2. If the addressed amplicons include more than one SNP, previous reporter probes and stabilizers need to be stripped by incubating the cartridge for 3 min with 0.1 *N* NaOH and washed five times with 150 µL H_2O and five times with 150 µL high salt buffer. The target DNA is now free for a new reporting.
3. Multiple hybridization protocols on the same chip with more than one set of probes can be performed (*see* **Note 7**).

3.5. Fluorescence Analysis

1. The cartridge is placed into the Reader.
2. To detach all fluorescent probes that are not perfectly matched to template DNA, a thermal stringency step is performed by raising the temperature inside the cartridge (*see* **Note 8**).
3. Hybridization is detected by fluorescence automated scanning.
4. The user creates a Reader protocol containing heating, washing, and scanning steps specifically developed for each SNP.
5. For appropriate fluorescence, signal acquisition, users can adjust the photomultiplier sensitivity by changing both the accumulation time (automatic/manual) and, for each of the two lasers, the tube gain (low/medium/high).
6. A standard reader protocol includes the following steps:
   a. Initial scanning in high salt buffer at room temperature, to check the hybridization in non stringent conditions.
   b. Heating the cartridge to SNP specific discrimination temperature.
   c. Washing the cartridge with low salt buffer.
   d. Scanning the cartridge.
7. For setting up a new assay, we suggest that the Reader protocol be performed with a gradient of increasing temperatures to asses the optimal thermal stringency condition.
8. The addressed cartridge can be stored at 4°C up to 1 mo. If the cartridge is not completely loaded, the addressing process can be repeated several times until all pads are filled.

3.6. Interpretation of Results

1. Quantitative analysis of the hybridization results is performed by a dedicated software provided on the workstation, which can be also accessed through the web from other PC stations.
2. The software automatically subtracts background control values from each pad signal. Only samples displaying a signal-to-noise ratio greater than five fold are considered.
3. Fluorescence values from the heterozygous control are used to normalize the Cy3/Cy5 signals to a value of 1, and fluorescence data for all samples are adjusted accordingly.
4. Alleles are assigned by evaluating the specific to nonspecific hybridization ratio. As default parameters, samples displaying a wild-type to mutant signal ratio of between 1:1 and 1:2 are assigned as heterozygotes; samples with a wild-type/mutant signal ratio of > 5:1 or > 1:5 are assigned as wild-type homozygotes or mutant homozygotes, respectively; samples with a wild-type/mutant signal ratio between 1:2 and 1:5 are not scored (no call).
5. If samples are addressed on more than one pad, the mean value is automatically calculated by the software for sample genotyping.
6. For standard SNP analysis, we recommend that these rules be followed; however, both black-to-gray and signal-to-noise ratio default threshold values can be changed for specific applications.
7. Row data are recorded and archived in specific files on the workstation and can be exported through a zip disk. These data can be analyzed time by time on selected samples by changing different parameters. Processed data are available in a useful format that includes genotype designation.

Example **Figure 4** shows an example of results for the β-globin IVS1-110 (G→A) mutation.

1. Figure 4A shows the histogram for 10 samples.
2. Figure 4B gives black and gray normalized signals, number of pads used for the analysis, black and gray ratios (B:G), and final genotyping for IVS1-110, for each sample. Sample 10 is used as an internal heterozygous control for normalization of wild-type and mutated signals.
3. Figure 4C gives more detailed and useful statistical information.

    a. For each sample the number of pads addressed (Count) is indicated.
    b. Background black and gray values are the means of the signals obtained from the pads addressed with histidine; this background is automatically subtracted from all the samples addressed during the same process.
    c. Signal-to-noise ratio mean (Mean SNR) is calculated on row data (without background subtraction). Mean corresponds to row fluorescence signals after background subtraction, standard deviation (STD) and correlation value percentage (%CV) are calculated among pads on the mean value.

A

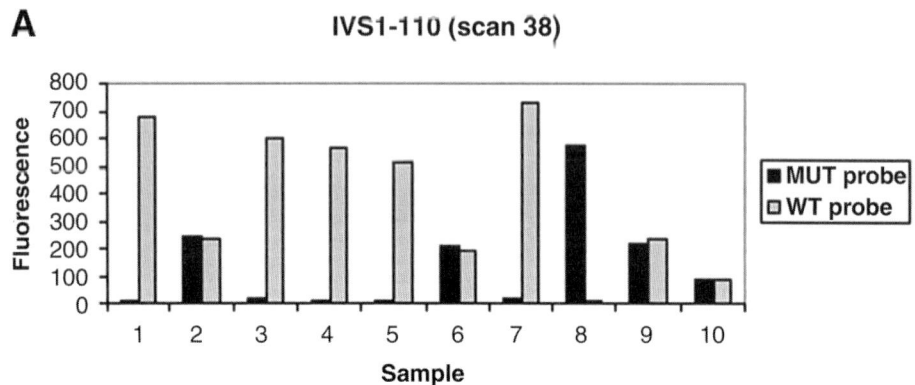

B

		Results for IVS1-110 (scan 38)			
Sample	Red	Green	Pads	Ratio (R:G)	Probe Designation
1	11.45	674.36	2	1:58.9	G/G
2	245.25	231.06	2	1.06:1	A/G
3	13.45	598.7	2	1:44.51	G/G
4	8.6	561.71	2	1:65.32	G/G
5	11.7	512.96	2	1:43.84	G/G
6	210.75	192.95	2	1.09:1	A/G
7	17.8	733.2	2	1:41.19	G/G
8	574.65	12.49	2	57.77:1	A/A
9	218.25	230.5	2	1:1.06	A/G
10	83.75	83.73	2	1:1	A/G

C

			Statistics for IVS1-110 (scan 38)						
Locus	Sample	Background	Mean SNR	Count	Mean	STD	%CV	Scale	Scaled Mean
IVS1-110	1	6.25	2.83 :: 1	2	11.45	0.85	7.41	1	11.45
		13.85	44.43 :: 1		601.45	48.79	8.11	1.12	674.36
IVS1-110	2	6.25	40.24 :: 1	2	245.25	0.71	0.29	1	245.25
		13.85	15.88 :: 1		206.15	7.07	3.43	1.12	231.06
IVS1-110	3	6.25	3.15 :: 1	2	13.45	7.92	58.88	1	13.45
		13.85	39.57 :: 1		534.15	31.11	5.82	1.12	598.7
IVS1-110	4	6.25	2.38 :: 1	2	8.6	0.35	4.11	1	8.6
		13.85	37.18 :: 1		501.15	80.61	16.09	1.12	561.71
IVS1-110	5	6.25	2.87 :: 1	2	11.7	0.64	5.44	1	11.7
		13.85	34.04 :: 1		457.65	133.64	29.2	1.12	512.96
IVS1-110	6	6.25	34.72 :: 1	2	210.75	32.53	15.43	1	210.75
		13.85	13.43 :: 1		172.15	9.9	5.75	1.12	192.95
IVS1-110	7	6.25	3.85 :: 1	2	17.8	5.02	28.2	1	17.8
		13.85	48.23 :: 1		654.15	46.67	7.13	1.12	733.2
IVS1-110	8	6.25	92.96 :: 1	2	574.75	118.09	20.55	1	574.75
		13.85	1.81 :: 1		11.15	3.11	27.9	1.12	12.49
IVS1-110	9	6.25	35.92 :: 1	2	218.25	38.89	17.82	1	218.25
		13.85	15.85 :: 1		205.65	26.16	12.72	1.12	230.5
IVS1-110	10	6.25	14.40 :: 1	2	83.75	4.53	5.4	1	83.75
		13.85	6.39 :: 1		74.7	3.04	4.07	1.12	83.73

d. Finally, the software calculates the normalization factor (Scale) on the basis of the value obtained for the heterozygous control (sample 10) and normalizes all the unknown samples by multiplying the mean by the scale value (Scaled Mean).

A different representation of results is available in which all the analyses performed on the same sample are displayed together. This format is particularly indicated for setting up a new assay. **Figure 5** shows the trend of fluorescence by increasing temperature steps for the IVS1-110 mutation. As expected, for the heterozygous sample, the same behavior for both wild-type (diamonds) and mutant (circles) signals are obtained. Conversely, in the homozygous wild-type sample, the mutant signal, corresponding to the mismatched probe, shows a low hybridization signal rapidly decreasing to background level by increasing thermal stringency. As the final testing temperature, the one at which matched and mismatched probes display the higher ratio and fluorescent signal is chosen.

4. Notes

1. Classical 5′ Cy3 and 5′ Cy5 labeling of wild-type and mutant reporter, respectively, is more advisable. In case a secondary structure is present at the 5′ end of a target strand, which can interfere with the hybridization process, the location of stabilizer and reporters can be inverted. In this situation labeling must be performed at the free 3′ end of reporters; in this case we suggest the use of different fluorophores, such as 6-TAMRA/Bodipy650/665 for wild-type and mutant reporter, respectively, owing to the low labeling efficiency of Cy3/Cy5 at the 3′ end.
2. The universal probe reporting system consists of a pair of random, nonhuman either Cy3 or Cy5 dye-tagged oligonucleotides (universal probes). The wild-type and mutated reporter oligos, called *discriminators*, are designed to contain a tail that specifically hybridizes to either the Cy3 or the Cy5 universal probe. This approach eliminates the need to purchase large numbers of expensive labeled reporters for SNP analysis.
3. Optimal fragment length can vary according to the base pair context surrounding the DNA variation of interest.
4. Samples can be directly resuspended into 130 µL of 50 mM histidine. A loading volume of 60 µL is enough to load samples twice. If the total volume of the PCR

Fig. 4. *(Figure on opposite page)* **(A)** Microchip analysis for the IVS1-110 β–globin gene mutation. The graph shows fluorescence signals for 10 samples at a stringency temperature of 38°C. Sample 10 was used as heterozygous control for normalization of gray (wild-type [WT]) and black (mutant [MUT]) signals. Result table **(B)** and statistical table **(C)** for IVS1-110 β–globin gene mutation. For each sample, background values, signal-to-noise ratio mean (Mean SNR), number of pads addressed with the same sample (Count), background subtracted mean (Mean), standard deviation (STD), correlation value percentage (%CV), normalization factor (Scale) and normalized fluorescence signals (Scaled Mean) are indicated. B:G, black/grey.

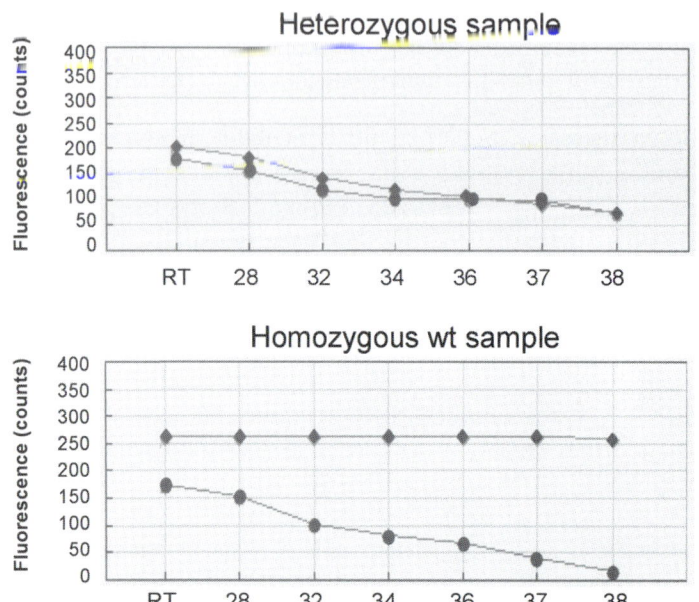

Fig. 5. Trending graph for the IVS1-110 β–globin gene mutation ranging from room temperature (RT) to 40°C. WT, wild type (diamonds); mutant (circles).

has to be loaded onto the NanoChip, samples can be resuspended in 70 µL of 50 m*M* histidine. Samples resuspended in histidine should be stored at −20°C, to keep conductivity within the optimal range.

5. The same chip can be sequentially addressed with PCR reactions amplifying different genomic region. Since previous amplicons cannot be removed from the chip, probes for the new assay should not hybridize to them. Before each addressing, stripping of the previous probes by incubating the chip with 0.1 *M* sodium hydroxide for 3 min is performed.

6. When stabilizer and reporters are partially complementary to each other, they should be separately and subsequently hybridized.

7. If different pads on the same chip are addressed, each with a different DNA fragment, the reporting process can be performed by hybridizing all pads simultaneously with a mixture containing several pairs of wild-type and mutant probes, each specific for only one mutation per amplicon. After passive multiple hybridization with each set of probes, thermal stringency specific for the probe with the lower melting temperature is applied, and fluorescence is detected at the corresponding pads. The scanning process is repeated in succession at the increasing temperature specific for the remaining sets of probes.

8. As an alternative to thermal stringency, the system is also designed to allow application of electronic stringency. To remove any unbound or nonspecifically bound DNA from each site, the charge polarity of the site can be reversed to negative,

thereby forcing any unbound or nonspecifically bound DNA back into solution. This approach is useful when coupled with multiple hybridization. When more than one set of probes is used, the system is able to modulate electronic stringency specifically to each sample. This allows one to analyze probes with different melting temperature simultaneously, an advantage that cannot be exploited by using thermal stringency.

Acknowledgments

This work was supported by Amplimedical S.p.A. Diagnostic Group, Italy and FIRB grant RBNE01SLRJ (to M. Ferrari).

References

1. Ferrari, M., Stenirri, S., Bonini, P., and Cremonesi, L. (2003) Molecular diagnostics by microelectronic microchips. *Clin. Chem. Lab. Med.* **41,** 462–467.
2. Edman, C. F., Raymond, D. E., Wu, D. J., et al. (1997) Electric field directed nucleic acid hybridization on microchips. *Nucleic Acids Res.* **25,** 4907–4914.
3. Sosnowski, R. G., Tu, E., Butler, W. F., O'Connell, J. P., and Heller, M. J. (1997) Rapid determination of single base mismatch mutations in DNA hybrids by direct electric field control. *Proc. Natl. Acad. Sci. USA* **94,** 1119–1123.
4. Gilles, P. N., Wu, D. J., Foster, C. B., Dillon, P. J., and Chanock, S. J. (1999) Single nucleotide polymorphic discrimination by an electronic dot blot assay on semiconductor microchips. *Nat. Biotechnol.* **17,** 365–370.
5. Heller, M. J., Forster, A. H., and Tu, E. (2000) Active microeletronic chip devices which utilize controlled electrophoretic fields for multiplex DNA hybridization and other genomic applications. *Electrophoresis* **21,** 157–164.
6. Santacroce, R., Ratti, A., Caroli, F., et al. (2002) Analysis of clinically relevant single-nucleotide polymorphisms by use of microelectronic array technology. *Clin. Chem.* **48,** 2124–2130.
7. Foglieni, B., Cremonesi, L., Travi, M., et al. (2004) Beta-thalassemia microelectronic chip: a fast and accurate method for mutation detection. *Clin. Chem.* **50,** 73–79.
8. Pollak, E. S., Feng, L., Ahadian, H., and Fortina, P. (2001) Microarray-based genetic analyses for studying susceptibility to arterial and venous thrombotic disorders. *Ital. Heart J.* **2,** 568–572.
9. Erali, M., Schmidt, B., Lyon, E., and Wittwer, C. (2003) Evaluation of electronic microarrays for genotyping factor V, factor II, and MTHFR. *Clin. Chem.* **49,** 732–739.
10. Thistlethwaite, W. A., Moses, L. M., Hoffbuhr, K. C., Devaney, J. M., and Hoffman, E. P. (2003) Rapid genotyping of common MeCP2 mutations with an electronic DNA microchip using serial differential hybridization. *J. Mol. Diagn.* **5,** 121–126.
11. Schrijver, I., Lay, M. J., and Zehnder, J. L. (2003) Diagnostic single nucleotide polymorphism analysis of factor V Leiden and prothrombin 20210G > A. A comparison of the Nanogen Electronic Microarray with restriction enzyme digestion and the Roche LightCycler. *Am. J. Clin. Pathol.* **119,** 490–496.

12. Stenirri, S., Foglieni B., Manitto, M.P., et al. (2002) Single nucleotide polymor phism and mutation identification by microelectronic chip technology. *Minerva Biotec.* **14,** 241–246.

13. Behrensdorf, H. A., Pignot, M., Windhab, N., and Kappel, A. (2002) Rapid parallel mutation scanning of gene fragments using a microelectronic protein-DNA chip format. *Nucleic Acids Res.* **30,** e64.

14. Radtkey, R., Feng, L., Muralhidar, M., et al. (2000) Rapid, high fidelity analysis of simple sequence repeats on an electronically active DNA microchip. *Nucleic Acids Res.* **28,** E17.

15. Westin, L., Xu, X., Miller, C., Wang, L., Edman, C. F., and Nerenberg, M. (2000) Anchored multiplex amplification on a microelectronic chip array. *Nat. Biotechnol.* **18,** 199–204.

16. Westin, L., Miller, C., Vollmer, D., et al. (2001) Antimicrobial resistance and bacterial identification utilizing a microelectronic chip array. *J. Clin. Microbiol.* **39,** 1097–1104.

17. Edman, C. F., Mehta, P., Press, R., Spargo, C. A., Walker, G. T., and Nerenberg, M. (2000) Pathogen analysis and genetic predisposition testing using microelectronic arrays and isothermal amplification. *J. Invest. Med.* **48,** 93–101.

18. Huang, Y., Joo, S., Duhon, M., Heller, M., Wallace, B., and Xu, X. (2002) Dielectrophoretic cell separation and gene expression profiling on microelectronic chip arrays. *Anal. Chem.* **74,** 3362–3371.

19. Weidenhammer, E. M., Kahl, B. F., Wang, L., et al. (2002) Multiplexed, targeted gene expression profiling and genetic analysis on electronic microarrays. *Clin. Chem.* **48,** 1873–1882.

20. Cheng, J., Sheldon, E. L., Wu, L., et al. (1998) Preparation and hybridization analysis of DNA/RNA from *E. coli* on microfabricated bioelectronic chips. *Nat. Biotechnol.* **16,** 541–546.

21. Ewalt, K. L., Haigis, R. W., Rooney, R., Ackley, D., and Krihak, M. (2001) Detection of biological toxins on an active electronic microchip. *Anal. Biochem.* **289,** 162–172.

22. Yang, J. M., Bell, J., Huang, Y., et al. (2002) An integrated, stacked microlaboratory for biological agent detection with DNA and immunoassays. *Biosens. Bioelectron.* **17,** 605–618.

23. Cooper, K. L. and Goering, R. V. (2003) Development of a universal probe for electronic microarray and its application in characterization of the *Staphylococcus aureus* polC gene. *J. Mol. Diagn.* **5,** 28–33.

6

Molecular Diagnostic Testing for Inherited Thrombophilia Using Invader®

Margaret A. Keller

Summary

Physicians in the United States and Europe began testing patients who had idiopathic thrombotic events for inherited risk factors in 1990s. The College of American Pathologists (CAP) offered proficiency testing for molecular genetic screening for thrombophilia in 1997. Today, a hypercoagulable workup including screening for inherited thrombophilia defects is becoming part of the standard of care in many parts of the world *(1)*. Who, what, and when to test continue to be controversial and challenging questions *(2)*; however, laboratories developing new or improved mutation detection methodologies have used the most commonly screened inherited thrombophilia polymorphism, factor V Leiden (R506Q), for many years *(3)*. In the most recent CAP survey (MGL-A 2003), the most commonly employed method used in one-third of all participating clinical laboratories testing for factor V Leiden and prothrombin G20210A was the non-PCR-based method called Invader®, developed by Third Wave Technologies. The remainder of the clinical laboratories reported testing for these variants using PCR-restriction fragment length polymorphism (RFLP), allele-specific PCR, and allele-specific hybridization. Emerging mutation detection technologies include DNA resequencing approaches such as pyrosequencing *(4)* fluorescence polarization detection *(5)*, genotyping on microelectronic DNA chips like Nanogen's nanochip *(6)*, and oligonucleotide hybridization with photocrosslinking *(7)*. Invader technology is currently a medium-throughput, 96-well plate format assay that is sufficient for most hospital clinical laboratories. Although this assay format is not currently performed in a microarray format, Invader is amenable to performance on a solid support; specifically, the reaction can be performed on the surface of microspheres and the resulting fluorescence measured using flow cytometry *(8)*. This chapter presents the method as well as some suggestions for utilizing the Invader system for mutation/polymorphism screening in general and for thrombophilia testing in particular.

Key Words: Thrombophilia; venous thrombosis; hypercoagulable; factor V Leiden; prothrombin; factor II; methylene tetrahydrofolate reductase (MTHFR); mutation detection; Invader.

From: *Methods in Molecular Medicine, Vol. 144, Microarrays in Clinical Diagnostics*
Edited by: T. Joos and P. Fortina © Humana Press Inc., Totowa, NJ

1. Introduction

Thrombophilia is defined as a predisposition to thrombosis or obstructive clot formation. Although there are acquired forms of thrombophilia, such as thrombosis associated with surgery, trauma, immobilization, or malignancy, in this context, we present molecular genetic screening for *inherited* thrombophilia. We will address only heritable risk factors for *venous* thrombosis, which most commonly manifests in the form of deep vein thrombosis (DVT) or pulmonary embolism (PE). Both the environmental and genetic risk factors for *arterial* thrombotic events, such as myocardial infarction and ischemic stroke, are different *(1)*. Often, when patients present with an idiopathic venous thrombotic event such as DVT or PE, their physician orders what is sometimes referred to as a hypercoagulable workup, several functional tests including measuring plasma levels of antithrombin III, protein C, and protein S, fasting total plasma homocysteine levels, and several genetic tests, including factor V Leiden and prothrombin 20210. Some laboratories also perform genetic tests to screen for polymorphisms in the methylene tetrahydrofolate reductase (MTHFR) gene, a rate-limiting enzyme involved in folate metabolism. These tests are also available for Invader assay.

1.1. Inherited Thrombophilia Risk Factors

1.1.1. Factor V Leiden

Factor V Leiden (FVL) results from a single base change in the coding region of the FV gene, causing substitution of arginine (R) for glutamine (Q) at amino acid 506. This change results in an FV protein that is resistant to inactivation by activated protein C *(9)*. Thus, once activated, plasma FVa levels persist, resulting in a prothrombotic environment *(3,10)*. Individuals homozygous for FVL are at an 80-fold increased risk of a thrombotic event over their lifetime *(11)*, and 20–50% of patients with venous thrombosis (VT) are FVL heterozygotes *(1)*. The risk of VT is affected by gene-gene interactions such as the inheritance of multiple, distinct prothrombotic polymorphisms, as is the case in patients compound heterozygous for FVL and PT20210, who have an 20-fold increased risk of experiencing VT over their lifetime. The risk of VT is also affected by gene-environment interactions such as heterozygosity for the FVL mutation combined with oral contraceptive use *(12)*. FVL heterozygote frequencies range from a high of 5.2% in Caucasians to a low of 0.45% in African Americans *(13)*.

1.1.2. Prothrombin G20210A

The PT20210 polymorphism is a G → A nucleotide change in the noncoding region of the gene encoding factor II, also known as prothrombin. Individuals with the PT20210 polymorphism have elevated levels of prothrombin in their plasma *(14)*. The mechanism for this is not clear; however, PT20210A may increase the efficiency of processing of messenger RNA transcribed from this gene *(15)* or increased translational efficiency of the message *(16)*. Carriers of PT20210 have an approximately threefold increased risk of VT *(14)*. The PT20210 G →A polymorphism is present in 1–6% of Caucasians and is nearly absent in individuals of Asian and African descent *(17,18)*.

1.1.3. MTHFR

The link between elevated homocysteine and increased risk of VT thrombosis is well documented *(19,20)*. Hyperhomocysteinemia has been linked to mutations in the enzymes that regulate folate metabolism, including cystathionine β-synthase and MTHFR *(21)*. A C to T polymorphism in the MTHFR gene leads to an alanine to valine substitution at codon 677 in the catalytic domain of the MTHFR protein, resulting in a variant protein that has reduced enzymatic activity and is heat sensitive in vitro *(22)*.

The association of elevated homocysteine and risk of thromboembolic event has been shown by several large studies and meta-analyses *(23)* and is probably affected by gene-environment interactions such as folate and vitamin B_{12} status. However, the influences of the C677T polymorphism that generates a thermolabile variant of the enzyme and that of another common polymorphism, A 1298 C (alanine to glutamate) *(24)*, are less clear *(20,25)*. In general, individuals homozygous for the C677T polymorphism have elevated homocysteine *(21)*, and some studies suggest that compound heterozygotes for C677T and A1298C have elevated levels *(24)*. Plasma homocysteine levels are significantly influenced by environmental factors such as diet and vitamin B_{12} status *(26,27)*. The MTHFR C677T variant allele is present in 65% of Japanese, 30% of Caucasians, and 10% of African Americans *(18,28,29)*. The MTHFR A1298C variant allele frequency is approx 2% in Asian and African Americans and approx 8% in Caucasians. Given these allele frequencies, approx 5% of Asians and African Americans and approx 15% of Caucasians would be expected to be compound heterozygotes for these two polymorphisms *(30)*.

An emerging challenge in inherited thrombophilia is how to interpret the inheritance of multiple thrombophilic polymorphisms *(31)*. Also, optimal clinical management of these individuals on long-term anticoagulation therapy is controversial *(1)*. The FVL and PT20210 variants are extremely rare in some ethnic groups, including individuals of African descent *(32)*. Recently, elevated levels of factor VIII were shown to confer risk for VT *(33)*, and a polymorphism in its receptor, LRP, has been associated with variation in Factor VIII levels *(34)*.

Given that thrombophilia testing is becoming the standard of care in many parts of the world, the testing methodology must be simple and reliable enough that it can be implemented by clinical laboratory staff not trained in molecular biology techniques. The Invader technology meets both requirements, in that there is minimal manipulation of DNA and no requirement for polymerase chain reaction (PCR) amplification, thereby eliminating accompanying optimization and contamination issues.

1.2. Molecular Basis of Polymorphism Detection Using Invader

The Invader assay detects biallelic polymorphisms through the use of allele-specific dye and quencher, dual-labeled probes, termed fluorescence resonance energy transfer (FRET) cassettes, and Cleavase, a naturally occurring "flap" endonuclease. Cleavase digestion separates the dye from the quencher when the FRET cassette forms a specific 3D structure with the 5′ portion of a polymorphism-specific oligonucleotide, termed the primary probe 5′ flap. This 5′ flap, which is not complementary to the genomic DNA and remains single stranded while the 3′ portion hybridizes to the target, is only available to interact with the FRET cassette when it has been generated in a Cleavase reaction involving the target genomic DNA, the primary probe, and an Invader oligonucleotide that is gene specific and overlaps with the primary probe.

In the first step of the reaction (**Fig. 1**), Cleavase recognizes and cleaves the unannealed or "flap" portion of one member of a trimolecular DNA complex. The Invader oligonucleotide hybridizes to the region directly next to the region occupied by the 3′ portion of the probe and overlaps with the probe at one nucleotide. It is this trimolecular structure of the target DNA with the primary probe's 3′ portion and the invader oligo that is recognized by the enzyme. The primary probe will only anneal if the mutant/polymorphic base is a perfect match with the genomic DNA. The melting temperature of the primary probe is close to the temperature of the reaction such that there is repeated hybridization and denaturation of the probe from its template. Thus, this process of primary probe and Invader oligo hybridization, Cleavase recognition, and primary probe 5′ "flap" release is repeated many times such that there is isothermal amplification of the cleaved 5′ flap during the incubation.

In the second step of the reaction, the 5′ flap hybridizes with a FRET cassette to form a trimolecular structure that is recognized by the Cleavase. The FRET cassette has areas of self-complementarity, such that a "trimolecular" structure can assemble when the 5′ flap anneals. The Invader assay detects both alleles in the same reaction, with two primary probes containing distinct 5′ sequences for each allele of a biallelic system; each is partnered with a specific FRET cassette. The two FRET cassettes have spectrally distinct fluors (F), one each for the normal and variant alleles, allowing for simultaneous detection of both alleles by the fluorescence plate reader. When intact, the FRET cassette gives off no fluorescence upon excitation, since the energy from the fluor is transferred to the neighboring quencher dye (Q). In this second step, the enzyme cleaves the

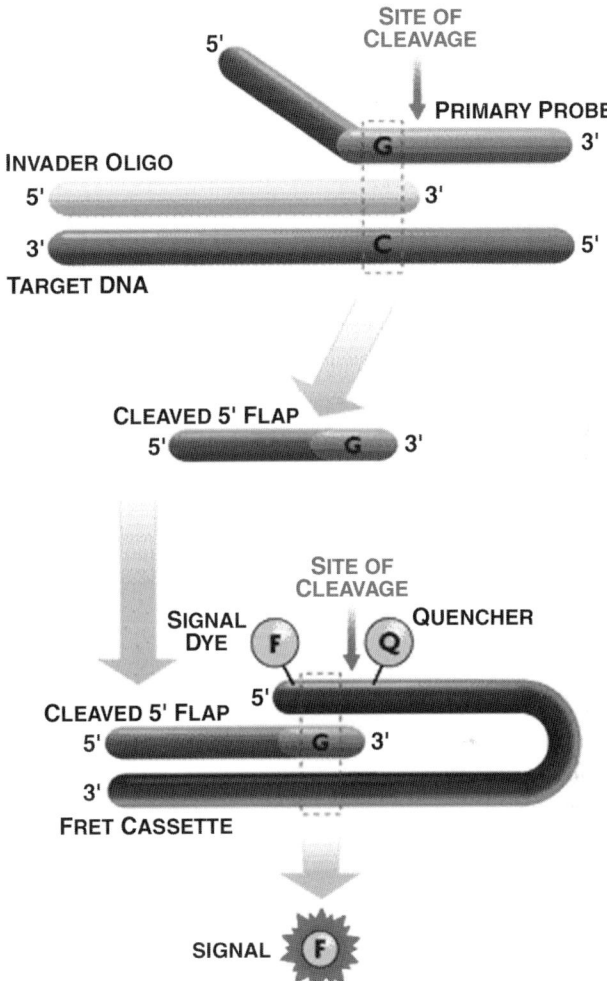

Fig. 1. Schematic representation of the biplexed Invader assay. The Cleavase enzyme recognizes the invasive structure formed by the probe and Invader oligonucleotides in the presence of the appropriate target. Use of two different fluorescence resonance energy transfer (FRET) cassettes allows simultaneous detection of wild-type and mutant targets in a single reaction vessel. (©Third Wave Technologies, Inc. 2004; figure and legend reproduced with permission from Third Wave Technologies, Inc.)

FRET cassette between these two dyes, resulting in release of the portion of the FRET cassette containing the fluor, thus yielding detectable fluorescence.

In the context of a biallelic mutation/polymorphism detection, there are two primary probes in the reaction, each with distinct 5′ "flap" sequences, and 3′ gene-specific regions with the mutant/polymorphic base at the junction of these two halves of the oligonucleotide. The assay obtains its gene and allele specificity from the sequence of the 3′ portion of the primary probes and from the Invader oligonucleotide. In contrast, the flap sequences are designed to hybridize to the Invader FRET cassettes, which in combination can be used in many Invader assays.

2. Materials

1. Thermal cycler such as Applied Biosystems 9700 (Foster City, CA).
2. Fluorescent plate reader such as Genios (Tecan, Maennedorf, Switzerland).
3. Chill-out liquid wax (clear; cat. no. CHO-1411, MJ Research, Waltham, MA).
4. Low Profile Polypropylene Microplate 96, (cat. no. MLL-9601, MJ Research).
5. Sealing Mat-M 96-well gasket (cat. no. MSM-1001, MJ Research).
6. Invader® reagents (Third Wave Technologies, Madison, WI):
   a. Biplex Reagents (cat. no. 97-004).
   b. DNA Reaction Buffer 1.
   c. 40 ng/mL Cleavase, cat. no. 97-004.
   d. Factor V Leiden (G1691A) oligonucleotides, cat. no. 95-311.
   e. Factor V Leiden (G1691A) DNA Assay Controls, cat. no. 98-310.
   f. PT20210 oligonucleotides, cat. no. 95-313.
   g. PT20210 DNA Assay controls, cat. no. 98-312.
   h. MTHFR C677T oligonucleotides, cat. no. 95-312.
   i. MTHFR C677T DNA assay controls, cat. no. 98-311.
   j. MTHFR A1298C oligonucleotides, cat. no. 95-314.
   k. MTHFR A1298C DNA assay controls, cat. no. 98-313.
   l. No Target Blank, cat. no. 97-005.

3. Methods

3.1. DNA Extraction

Most laboratories start with a vacutainer tube of anticoagulated blood collected in EDTA. Several commercial products based on differential hypotonic lysis of red blood cells followed by lysis and isolation of genomic DNA from leukocytes can be used. A commonly used product is Gentra Systems (Minneapolis, MN) Puregene Genomic DNA Isolation kit, which can accommodate 300 μL or 3 mL of whole blood for the isolation. The 3-mL preparation generally yields sufficient material for several Invader assays such as might be ordered for a hypercoagulable workup (*see* **Note 1**).

3.2. Invader Procedure

3.2.1. Mastermix Preparation

1. Take reagents out of the refrigerator, and allow 30 min to come to room temperature.
2. Use the Microsoft Excel worksheet with macros supplied by Third Wave technologies. This quickly adjusts the volumes of reagents based on the number of samples to be tested.

3. Since the reaction is done in a thermal cycler (although a well-calibrated digital dry block would be sufficient) and read in a plate reader, the format is based on the 96-well plate, with four controls.
4. We suggest including at least one genomic DNA control (*see* **Note 2**) such that there are 91 wells per run available for patient samples.

3.2.2. Plate Setup

1. Heat the stock tubes of genomic DNA at 55°C for 15 min (*see* **Note 3**) before adding 10 µL of each per well.
2. Add 20 µL of mineral oil to the wells of the empty plate, followed by addition of the 10 µL of preheated genomic DNA, being careful to pipet it below the oil.

3.2.3. Running Plate

1. The thermal cycler is programmed as follows: 95°C/5 min, 63°C/4 h, and 10°C hold forever (*see* **Note 4**).
2. The plate is topped with an autoclavable rubber gasket (*see* **Note 5**) and put in the thermal cycler for the initial denaturation of 5 min at 95°C.
3. The machine is paused while the 10 µL of master mix is added below the mineral oil and the liquid is pipeted up and down three times.
4. The gasket is replaced and the run continued.

3.2.4. Reading the Plate on a Fluorescence Plate Reader

1. The plate can be read immediately after the run is complete or stored at 4°C in the dark for several days before being read.
2. It is important that the reactions liquify before you place the plate in the reader (*see* **Note 4**). This can be accomplished simply by holding the plate in the palm of your hand for several minutes.
3. Before putting the plate in the reader, release any air bubbles from the wells by gently tapping the side of the plate or spinning briefly in a table-top centrifuge equipped with plate adaptors.
4. Follow the plate reader instructions to set up the machine with the appropriate filter sets (485/530 nm and 560/620 nm).
5. Adjust the gain setting for each scan to give No Target Blank signal values between 150 and 200 AFUs for each scan.
6. Repeat until both NTCs are acceptable.
7. Confirm that the other controls (WT, HET, MUT) were all genotyped correctly.
8. Read the plate at both wavelength settings, and cut and paste the readings into the Microsoft Excel macro supplied by Third Wave Technologies.

3.3. Data Analysis

If you are not using a Microsoft Excel spreadsheet with customized macros supplied by Third Wave Technologies, calculations can be performed manually using the following formulas. To determine whether the assay is valid, the raw

Table 1
FOZ Ranges for Invader MTHFR 677 Assay Controls

Controls[a]	Major allele FOZ[b]	Minor allele FOZ[c]	Allele ratio
Homozygous Major	≥1.5	≥0.2	≥5.0
Heterozygous	≥1.5	≥1.5	≥0.5 to ≤2.0
Homozygous Minor	≥0.2	≥1.5	≤0.2

[a] Signal of the NTC must be $100 \leq x \leq 200$.
[b] In the case of FVL and PT20210, the major allele is detected by the 560/620-nm filter set.
[c] In the case of FVL and PT20210, the minor allele is detected by the 485/530-nm filter set.
Adapted from Invader package insert; Third Wave Technologies.

Sample results must meet the ratio requirements in **Table 1** and also meet the FOZ values given in **Table 2**. If the ratio falls between the three ratio ranges, the result is equivocal and must be repeated (*see* **Note 1**).

signals of both filter sets for the control and the NTC samples are used to make a ratio in which

$$FOZ = \frac{\text{raw signal of control or sample}}{\text{raw signal of No Target Blank}}$$

Next, for both samples and controls for both filter sets, generate a net FOZ value by subtracting 1 from the FOZ value. When the net FOZ is less than 0.04 or a negative number, substitute 0.04 as the net FOZ for ratio calculations. These adjusted signals are used in a ratio in which the major allele signal is in the numerator and the minor allele is in the denominator:

$$\text{Ratio} = \frac{\text{net major allele FOZ}}{\text{net minor allele FOZ}}$$

3.4. Data Interpretation

The ratios calculated above are used to determine whether the run is acceptable. For this to be the case, the three controls that represent the three potential genotypes must yield ratios within the ranges shown in **Table 1**.

In the sample data given in **Table 3**, MTHFR 677 is the target with the major allele (677C) detected by the 485-nm filter and the minor allele (677T) detected by the 560-nm filter. First we confirm that the NTC signal is between 150 and 200. Then we divide the control values by the NTC values to obtain the FOZ and subtract 1 from the FOZ to obtain the net FOZ, remembering to substitute 0.04 if the value is less than 0.04 or a negative number. Finally, we make a ratio of the net FOZ for the major allele over the net FOZ for the minor allele and determine whether the generated value conforms with the requirements in **Table 1**.

Table 2
FOZ Ranges for Invader MTHFR 677 Patient Samples

Sample ratio	Major allele FOZ[a]	Minor allele FOZ[b]	Allele genotype
≥5.0	≥2.0	≥0.2	Homozygous Major
≥0.5 to ≥2.0	≥2.0	≥2.0	Heterozygous
≤0.2	≥0.2	≥2.0	Homozygous Minor

[a] In the case of FVL and PT20210, the major allele is detected by the 560/620-nm filter set.
[b] In the case of FVL and PT20210, the minor allele is detected by the 485/530-nm filter set.
Adapted from Invader package insert, Third Wave Technologies.

Table 3
Invader Sample Dataset

Sample	F signal	R signal	F signal FOZ	R signal	Net F signal FOZ	FOZ	Net R signal FOZ	Ratio
Control 1	999	188						
Control 2	635	573						
Control 3	170	1005						
Control 4 (NTC)	162	171						
Sample 1	1102	706						
Sample 2	1672	303						
Sample 3	253	1410						

1. Control homozygous major.
   a. 485 nm: FOZ = 999/162 = 6.17; net FOZ = 6.17 − 1 = 5.17.
   b. 560 nm: FOZ = 188/171 = 1.10; net FOZ = 1.10 − 1 = 0.1.
   c. Ratio: net FOZ (major)/net FOZ (minor) = 5.17/0.1 = 51.7. Since the ratio is ≥5.0, it is a valid result.

2. Control heterozygous.
   a. 485 nm: FOZ = 635/162 = 3.92; net FOZ = 3.92 − 1 = 2.92.
   b. 560 nm: FOZ = 573/171 = 3.35; net FOZ = 3.35 − 1 = 2.35.
   c. Ratio: net FOZ (major)/net FOZ (minor) = 2.92/2.35 = 1.24. Since this ratio meets the criteria ≥0.5 to ≤2.0, it is a valid result.

3. Control homozygous minor.
   a. 485 nm: FOZ = 170/162 = 1.05; net FOZ = 1.05 − 1 = 0.05.
   b. 560 nm: FOZ = 1005/171 = 5.88; net FOZ = 5.88 − 1.0 = 4.88.
   c. Ratio: net FOZ (major)/net FOZ (minor) = 0.05/4.88 = 0.01. Since the ratio is ≤0.2, it is a valid result.

Since the three genotype controls and the NTC meet the criteria, the assay is valid, and we can proceed with the sample calculations below:

1. Sample 1.
    a. 485 nm: FOZ = 1102/162 = 6.80; net FOZ = 6.80 − 1 = 5.80,
    b. 560 nm: FOZ − 706/171 = 4.13; net FOZ = 4.13 − 1 = 3.13.
    c. Ratio: net FOZ (major)/net FOZ (minor) = 5.80/3.13 = 1.85. Since the ratio is ≥0.5 to ≤2.0, following the criteria in **Table 2**, the genotype of this sample is MTHFR 677 heterozygous.

2. Sample 2.
    a. 485 nm: FOZ = 1672/162 = 10.32; net FOZ = 10.32 − 1 = 9.32.
    b. 560 nm: FOZ = 303/171 = 1.77; net FOZ = 1.77 − 1 = 0.77.
    c. Ratio: net FOZ (major)/net FOZ (minor) = 9.32/0.77 = 12.10. Since the ratio is ≥5.0, following the criteria in **Table 2**, the genotype of this sample is MTHFR 677C homozygote.

3. Sample 3.
    a. 485 nm: FOZ = 253/162 = 1.56; net FOZ = 1.56 − 1 = 0.56.
    b. 560 nm: FOZ = 1410/171 = 8.24; net FOZ = 8.24 − 1.0 = 7.24.
    c. Ratio: net FOZ (major)/net FOZ (minor) = 0.56/4.88 = 0.08. Since the ratio is ≤0.2, following the criteria in **Table 2**, the genotype of this sample is MTHFR 677T homozygote.

4. Notes

1. One advantage of using Invader in a clinical laboratory setting is that since the sensitivity of the assay is fairly high, the protocol does not call for quantifying the concentration of the genomic DNA generated from peripheral blood leukocytes, thus saving time. However, we have a 2.5% failure rate, with most of these samples yielding low signals by both the 485- and 560-nm filters. Assuming that these low signals are caused by a concentration of genomic DNA below that required for the assay (100 ng/μL, according to the manufacturer), the existing sample can be concentrated to half the original volume by vacuum centrifugation or precipitated in ethanol with sodium acetate following a wash with 70% ethanol and resuspension in half the original volume. Either of these approaches, in our hands, yields a valid genotyping result. Failure of a specimen owing to insufficient material may correlate with samples derived from patients with low leukocyte counts, such that an equivalent volume of peripheral blood when used for genomic DNA extraction results in a lower yield of genomic DNA.

2. The manufacturer of the Invader reagents sells genotype controls for each assay (*see* **Subheading 2., Materials**); however, these are synthetic controls, not human genomic DNA of known genotype. We feel that the assay is strengthened by including genomic DNA from an individual that is shown by another method, for instance PCR-RFLP, to be heterozygous for the polymorphism or mutation of interest. Such a genomic DNA control sample is included with every assay, and the acceptability

of the assay is dependent on correct genotyping of this sample, as well as the controls supplied by the manufacturer.

3. To avoid equivocal results owing to sampling variability from concentrated non-homogeneous solutions of genomic DNA, we heat the stock tubes of genomic DNA in a dry block for 5–10 min at 55°C. After gentle tapping of the tube to mix and brief centrifugation to pellet condensation, an aliquot is removed for genotyping.

4. Upon completion of the thermal cycler incubation, the plate can be scanned in the fluorescence plate reader immediately. However, when the samples are not going to be removed from the thermal cycler for several hours or until the next day, the instrument can be programmed to hold the plate at 10°C. This allows for preservation of the samples below room temperature while keeping them in the liquid state. This is in contrast to programming the thermal cycler to hold the samples at 4°C (as in a standard PCR reaction), such that when the plate is removed, it must be held for several minutes to liquefy the mineral oil in the samples prior to reading in the fluorescence plate reader.

5. The assay protocol utilizes PCR cycle plates and does not call for covering the plates while they are in the thermal cycler. To avoid contamination of the instrument's heated lid with genomic DNA, we use autoclavable rubber gaskets (*see* **Subheading 2.**, **Materials**) to cover the plate while it is in the thermal cycler. The gasket is autoclaved after use. If you are using 24-well cycle plates, the gasket sold for the 96-well plate can be cut to size with scissors.

References

1. Bauer, K. A., Rosendaal, F. R., and Heit, J. A. (2002) Hypercoagulability: too many tests, too much conflicting data. *Hematology (Am Soc Hematol Educ Program)*, 353–368.
2. Jennings, I., and Cooper, P. (2003) Screening for thrombophilia: a laboratory perspective. *Br. J. Biomed Sci.* **60**, 39–51.
3. Dahlback, B. (2003). The discovery of activated protein C resistance. *J. Thromb. Haemost.* **1**, 3–9.
4. Fakhrai-Rad, H., Pourmand, N., and Ronaghi, M. (2002) Pyrosequencing: an accurate detection platform for single nucleotide polymorphisms. *Hum. Mutat.* **19**, 479–485.
5. Hsu, T. M., and Kwok, P. Y. (2003). Homogeneous primer extension assay with fluorescence polarization detection. *Methods Mol. Biol.* **212**, 177–187.
6. Santacroce, R., Ratti, A., Caroli, F., et al. (2002). Analysis of clinically relevant single-nucleotide polymorphisms by use of microelectronic array technology. *Clin. Chem.* **48**, 2124–2130.
7. French, C. L. C., Strom, C, Sun, W, Van Atta, R, Gonzalez, B, Wood, M. (2004). Detection of the factor V Leiden mutation by a modified photo-cross-linking oligonucleotide hybridization assay. *Clin. Chem.* **50**, 296–305.
8. Rao, K. V., Stevens, P. W., Hall, J. G., Lyamichev, V., Neri, B. P., Kelso, D. M. (2003). Genotyping single nucleotide polymorphisms directly from genomic DNA by invasive cleavage reaction on microspheres. *Nucleic Acids. Res* **31**, e66.

9. Bertina, R. M., Koster, T., F. Rosendaal, F. R., et al. (1994) Mutation in blood coagulation factor V associated with resistance to activated protein C. *Nature* **369**, 64–67.

10. Bertina, R. M., Reitsma, P. H., Rosendaal, F. R., and Vandenbroucke, J. P. (1995). Resistance to activated protein C and factor V Leiden as risk factors for venous thrombosis *Thromb. Haemost.* **74,** 449–453.

11. Rosendaal, F. R., Koster, T., Vandenbroucke, J. P., and Reitsma, P. H. (1995). High risk of thrombosis in patients homozygous for factor V Leiden (activated protein C resistance). *Blood* **85,** 1504–1508.

12. Vandenbroucke, J. P., Koster, T., Briet, E., Reitsma, P. H., Bertina, R. M., and Rosendaal, F. R. (1994). Increased risk of venous thrombosis in oral-contraceptive users who are carriers of factor V Leiden mutation. *Lancet* **344,** 1453–1457.

13. Ridker, P. M., Miletich, J. P., Hennekens, C. H., and Buring, J. E. (1997). Ethnic distribution of factor V leiden in 4047 men and women: implications for venous thromboembolism screening. *JAMA* **277,** 1305–1307.

14. Poort, S. R., Rosendaal, F. R., Reitsma, P. H., and Bertina, R. M. (1996). A common genetic variation in the 3'-untranslated region of the prothrombin gene is associated with elevated plasma prothrombin levels and an increase in venous thrombosis. *Blood* **88,** 3698–3703.

15. Ceelie, H., Riel, C. C. S.-v., Bertina, R. M., & Vos, H. L. (2004). G20210A is a functional mutation in the prothrombin gene; effect on protein levels and 3'-end formation. *J. Thromb. Haemost.* **2,** 119–127.

16. Pollak, E. S., Lam, H. S., Russell, J. E. (2002). The G20210A mutation does not affect the stability of prothrombin mRNA in vivo. *Blood* **100,** 359–362.

17. Ferraresi, P., Marchetti, G., Legnani, C., et al. (1997) The heterozygous 20210 G/A prothrombin genotype is associated with early venous thrombosis in inherited thrombophilias and is not increased in frequency in artery disease. *Arterioscler. Thromb. Vasc. Biol.* **17,** 2418–2422.

18. Franco, R. F., Santos, S. E., Elion, J., Tavella, M. H., & Zago, M. A. (1998). Prevalence of the G20210A polymorphism in the 3'-untranslated region of the prothrombin gene in different human populations. *Acta Haematol.* **100,** 9–12.

19. den Heijer, M., Koster, T. and Blom, H. J. P. B., et al. (1996). Hyperhomocysteinemia as a risk factor for deep-vein thrombosis. *New England J Med* **334,** 759–762.

20. den Heijer, M. (2003). Hyperhomocysteinaemia as a risk factor for venous thrombosis: an update of the current evidence. *Clin. Chem. Lab. Med.* **41,** 1404–1407.

21. Tsai, M. Y., Bignell, M., Yang, F., Welge, B. G., Graham, K. J., & Hanson, N. Q. (2000). Polygenic influence on plasma homocysteine: association of two prevalent mutations, the 844ins68 of cystathionine beta-synthase and A(2756)G of methionine synthase, with lowered plasma homocysteine levels. *Atherosclerosis* **149,** 131–137.

22. Frosst, P., Blom, H. J., Milos, R. et al. (1995). A candidate genetic risk factor for vascular disease: a common mutation in methylenetetrahydrofolate reductase. *Nat. Genet.* **10,** 111–113.

23. Tsai, A. W., Cushman, M., Tsai, M. Y., et al. (2003). Serum homocysteine, thermolabile variant of methylene tetrahydrofolate reductase (MTHFR), and venous thromboembolism: longitudinal investigation of thromboembolism etiology (LITE). *Am. J. Hematol.* **72,** 192–200.

24. van der put, N. M. J., Gabreels, F., Erik, M. B., et al. (1998). A second common mutation in the methylenetetrahydrofolate reductase gene: an additional risk factor for neural-tube defects? *Am. J. Hum. Genet.* **62,** 1044–1051.

25. Kluijtmans, L. A., and Whitehead, A. S. (2001). Methylenetetrahydrofolate reductase genotypes and predisposition to atherothrombotic disease; evidence that all three MTHFR C677T genotypes confer different levels of risk. *Eur. Heart J.* **22,** 294–299.

26. Girelli, D., Martinelli, N., Pizzolo, F., et al. (2003). The interaction between MTHFR 677 C → T genotype and folate status is a determinant of coronary atherosclerosis risk. *J. Nutr.* **133,** 1281–1285.

27. Geisel, J., Hubner, U., Bodis, M., et al. (2003). The role of genetic factors in the development of hyperhomocysteinemia. *Clin. Chem. Lab. Med.* **41,** 1427–1434.

28. Nishio, H., Lee, M. J., Fujii, M., et al. (1996). A common mutation in methylenetetrahydrofolate reductase gene among the Japanese population. *Jpn. J. Hum. Genet.* **41,** 247–251.

29. McAndrew, P. E., Brandt, J. T., Pearl, D. K., and Prior, T. W. (1996). The incidence of the gene for thermolabile methylene tetrahydrofolate reductase in African Americans. *Thromb. Res.* **83,** 195–198.

30. Esfahani, S. T., Cogger, E. Z., & Caudill, M. A. (2003). Heterogeneity in the prevalence of methylenetetrahydrofoalte reductase gene polymorphisms in women of different ethnic groups. *J. Am. Diet. Assoc.* **103,** 200–207.

31. Salomon, O. S. D., Zivelin A., Gitel, S., et al. (1999). Single and combined prothrombotic factors in patients with idiopathic venous thromboembolism: prevalence and risk assessment. *Arterioscler. Thromb. Vasc. Biol.* **34,** 1821–1826.

32. Jerrard-Dunne, P. E. A., McGovern, R, Hajat, C., et al. (2003). Ethnic differences in markers of thrombophilia: implications for the investigation of ischemic stroke in multiethnic populations: the South London Ethnicity and Stroke Study. *Stroke* **34,** 1821–1826.

33. Patel, R. K., Ford, E., Thumpston, J., & Arya, R. (2003). Risk factors for venous thrombosis in the black population. *Thromb. Haemost.* **90,** 835–838.

34. Morange, P. E., Tregouet, D. A., Frere, C., et al. (2005). Biological and genetic factor's influencing plasma factor VIII levels in a healthy family population: results from the Stanislas cohort. *Br. J. Haematol.* **128,** 91–99.

7

Controlled Agitation During Hybridization

Surface Acoustic Waves Are Shaking Up Microarray Technology

Achim Wixforth

Summary

Microarray hybridization experiments are mostly based on quite small sample volumes being confined between the microarray itself and a cover slip or lifter slip on top of the narrow fluid layer. Under such conditions, the system is governed by the rules of microfluidics, i.e., by the regime of small Reynold's numbers. Here, diffusion is the only source for moving sample molecules toward their target spots. However, for a typical macromolecule such as that used in microarray hybridization experiments, the diffusion constant is very small. Hence, because they are driven by diffusion only, traveling over typical distances on a microarray may take them a very long time. Additionally, the slow time constants associated with the diffusion limit lead to pronounced depletion effects, which strongly influence the dynamics of a hybridization assay. In this report, we describe a novel technique to overcome the diffusion limit in microarray hybridization experiments. Surface acoustic waves on a piezoelectric substrate are coupled with the sample fluid on a microarray, where they act as a highly efficient agitation source. We demonstrate that the diffusion limit can be overcome in this fashion, leading to a remarkable increase in signal intensity and homogeneity in fluorescence-labeled microarray assays.

Key Words: Hybridization; microarray; agitation; surface acoustic wave; SAW; ArrayBooster; diffusion; microfluidics.

1. Introduction

For microarray production, large sets of genes are arrayed on small surface areas using high-throughput robotic platforms. In most cases, the substrates for DNA microarrays are conventional microscope slides with dimensions of about 75×25 mm. Up to several thousand spots of oligonucleotides or cDNA probes *(1)* with known identities cover the slide in a checkerboard pattern. In gene expression profiling assays, the ratio of binding of complementary nucleic acids

From: *Methods in Molecular Medicine, Vol. 144, Microarrays in Clinical Diagnostics*
Edited by: T. Joos and P. Fortina © Humana Press Inc., Totowa, NJ

from test and control samples is determined. This allows a parallel, semiquantitative analysis of transcription levels in a single experiment. In a standard microarray experiment, the sample solution is sandwiched between the DNA microarray and a cover slip, forming a capillary gap of about 20–100 µm in thickness and several centimeters in width. Especially for low concentrated cDNA molecules representing the low expressed genes, the immediate vicinity of the corresponding probe spot will be quickly depleted. Without active agitation, diffusion is the only mechanism for the DNA strands to be transported to their complementary spots. However, on the scale of several centimeters, diffusion is a notoriously slow process for molecules that are the size of the DNA strands discussed here. It has been estimated that it would require weeks for the hybridization reaction to reach equilibrium *(2,3)*. Here we report on a novel agitation mechanism approach using surface acoustic wave (SAW)-based technology to increase the results in standard microarray experiments.

2. Materials

We present a novel technique for microagitation of the sample solution during microarray hybridization experiments. The biological experiments presented here are meant to be representative only for some microarray hybridization experiments, including DNA, protein, and tissue microarrays for various applications.

1. $LiNbO_3$ piezoelectric substrates for the SAW devices.
2. Kidney and liver tissue of adult female Wistar-Han rats (Charles River Laboratories, Sulzfeld, Germany).
3. RNeasy Maxi Kit including column DNase digestion (Qiagen, Hilden, Germany).
4. Label Star Array Kit (Qiagen).
5. Cyanine-3-dCTP for rat kidney (Perkin Elmer Life Sciences, Freiburg, Germany).
6. Cyanine-5-dCTP for rat liver cDNA (Perkin-Elmer).
7. ArrayDesigner 2 (Premierbiosoft, Palo Alto, CA).
8. Gene Machines OmniGrid contact printer.
9. SMP3 split pin (Telechem, Sunnyvale, CA).
10. ddH$_2$O.
11. 4X Standard saline citrate (SSC) 1% (w/v).
12. Sodium dodecyl sulfate (SDS)/1% (w/v).
13. Bovine serum albumin (BSA; Sigma).
14. ArrayBooster™ Hybridization station (Advalytix, Brunnthal, Germany).
15. AdvaCard™ micro mixer card (Advalytix).
16. 50% Formamide, 6X SSC, 0.5% (w/v) SDS, and 5X Denhardt's solution (Sigma).
17. 22 × 22-mm Cover glasses (Sigma).
18. Fixogum glue (Marabu, Tamm, Germany).
19. 2X SSC/0.1% Tween 20.
20. GenePix4000B Scanner (Axon Instruments, Union City, CA).

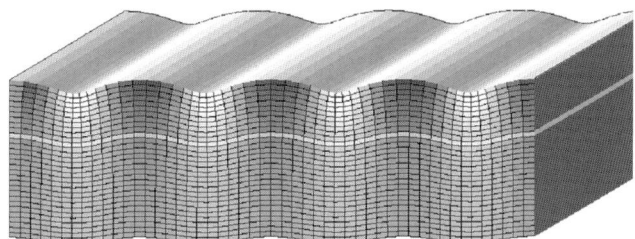

Fig. 1. Sketch of a surface acoustic wave (Rayleigh mode) propagating at the surface of an elastic and piezoelectric solid. The particle displacements, as well as the piezoelectric fields associated with the wave, decay in an exponential manner with depth into the substrate.

21. GenePix Pro 4.0 Software (Axon Instruments).
22. QIAquick PCR Purification Kit (Qiagen).
23. Label Star Array Kit (Qiagen).

3. Methods

3.1. Surface Acoustic Waves

SAWs were first described in combination with earthquakes *(4)*. Reduced to a significantly smaller nanoscale, they found their way into friendlier fields: SAW devices are widely used for radio frequency (RF) signal processing and filter applications and also play a large part in mobile communication. SAW devices have been around for years in communication circuitry—every cell phone has filters using the effect. An electrical signal fed into so-called transducers on the surface of a piezoelectric chip is converted into a deformation of the crystal underneath. Given the right frequency of the signal, a mechanical wave is launched across the chip. In **Fig. 1**, we sketch a snapshot of a SAW propagating on a solid. A SAW is characterized by subsequent regions of compressed and expanded material, as indicated in grayscale. All physical quantities related to the SAW-like material deformation, stress, and strain, (as well as the piezoelectric fields and potentials associated with the mechanical part of the wave) decay in an exponential manner with the substrate and vanish at approximately one wavelength depth. Typical wavelengths of technically exploited SAWs range from about $\lambda = 30 \mu m$ at $f \approx 100$ MHz down to $\lambda = 0.5$ μm at $f \approx 6$ GHz.

It is especially convenient to excite SAWs on piezoelectric substrates. Such materials undergo a defined deformation if they are subjected to an electric field, and, vice versa, they produce an electric field if they are deformed. The reason is the lack of inversion symmetry of the respective crystal structures.

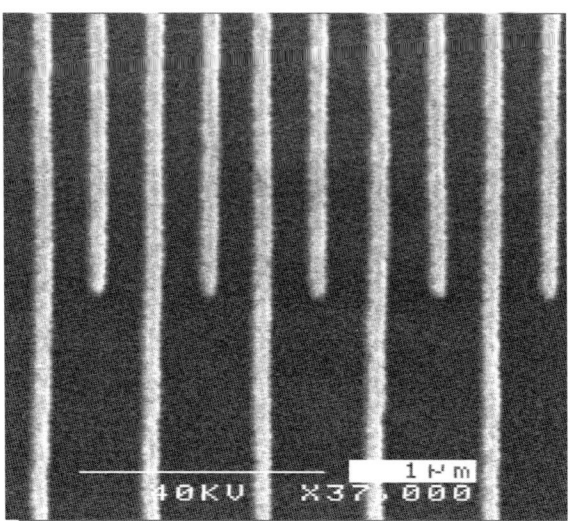

Fig. 2. Detail of an interdigitated transducer (IDT) for exciting a surface acoustic
wave on a piezoelectric substrate. In its simplest form, it consists of two comb-like
metallic electrodes on top of the crystal. A high-frequency signal applied to the transducer is effectively converted into a surface wave once the resonance condition is met,
as explained in the text.

The piezoelectric effect is well known from cigarette lighters, or, more high
tech, from actuators in scanning probe microscopes. A SAW of a well-defined
wavelength and frequency can be excited on such a piezoelectric substrate if a
specially formed pair of metal electrodes is deposited on top of the substrate.
Such electrodes are usually referred to as interdigitated transducers (IDTs), as
shown in **Fig. 2**. A high-frequency signal applied to such an IDT is then converted into a periodic crystal deformation, and, if fed with the right frequency f
$= v/\lambda$, a monochromatic and coherent SAW is launched. Here, v denotes the
sound velocity of the respective substrate, and the wavelength λ is given by the
lithographically defined periodicity of the IDT. If a second IDT is placed downstream from the substrate surface, a so-called delay line is formed. An electrical high-frequency signal fed into one of the transducers is converted into a surface sound wave, which travels along the substrate surface until it reaches the
receiving transducer. There, it is reconverted into an electrical high-frequency
signal. Both transducers, their design, and the substrate properties thus act as a
high-frequency filter with a predetermined frequency response. They are light
weight, relatively simple, and low cost and can be produced very reproducibly,
which explains their massive use in high-frequency signal processing like
mobile telephony.

In the recent past, however, SAWs have also been used in a completely different way than for filtering and signal processing by converting electrical signals into mechanical vibrations and vice versa. Excited on piezoelectric substrates, they are accompanied by large electric fields. These electric fields travel at the speed of sound of the substrate (~3000 m/s), having the same spatial and temporal periodicity as their mechanical companions. Charges at or close to the surface couple to these electric fields, and currents are induced withing a conducting layer.

Nearly 20 yr ago, we introduced SAWs to study the dynamic conductivity $\sigma(\omega, k)$ of low-dimensional electron systems (LDES) in high magnetic fields and at low temperatures. It turns out that the interaction between a SAW and the mobile charges in a semiconductor is strongest for very low sheet conductivities, such as those observed, e.g., in the regime of the quantum Hall effect *(5)*. However, SAWs can be used not only to probe the properties of quantum systems but also to deliberately alter some of them. SAWs represent a spatially modulated strain and stress field accompanied by strong electric fields in a solid, propagating at the speed of sound. Such an interaction between SAWs and the optical properties of a semiconductor quantum well led us to the discovery that photogenerated electron hole pairs in a semiconductor quantum well can be spatially separated under the influence of a SAW-mediated electric field. This, in turn, has an enormous impact on the photoluminescence (PL) of the semiconductor. We were able to show that the PL not only is quenched under the influence of a SAW, but also can be reestablished at will at a remote location on the sample and after a certain delay time *(6)*. Further studies include the acoustic charge transport and the creation of dynamically induced electron wires *(7)*, as well as the study of nonlinear acoustic interaction with low-dimensional electron systems in semiconductors *(8)*.

However, the piezoelectric effect is usually only a small contribution to the elastic properties of a solid: most of the energy propagating in a SAW (usually more than 95%) is of a mechanical nature. Hence, not only electrical interactions as described above, but also mechanical interactions present possibilities for experimental investigations. Because they have wavelengths of a few microns and amplitudes of about only a nanometer, however, the forces and electric fields within the SAW "nanoquake" are sufficient to have a macroscopic effect. Any piece of matter at the surface along the way of a SAW experiences their vibrating force: viscous materials like liquids absorb a lot of their energy. It turns out that the interaction between a SAW and a liquid on top of the substrate surface induces an internal streaming, and, as we point out in **Subheading 3.2.** below, at large SAW amplitudes this can even lead to a movement of the liquid as a whole.

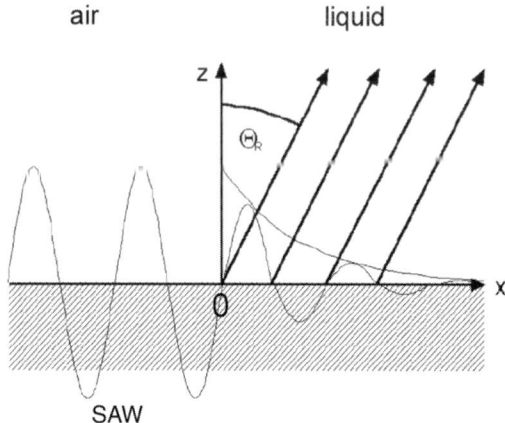

Fig. 3. Schematic representation of the diffraction of a surface acoustic wave on an elastic solid into a fluid layer on top of the substrate. As the sound velocities in both materials differ considerably, a wave is launched under an angle into the fluid.

3.2. Acoustic Streaming

The origin of such an acoustically induced internal streaming is depicted schematically in **Fig. 3**: a SAW is propagating from left to right along the x-axis. At $x = 0$, it reaches the boundary of a liquid at the surface of the substrate. A SAW with a nonvanishing amplitude in the z-direction, i.e., normal to the surface of the substrate, is then strongly absorbed by the fluid, as indicated by the decaying amplitude for positive x values. Moreover, it creates a small but finite pressure difference $2\Delta p$ in the fluid between the ridges and the wells of the wave, which transforms into a small but finite difference $2\Delta \rho$ in the liquid density. Both quantities then spatially and temporally oscillate around their equilibrium value p_0 and ρ_0, respectively. The pressure difference directly above the surface of the substrate leads to the excitation of a longitudinal sound wave into the liquid. As the sound velocities for the liquid and the solid substrate are in general not equal, this wave is launched under a diffraction angle Θ_R, given by:

$$\Theta_R = \arcsin\left(\frac{v_s}{v_f} \right)$$

Here, v_s and v_f denote the sound velocities of the substrate and the fluid, respectively. In addition, the SAW is responsible for the buildup of an acoustic radiation pressure

$$P_s = \rho_0 v_s^2 \left(\frac{\Delta \rho}{\rho_0} \right)$$

Interdigital transducer

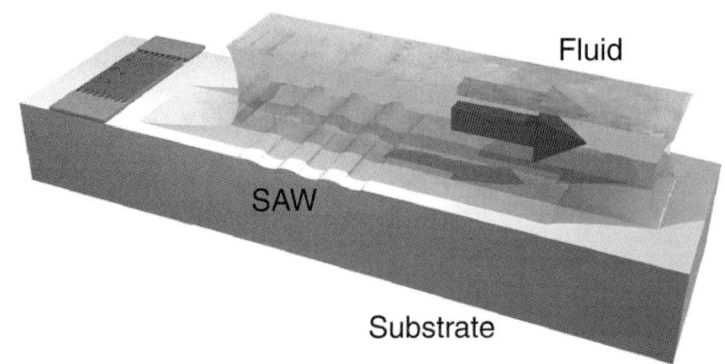

Fig. 4. Cartoon visualizing the effect of acoustic streaming. A surface acoustic wave (SAW) is generated by an interdigital transducer and then generates an internal streaming in the fluid.

in the direction of the sound propagation in the fluid, depicted schematically in **Fig. 4**.

Such internal SAW-driven streaming in a small droplet can be nicely visualized by dissolving a dried fluorescent dye under the influence of a SAW. In **Fig. 5**, we depict two snapshots of such a fluorescence experiment, taken approx 1 s apart from each other. Similar streaming patterns can be achieved in a closed fluid volume, for instance in the capillary fluid layer between a glass slide (microarray), and a cover slip. To illucidate further the streaming pattern and the velocity profile of such a thin fluid layer, we have performed visualization experiments using ink in buffer solution. Although not exactly applicable to the biological assays discussed here, the ink experiment gives important insight into the physics of acoustic streaming in narrow fluid channels. The result of such an experiment is shown in **Fig. 6**. Here we show a bird's-eye view micrograph of the dye distribution in a capillary fluid layer of approx 200 μm thickness. The position of the center of the IDT acting as a nanopump is indicated as a black bar in the leftmost picture. The three images were recorded a few seconds after each other. Clearly one observes the pumping action as a spreading of the dye into a direction perpendicular to the transducer axis.

A detailed analysis of the SAW-induced streaming pattern *(9)* yields an explicit velocity profile as a function of the distance from the transducer, which enables the engineer to carefully design a mixer chip to efficiently actuate the thin fluid layer, for example, between a microarray and a cover lid in a hybridization assay. In **Fig. 7**, we depict such a typical arrangement consisting of a microarray (A), the cover slide (B), and the sample fluid (C) in between.

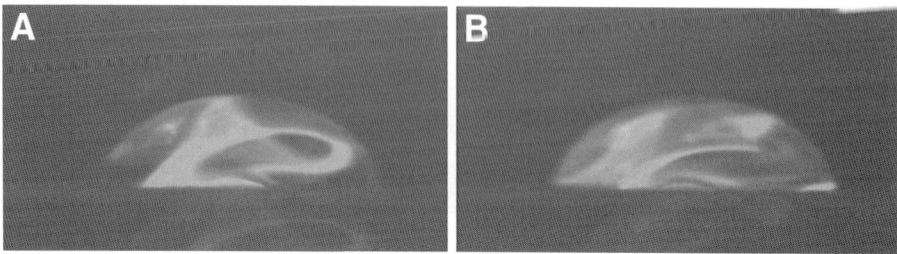

Fig. 5. Visualization of the acoustically induced internal streaming in a small droplet (50 nL) on top of a SAW substrate. The SAW induces internal streaming within the droplet, which in this case is used to dissolve a small amount of dried fluorescent dye. The photograph on the right (**B**) was taken approx 1 s later than (**A**).

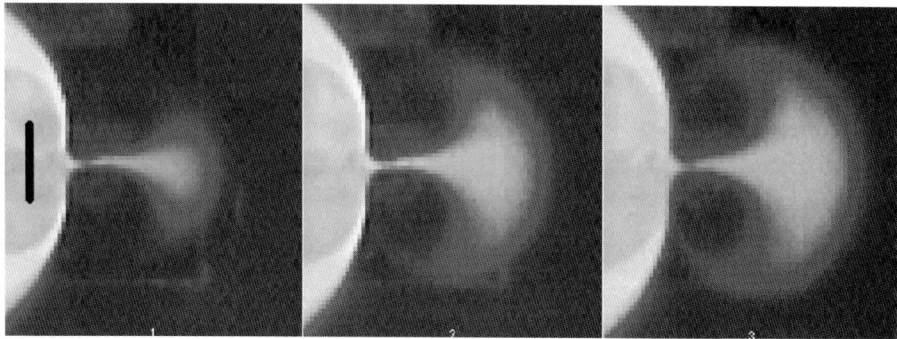

Fig. 6. Top view of acoustically induced streaming in a thin fluid layer sandwiched between two glass slides. To visualize the streaming pattern, we used fluorescent dye, which spreads across the image area after the SAW has been excited by the IDT (indicated as a black line in the left panel).

3.3. SAW-Driven Agitation of Hybridization Assays

The SAW streaming technology has proved to be a very attractive tool for the controlled agitation of sample fluids in microarray assays. As has been pointed out above, in such experiments under conventional conditions, diffusion is the only source for transporting a sample molecule to the target spot. To give a rough estimate over the typical time scales involved in diffusion-driven microarray assys, we have calculated the typical diffusion times for three different particles of different sizes at room temperature in **Table 1**. Even intuitively, it is clear that agitation of the sample fluid during hybridization would tremendously speed up the hybridization process. However, it should be pointed out again that the biochemical process of the binding itself is not the limiting factor, but rather the "refilling" of the depletion zones around the target spots. Such depletion

Table 1
**Diffusion Times at Room Temperature for Three Different Particles
of Different Sizes[a]**

Diffusion length (μm)	Potassium ion (0.2 nm)	Oligonucleotide (6 nm)	PCR product (100 nm)
1	0.2 ms	6 ms	100 ms
10	20 ms	600 ms	10 s
100	2 s	60 s	20 min
1000	200 s	100 min	30 h

[a] To overcome a distance of a mere millimeter, large molecules like PCR products may need more than a day. The diffusion limit, together with the related depletion effects around a target spot, makes microarray experiments a time-consuming procedure, unless very high sample concentrations are used.

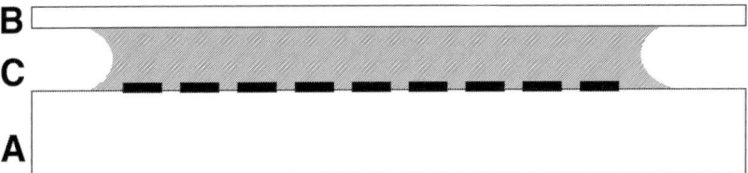

Fig. 7. Schematic of a microarray hybridization setup as described in the text. An array of functionalized spots (indicated by the small black rectangles) containing target molecules is deposited on a slide (**A**). A thin fluid layer (**C**) containing the sample molecules is spread over the spotted area of the microarray. A cover slide (**B**) is used to define the height of the fluid layer and to confine the sample volume.

zones occur if (as in most practical cases) the target molecule concentration in the spots is much higher than the one of sample molecules in the buffer solution.

Three different methods to apply the SAW agitation technique to microarray hybridization assays have been successfully demonstrated so far. First, in the so-called ArrayBooster geometry (*see* **Acknowledgments**), as schematically depicted in **Fig. 8**, a SAW mixer chip embedded in a glass or epoxy slide is submerged in the sample fluid from above. The SAW chip and the supporting glass or epoxy slide thus form a "mixer card" and at the same time serve as the cover slip. Spacers between the mixer card and the slide containing the microarray define a well-controllable fluid layer, if the sandwich is filled with sample liquid. To visualize the mixing process, in the experiment shown in **Fig. 9**, we have drilled four holes into the mixer card, through which we have pipeted small amounts of ink into the buffer solution confined between the slide and the

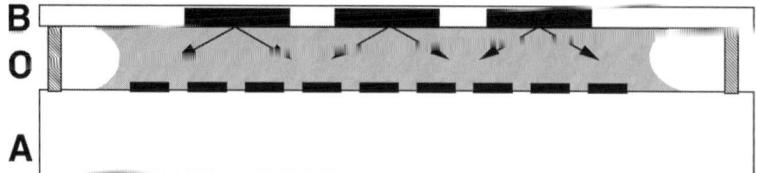

Fig. 8. Microarray experiment employing a SAW mixer card instead of a cover slip (ArrayBooster geometry). The slide containing the microarray (**A**) and the mixer card (AdvaCard) containing three SAW mixer chips (**C**) are separated by spacers (shaded rectangles). This sandwich is filled with bubble-free sample fluid (**B**) using the capillary effect. If connected to a high-frequency signal source, the SAW mixer chips induce acoustic streaming within the sample fluid.

mixer card. In the left parts of the subsequents photographs, the progress of mixing can be observed, once the SAW-mediated acoustic streaming takes place. In the right parts, the SAW had not been switched on. Hence, diffusion is the only source acting on the ink, and hardly any spreading of the dye is observed.

A second method to employ SAW-induced acoustic streaming to a microarray assay is depicted in **Fig. 10**. Here we use a specially designed SAW mixer chip whereby the ultrasonic energy is coupled through an index-matching fluid and through the slide hosting the microarray into the sample fluid on top of it. Even without direct contact between the sample hybridization solution and the mixer chip, the SAW-mediated agitation can be applied to the hybridization assay. This SlideBooster geometry has proved to be particularly sucessful for protein and tissue microarrays in immunohistochemistry applications. In both cases, the signal intensity as well as the homogeneity has been remarkably enhanced compared with conventional methods.

The third technique relies on the fact that mixing of a sample fluid in thin liquid layers is particularly effective if the sources of the ultrasonic energy are distributed randomly across the hybridization area on the microarray. Although both the ArrayBooster and SlideBooster geometries make use of a distributed source mixing by using multiple mixer chips containing multiple SAW transducers operating at different frequencies (multiplexing), the distribution of SAW sources in these cases is spatially fixed and given by the layout of the mixer chips used. The frequency multiplexing distributes the sources of ultrasonic energy in time, following a predetermined protocol. In other words, a set of transducers is powered at a given SAW intensity over a certain period and hence induces a specific streaming pattern, for instance that shown in **Fig. 9**. After a certain dwell time, a second set of transducers with a second streaming pattern is powered, and so on.

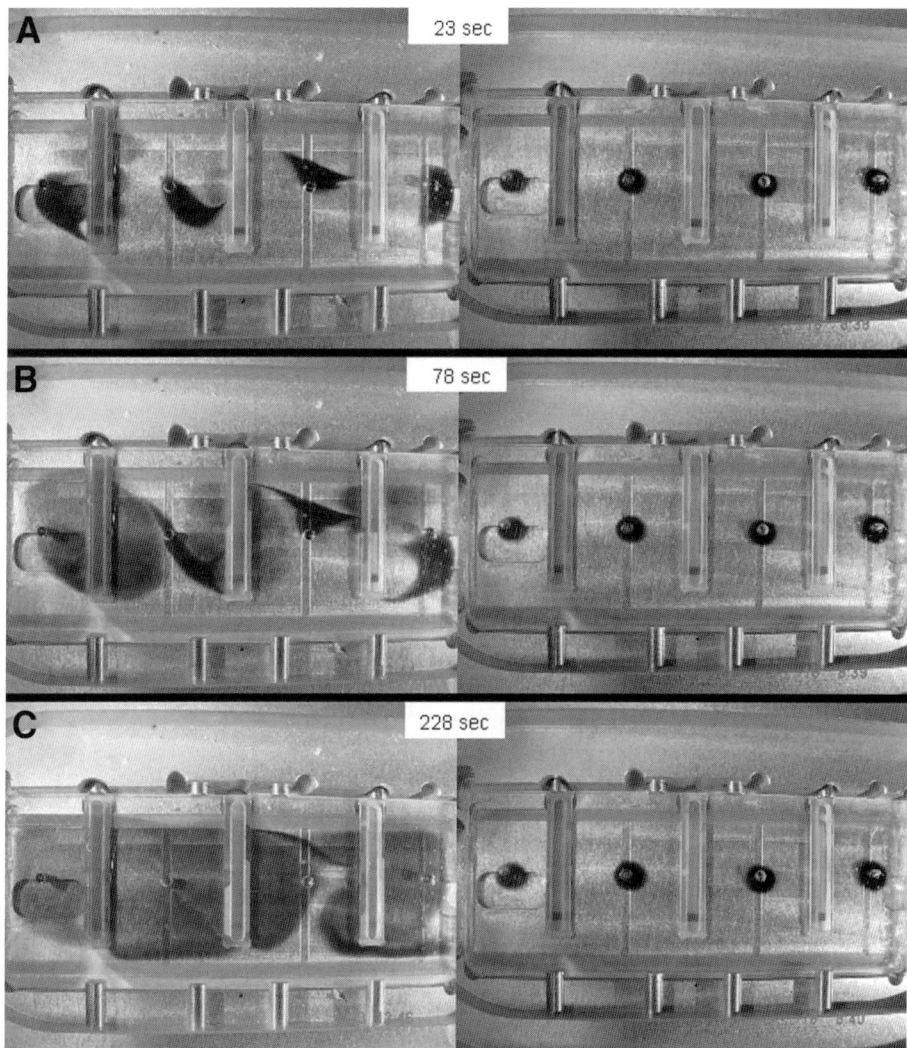

Fig. 9. (A–C) Visualization of SAW-mediated agitation of a microarray sandwich as sketched in **Fig. 8**. This top view into the hybridization chamber shows a mixer card with three SAW chips, which is used to confine a fluid layer of $d = 60$ µm on top of a glass slide containing an oligonucleotide microarray. Ink is pipeted through four deliberately drilled holes to visualize the agitation process. On the left side the mixer chips are activated, i.e., connected to a high-frequency signal source. On the right side the SAW is switched off, thus showing the effect of diffusion alone. After a few minutes, the SAW-induced acoustic streaming led to a complete homogenization of the ink, indicating a highly efficient mixing effect, whereas the ink on the right side (diffusion only) has not spread at all. The insets depict the time after which the photographs were taken.

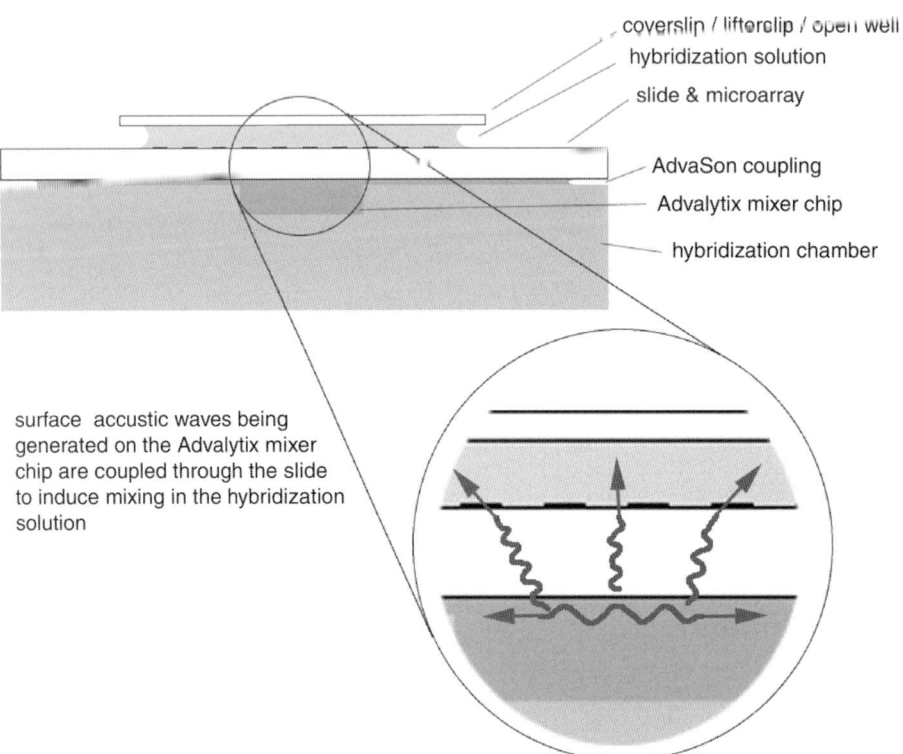

coverslip / lifterclip / open well
hybridization solution
slide & microarray

AdvaSon coupling
Advalytix mixer chip
hybridization chamber

surface accustic waves being
generated on the Advalytix mixer
chip are coupled through the slide
to induce mixing in the hybridization
solution

Fig. 10. Principle of the SlideBooster hybridization chamber. The SAW mixer chip in this case is mounted directly onto the chamber bottom, which can be heated to the desired hybridization temperature. An index matching fluid couples the sound wave through the slide hosting the microarray into the sample fluid on top. This way, the SAW mediated sample agitation can be employed in a variety of different assays: cover slip, lifter slip, or open well assays can be agitated with the same efficiency.

Recent experiments, however, have shown that source distribution in both time and space is even more effective for the hybridization process. Here we make use of the fact that SAW transducers not only excite a surface acoustic wave with a wave vector in the plane of the substrate, but also can be used to launch waves under an oblique angle into the substrate. This angle depends among other parameters on the frequency applied to the transducer. Varying the input frequency thus changes the angle of sound propagation and hence determines the point of impingement of the sound wave at the subtrate surface. This way, a "virtual" ultrasonic source is created at the interface between the sample solution and SAW substrate, the position of which can be changed during the hybridization process. The idea behind this approach is depicted in **Fig. 11**. To demonstrate

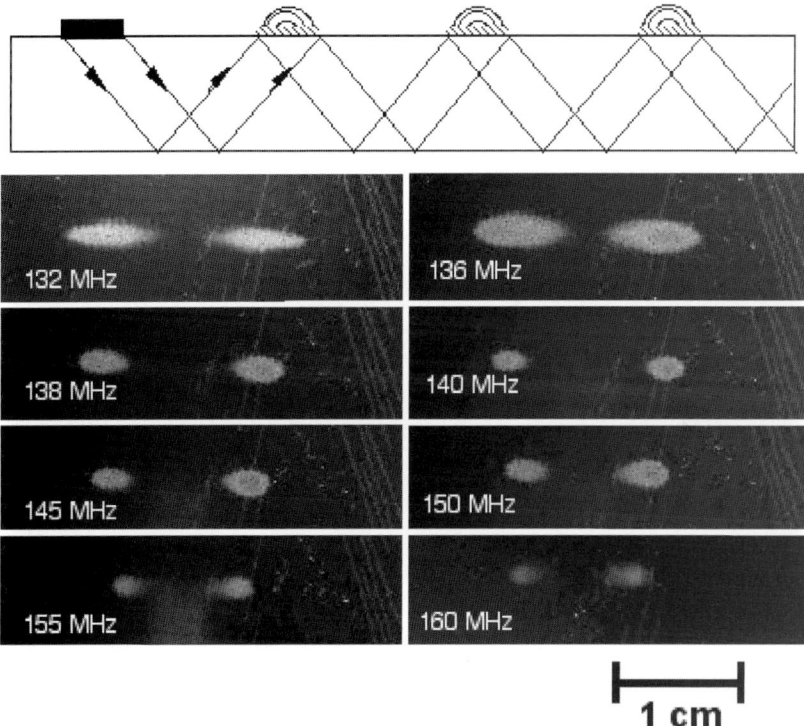

Fig. 11. (Top) Principle of the distributed source mixing as explained in the text. A SAW transducer (black rectangle on the left) is used to launch ultrasonic waves under an angle into the SAW substrate. Sound reflections at opposite surfaces of the substrate create sources of ultrasound at the boundary between the substrate and the sample solution. As the angle of incidence can be varied by varying the frequency applied to the transducer, the positions of the virtual sources can be varied also (bottom). This way, both temporal and spatial distributions of the sources of ultrasonic energy can be varied at will. The virtual sources in the lower part of the figure were visualized using a heat-sensitive liquid crystal probe on top of the SAW substrate.

that the position of the the the sources of acoustic energy can indeed be changed by changing the input frequency of the special sound transducer, we have visualized them by applying a heat-sensitive liquid crystal layer on top of the SAW substrate. The loci of ultrasonic energy distribution on the chip show up as bright spots in the lower part of the figure. For this experiment, an unusual high SAW intensity, however, has been used.

3.4. Biological Results

Several different assays have been performed in cooperation with A. Tögl, Ch. Gauer, and R. Kirchner (Advalytix) to demonstrate the usefulness and power of the SAW-mediated sample fluid agitation during hybridization processes. Two such experiments are presented here as representatives of a variety of different assays that have proved to result in a higher signal intensity and improved homogeneity compared with conventional methods.

3.4.1. Rat-Specific Oligonocleotide Microarray Hybridization

As an sample experiment *(10)*, we chose rat-specific microarrays representing the main model system for drug discovery and toxicology in humans *(11)*. Fifty-mer rat liver- and kidney-specific oligonucleotides ensure high specificity and a low rate of crossreactions *(12)*. Similar systems are widely used in research facilities and are available commercially as catalog arrays. To ensure that the SAW-mediated acoustic agitation during the hybridization process produces validated results, we performed in parallel a set of conventional cover glass control experiments using a water bath.

3.4.1.1. RNA ISOLATION

1. The first step is total RNA isolation from kidney and liver tissue of adult female Wistar-Han rats using the RNeasy Maxi Kit Protocol including the on column DNase digestion.
2. For reverse transcription and cDNA labeling, the Label Star Array Kit was used according to the manual enclosed. During reverse transcription, the tissue-specific total RNA is converted into first-strand cDNA using oligo dT primers. During this process, the cDNA is directly labeled with cyanine-3-dCTP (Perkin Elmer Life Sciences) for rat kidney and cyanine-5-dCTP for rat liver cDNA.
3. For each reaction, 50 µg of total RNA were used. The resulting cDNA products are sufficient for two hybridization reactions.

3.4.1.2. DESIGN AND FABRICATION OF MICROARRAYS

1. Fifty rat-specific oligonucleotides (Metabion, Martinsried, Germany) were designed according to NCBI GenBank sequences with ArrayDesigner 2 (Premierbiosoft) and spotted onto epoxy-coated glass slides (Advalytix) with the Gene Machines OmniGrid contact printer using a single SMP3 split pin (Telechem) to avoid variation in spot quality and morphology because of different pins.
2. The array consists of four identical subarrays with 100 spots each. Each gene on these subarrays is represented by a 50-mer oligonucleotide spotted in duplicate covering an overall area of 10×10 mm.
3. Postspotting slide processing was performed according to the protocol provided. Slides were incubated for 30 min in a humidity chamber to immobilize the spotted oligonucleotides followed by subsequent washing steps at room temperature for 5 min in 0.1% Triton X-100 (Sigma) 22 min in 0.004% HCl, 10 min in 100 mM KCl solution, and a final washing step in ddH$_2$O for 1 min.
4. Such slides can be stored up to 3 mo at room temperature after immobilization.

5. Prior to hybridization, however, the slides were blocked for 30 min in a solution of 4X SSC/0.1% (w/v) SDS/1% (w/v) BSA (Sigma) at 42°C followed by five washing steps with ddH$_2$O at room temperature.
6. The slides were dried in a nitrogen stream and used immediately after blocking.

3.4.1.3. HYBRIDIZATION

1. Hybridization was carried out in the commercially available ArrayBooster, a hybridization station with integrated microagitation for use with standard microarray slides.
2. The sample solution is sandwiched between the spotted glass slide and the AdvaCard as shown in **Fig. 8**. The AdvaCard contains a spacer 60 µm high. The hybridization solution is drawn into this sandwich by capillary action. Vent holes prevent bubble formation while one is loading the hybridization solution with a standard pipete tip.
3. Upon the start of the mixing program, an RF voltage is fed to the IDTs, resulting in a quasi-chaotic streaming in the solution. The instrument software allows the user to assign specific streaming patterns and temperatures to each experiment.
4. Hybridization was performed under stringent conditions in a buffer containing 50% formamide, 6X SSC, 0.5% (w/v) SDS, and 5X Denhardt's solution (Sigma) at 42°C for 16 h after a denaturation step for 3 min at 95°C.
5. The liquid volume under a cover glass was approx 20 µl, resulting in a calculative fluid layer thickness of about 25 µm. The liquid volumes in the ArrayBooster experiments using AdvaCards with an active area of 22 × 23-mm and 60-µm spacers were about 40 µl. This volume can be adjusted by choosing the appropriate AdvaCard and the individually preferred spacer thickness.
6. The control hybridization assays under cover glasses were performed in self-designed hybridization chambers using a 42°C water bath. To ensure similar geometries, we used 22 × 22-mm cover glasses (Sigma) and sealed them with Fixogum glue (Marabu) to avoid evaporation of the solution.
7. Subsequent washing steps were performed after overnight hybridization at 42°C using 2X SSC/0.1% Tween-20, 1X SSC, and 0.2X SSC for 5 min.
8. After washing, the slides were blown dry in a nitrogen stream and stored in a cool dark place until scanning.
9. For microarray scanning and image analysis we used the GenePix4000B Scanner and GenePix Pro 4.0 Software (Axon Instruments). Settings were as follows: laser power 100%, PMT gain 600.

3.4.1.4. DISCUSSION

All experiments were carried out using the same amount of target molecules. This leads to different concentrations of target molecules in the hybridization solution depending on the required volume. The concentration of target molecules in the cover glass experiments was twofold higher than in the microagitated experiments. Five cover glass reference experiments and five experiments using

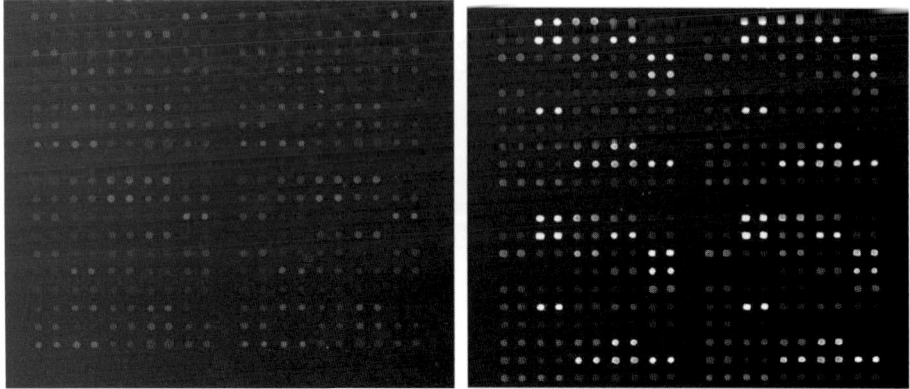

Fig. 12. Comparison of cyanine-5 (rat kidney cDNA) signal intensities of cover glass (left) and microagitated experiments (right) shows an increase in mean signal intensity by a factor of 6. Background values are identical.

micromixer cards in the ArrayBooster were performed for statistical reasons. When we compared the background-corrected signal intensities of the cover glass experiments with the ArrayBooster processed slides, we detected a nearly sixfold increase in the latter.

In **Fig. 12**, we show two examples of our hybridized slides. The background intensities do not differ between the two methods. In the case of the cover glass experiments, however, it was necessary to perform eight individual experiments, as three of the slides showed an extreme gradient across the array, possibly resulting from different fluid layer thickness or evaporation effects. Reportedly, however, even slides considered as highly reproducible show an on-slide variance of about 10% *(13)*. As such gradients (intensity, background, and so on) represent a severe problem in hybridization assays, we also analyzed the variation coefficient of the two different methods. In **Fig. 13**, we depict the variation coefficient of the cover glass experiments compared with the microagitated experiments in the ArrayBooster. In addition to higher signal intensities, microagitation results in higher on-slide reproducibility (CV 8.5%). The variation coefficient for the cover glass experiments was determined after manually sorting out the three slides exhibiting very large gradients. The remaining five slides used for signal intensity analysis showed a 16.5% CV.

To ensure that the signal amplification by microagitation had no effect on the gene expression patterns, we carried out two cohybridization experiments under cover glasses and two microagitated experiments. The ratios for cyanine-5-labeled cDNA from rat kidney vs cyanine-3-labeled rat liver cDNA were not significantly different between the two methods, as shown in **Fig. 14**. In addition to a remarkable signal enhancement, microagitation allows for reduction in

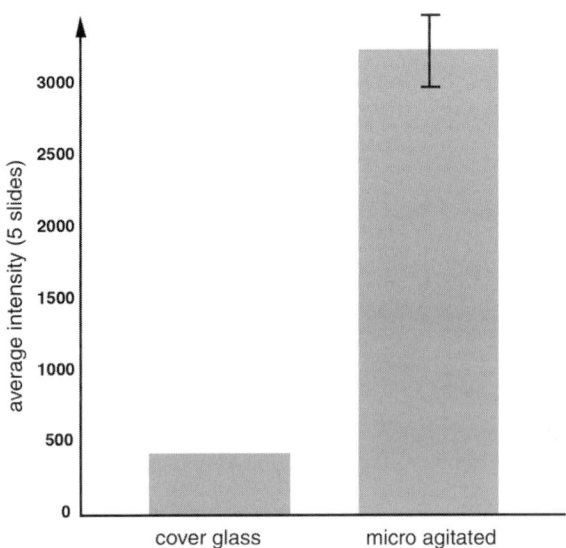

Fig. 13. Mean signal intensities of five slides using a conventional cover glass (left) and five microagitated slides. Experimental time was 16 h.

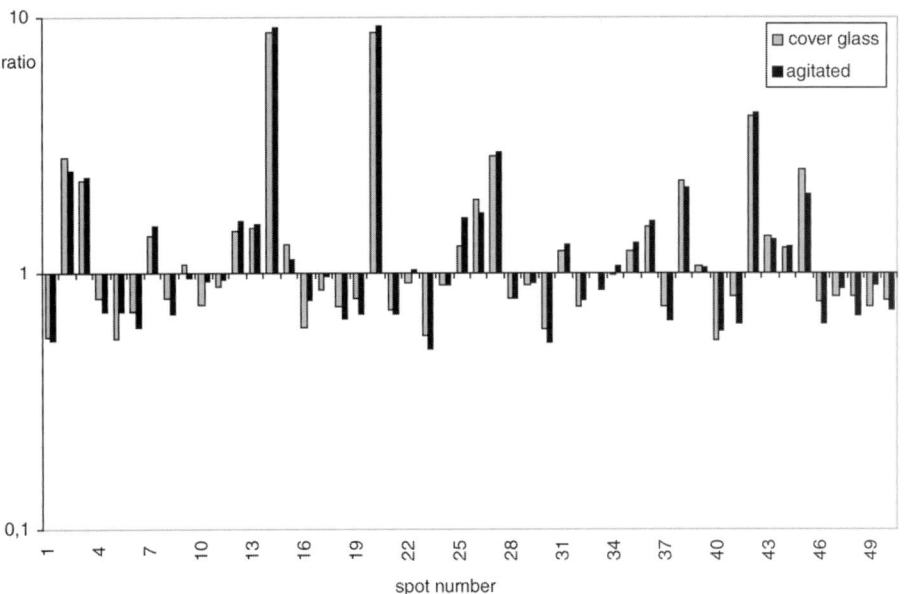

Fig. 14. Gene expression ratios of cyanine-5-labeled rat kidney cDNA vs cyanine-3-labeled rat liver cDNA comparing cover glass with hybridization with microagitation. As expected, is no significant change was seen in the gene expression pattern.

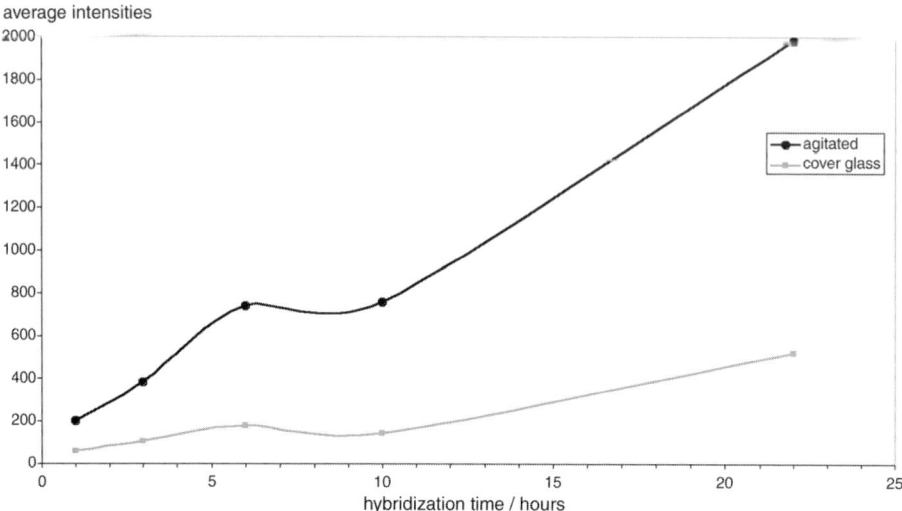

average intensities

Fig. 15. Reaction kinetics comparing cover glass (light gray) with microagitated experiments (black). The average signal intensities of two slides with four replicas each are background corrected.

incubation time by a factor of about four in microarray experiments without loss of signal intensities. To prove this additional benefit, we carried out some time series experiments using from 1 to 22 h of incubation time **(Fig. 15)**. The experimental conditions were similar to the ones above. However, we only used 10 µg of cyanine-5-labeled rat liver cDNA per slide, which resulted in somewhat lower overall signal intensities. Each time-point experiment was done in duplicate, and results were compared under 22 × 22-mm cover glasses (probe volume 40 µL) with hybridizations in the ArrayBooster using the standard agitation protocol (probe volume 40 µL). To ensure comparability between the 20 experiments, we prepared 10-µL aliquots of the pooled labeling reactions, which were filled up to the required volume with formamide/SSC hybridization buffer. Again, the microagitated slides showed higher overall signal intensities compared with the cover glass experiments. Furthermore, the increase in signal intensity was faster than without agitation so that saturation of the spots was reached earlier. For the agitated experiments, saturation was reached after 22 h. Based on diffusion kinetics, it is estimated that both lines will cross after about a week. The bump in intensity after about 6 h of incubation time was also observed in further experiments using the same model system and standard oligo-to-oligo hybridizations. The reason for such intermediate saturation is still unclear and will be the target of further investigations.

From these experiments, it becomes evident that migroagitation makes it possible to watch the reaction kinetics of a specific experiment. As most of the hybridizations to date have been done as so-called end-point assays, which are terminated after a more or less guessed at period, measurement of the reaction kinetics allows for precise control of assay time. Moreover, the time reponse of a specific hybridization protocol might provide the researcher with an additional important parameter for interpretation of the results, which remain unresolved if the hybridization reaction is duffusion limited.

3.4.2. Hybridization Efficiency of PCR-Product vs Oligonucleotide Microarrays

Microarrays for gene expression studies fall into two categories with respect to the type of probe molecules spotted: polymerase chain reaction (PCR) products amplified from cDNA clones or presynthesized oligonucleotides. Both systems have their specific advantages regarding biological functionality and cost. In this study, we compared the signal-to-noise ratios of cDNA hybridized simultaneously to spotted PCR products and oligonucleotides. To exclude any variations induced by sample preparation, hybridization, and washing conditions, we used hybrid slides, i.e., the PCR products and oligonucleotides were spotted on the same slide.

3.4.2.1. SAMPLE PREPARATION

1. Ten PCR products were amplified using rat liver genomic DNA as a template and checked for sequence consistency after purification with QIAquick PCR Purification Kit (Qiagen).
2. The PCR products were then spotted in one 10×10 subarray with 10 replicate spots in a row. Another 10×10 subarray consisted of the corresponding 50-mer *(14)* oligonucleotides (10 replicates each) specific for the same genes.
3. Each oligonucleotide had been designed to represent a part of the PCR product sequence (mostly at the 5′-end) to ensure comparability between both types of the spotted molecules. All probes, oligos, and PCR products, were designed with a GC content of 50%. We used QMT (Quantifoil) epoxy-coated slides for spotting the PCR products and oligonucleotides, again using a contact printing system with Telechem SMP3 split pins.
4. The spotting concentration for the PCR products was 0.5 pmol/µL, and that for the oligonucleotides 50 pmol/µL in a spotting buffer containing dimethyl sulfoxide (DMSO) to ensure that the PCR products were basically single stranded while binding to the slide surface.
5. Spotting concentrations were optimized separately such that the signal intensities reached their maximum in an overnight hybridization. Higher spotting concentrations showed no further effect on signal intensity.
6. Postspotting processing was carried out according to the slide manufacturer's manual for PCR products including a denaturation step. In **Table 2**, we show the gene ID and the lengths of the spotted PCR products.

Table 2
Gene ID and Fragment Length of the Rat PCR Products

No.	Gene	Length of PCR product(bp)
1	af 106860	292
2	x94242	297
3	j02742	143
4	x68282	324
5	m13501	596
6	m18053	155
7	l10339	193
8	x62145	428
9	x66370	337
10	x06483	348

3.4.2.2. HYBRIDIZATION

1. Hybridization was carried out using Cy3 directly labeled cDNA from 5 μg of rat liver total RNA using the Qiagen Label Star Array Kit.
2. The slides were prehybridized in a 1% BSA, 4X SSC, and 0.5% SDS solution at 42°C. Stringent washing after overnight hybridization was carried out in 2X SSC/0.1% SDS, 1X SSC, and 0.5X SSC for 10 min each.
3. All hybridizations were again done in the ArrayBooster with no mixing and with mixing activated (default program "42°C DNA-high").
4. The hybridization volume was 40 μL (AdvaCard AC1a) in all cases, i.e., we worked with the the same amount and concentration of target molecules by using a pooled aliquot. The only parameters changed were mixing and incubation time.
5. A time series with 8, 20, 32, 48, and 72 h of incubation time was carried out.

3.4.2.3. ANALYSIS

1. For microarray scanning and image analysis, we used again the Gene Pix 4000B scanner and the Gene Pix Pro 4.1 Software (Axon Instruments) for raw data generation.
2. All signal intensities were corrected for the local background. We used the background-corrected signal intensity rather than the signal-to-noise ratio, as the signals were generally at least one order of magnitude higher than the background.
3. Instruments settings were laser power 100% and PMT 600 using the rainbow 2 false color table for visualization.

3.4.2.3. DISCUSSION

In **Fig. 16A**, we show an image of a static (no agitation) hybridization with a 72-h incubation time. The PCR probes resulted in a higher signal intensity than their oligonucleotide counterparts. The gain in signal intensity achieved by using PCR products compared with oligonucleotides varied from 1.1 (row 9) to 10.2 (row 4). The PCR spots were generally smaller than their oligonucleotide counterparts, so this difference may be partly caused by cDNA molecules binding to a larger area.

Active microagitation of the hybridization solution improved the signal intensities for both oligonucleotides and PCR products, as can be seen in **Fig. 16 B**. However, the average amplification factor differed substantially. The gain in signal intensity was about 3-fold for PCR products and about 16-fold for oligonucleotides at an incubation time of 72 h. This pronounced amplification of oligonucleotide signal intensities by microagitation has been found for all incubation times (**Fig. 17**). As a result, agitation reversed the relative signal intensities of PCR products and oligonucleotides. With agitation, oligo probes gave higher intensities for all experimental conditions investigated, whereas PCR products resulted in higher signals without agitation. It is worth noting that the 72-h signal intensities are lower than the values at 48 h of incubation. Further studies are under way to corroborate this finding.

Under static hybridization conditions (no agitation), we assume that the number of target molecules within the diffusion radius limits the signal intensity for both spot types. In this case, the spotted PCR products exhibit a higher signal intensity compared with their corresponding oligonucleotides, as their binding constant is larger. With microagitation, the signal intensity is no longer determined by the number of target molecules in the immediate vicinity of the spots. Under these conditions, the larger number of binding sites on the oligo spots might overcompensate for the difference in the binding constant. For this specific array, we estimate the number of binding sites on the oligonucleotide spots to be at least two orders of magnitude larger than for PCR products. In a separate experiment, we found that almost all oligo molecules spotted on the chip act as binding sites. As the concentration of PCR products in the spotting buffer was about two orders of magnitudes lower than for oligonucleotides, we conclude that the number of available binding sites is also lower by the same factor. Furthermore, renaturation of the PCR product double-strand sections may compete with hybridization of the target molecules, which could result in a further reduction of the number of binding sites available.

3.5. Summary

We have shown that microagitation during microarray hybridization has an enormous impact on signal intensity and microarray homogeneity. For a given sample concentration, the hybridization time can be significantly reduced compared with conventional methods, vice versa, for a given hybridization time, the experiments can be done at significantly smaller sample concentrations without losing signal intensity.

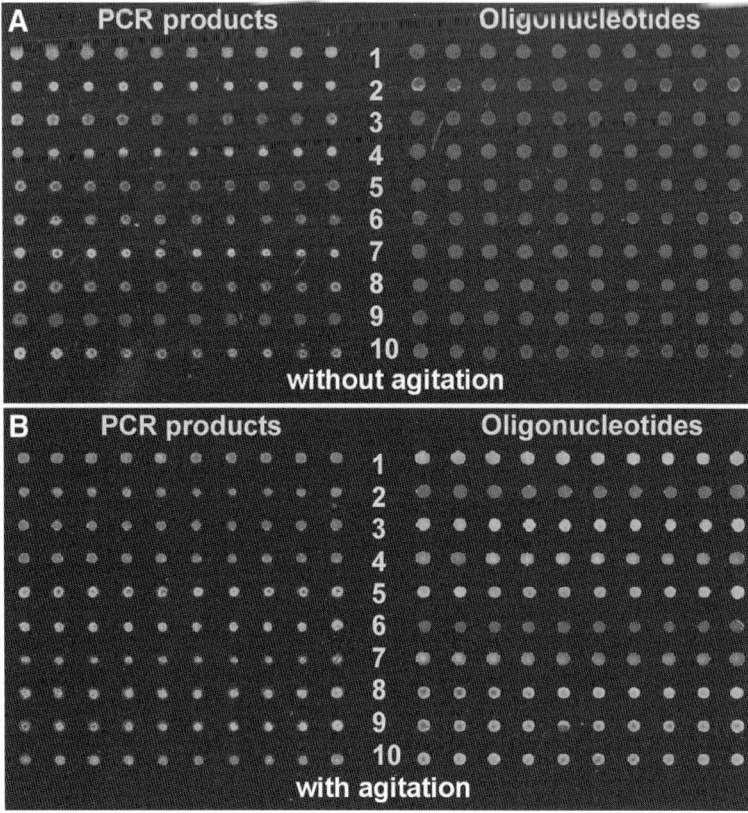

Fig. 16. (A) Static experiment without microagitation at 42°C; 72-h hybridization time. **(B)** Microagitated hybridization experiment under the same conditions.

Microagitation using SAWs has been successfully applied to DNA, oligonucleotide, PCR-product, protein, and tissue microarrays. Depending on the actual application, microagitation can be employed in either ArrayBooster or SlideBooster geometry. Novel mixing geometries like distributed source mixing or a special SAW-induced mixer for microtiter plates are being evaluated, and will be discussed elsewhere. Also, we are presently exploring the basics of spot-by-spot hybridization of microarrays. The SAW technology also allows for precise activation of small amounts of fluids as a whole, i.e., in the form of little droplets. This way, minimal amounts of sample solution can be moved to a specific spot, or a group of spots, where they can induce hybridization. Real-time monitoring of the hybridization progress then allows for an experimental strategy, another attractive feature of hybridization assays.

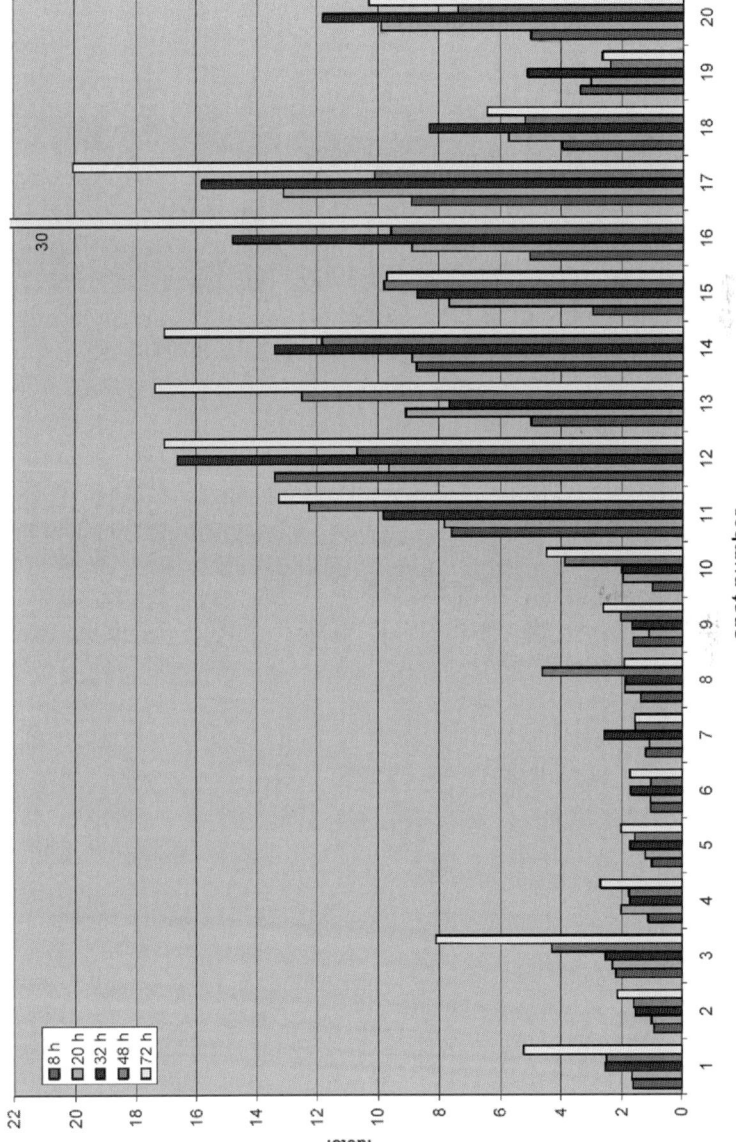

Fig. 17. Amplification factor from microagitation on a PCR product/oligonucleotide hybrid slide at different hybridization times. For all spots and all hybridization times, a significant amplification of the signal intensities was obtained. However, the amplification factor differs from about 1.1 up to more than 30 for different spots.

Acknowledgments

I gratefully acknowledge countless fruitful discussions with my collegues J. Scriba, Ch. Gauer, E. Neuhaus, A. Rathgeber, Ch. Keller, M. Wassermeier, S. Kraus, A. Tögl, R. Kirchner, M. Kantlehner, M. Riepl, Z. v. Guttenberg, A. Geisbauer, and H. Habermüller. Without their efforts, much of the work described here would have been impossible to realize.

This work was supported in part by the Bayerische Forschungsstiftung (FORNANO), the BMBF, VDI/VDE, the Bavarian Ministry of Economics, and the Deutsche Forschungsgemeinschaft under the program SFB 486.

* ArrayBooster, and SlideBooster are trademarks of the company Advalytix AG, Brunnthal, Germany (http://Advalytix.de), which is commercializing the microagitation techniques reported here in the form of different hybridization stations for DNA, protein, and tissue microarrays.

References

1. Hegde, P., Qi, R., Abernathy, K., et al. (2000) A concise guide to cDNA microarray analysis. *Biotechniques* **29,** 548–550, 552–554, 556.
2. Chan, V., Graves, D. J., and Mc Kenzie, S. E. (1995) The biophysics of DNA hybridization with immobilized oligonucleotide probes. *Biophys. J.* **69,** 2243–2255.
3. Allison, S. A., Northrup, S. H., and McCammon, JA. (1986) Simulation of biomolecular diffusion and complex formation. *Biophys. J.* **49,** 167–175.
4. Lord, Rayleigh, (1905) Surface acoustic waves. *Phil. Mag.* **10,** 364–374.
5. Wixforth, A., Kotthaus, J. P., and Weimann. G. (1986) Quantum oscillations in the surface acoustic wave attenuation caused by a two-dimensional electron system. *Phys. Rev. Lett.* **56,** 2104.
6. Rocke, C., Zimmermann, S., Wixforth, A., Kotthaus, J. P., Böhm, G., and Weimann, G. (1997) Acoustically induced storage of light in a solid. *Phys. Rev. Lett.* **78,** 4099.
7. Rotter, M., Kalameiscv, A. V., Govorov, A. O., Ruile, W., and Wixforth, A. (1999) Charge conveyance and nonlinear acoustoelectric phenomena for intense surface sound waves on a semiconductor quantum well. *Phys. Rev. Lett.* **82,** 2171.
8. Rotter, M., Wixforth, A., Govorov, A.O., Ruile, W., Bernklau, D., and Riechert, H. (1999) Nonlinear acoustoelectric interactions in GaAs/LiNbO$_3$ structures *Appl. Phys. Lett.* **75,** 965.
9. Rathgeber, A., v., Guttenberg, Z., Talkner, P., Rädler, J., and Wixforth, A. (2004) *Phy. Rev.* **E70,** 056311.
10. Tögl, A., Kirchner, R., Gauer, C., and Wixforth, A. (2002) Enhancing results of microarray hybridization by microagitation. *J. Biomole. Techn.* **14,** 197–204.
11. Bulera, S. J., Eddy, S. M., Ferguson, E., et al. (2001) RNA expression in the early characterization of hepatotoxicants in Wistar rats by high-density DNA microarrays. *Hepatology* **33,** 1239.
12. Kane, M. D., Jatkoe, T. A., Stumpf, C. R., Lu, J., Thomas, J. D., and Madore, S. J. (2000) Assessment of the sensitivity and specificity of oligonucleotide (50mer) microarrays. *Nucleic Acids Res.* **28,** 4552.

13. Diehl, F., Grahlmann, S., Beier, M., and Hoheisel, J. D. (2001), Manufacturing DNA microarrays of high spot homogeneity and reduced background signal. *Nucleic Acids Res.* **29,** e38.

14. Kane, M. D., Jatkoe, T. A., Stumpf, C. R., et al. (2000) Assessment of the sensitivity and specificity of oligonucleotide (50mer) microarrays. *Nucleic Acids Res.* **28,** 4552.

8

Rapid Screening for 31 Mutations and Polymorphisms in the Cystic Fibrosis Transmembrane Conductance Regulator Gene by Luminex® xMAP™ Suspension Array

Sherry A. Dunbar and James W. Jacobson

Summary

A suspension array hybridization assay is described for the detection of 31 mutations and polymorphisms in the cystic fibrosis transmembrane conductance regulator (CFTR) gene using Luminex® xMAP™ technology. The Luminex xMAP system allows simultaneous detection of up to 100 different targets in a single multiplexed reaction. Included in the method are the procedures for design of oligonucleotide capture probes and PCR amplification primers, coupling oligonucleotide capture probes to carboxylated microspheres, hybridization of coupled microspheres to oligonucleotide targets, production of targets from DNA samples by multiplexed PCR amplification, and detection of PCR-amplified targets by direct hybridization to probe-coupled microspheres. Mutation screening with the system is rapid, requires relatively few sample manipulations, and provides adequate resolution to reliably genotype the 25 CFTR mutations and 6 CFTR polymorphisms contained in the ACMG/ACOG/NIH-recommended core mutation panel for general population CF carrier screening.

Key Words: Cystic fibrosis; genetic screening; Luminex; xMAP technology; suspension array.

1. Introduction

Cystic fibrosis (CF) is a chronic, debilitating, autosomal recessive disorder that affects the respiratory, gastrointestinal, and reproductive systems and occurs in about 1 in 2500 persons in the United States *(1,2)*. The pulmonary disease is characterized by viscous, purulent secretions, persistent colonization and infection with common bacterial and viral pathogens, and an intense, chronic, neutrophil-dominated inflammatory response. Other CF-associated abnormalities include pancreatic insufficiency with secondary nutritional and

From: *Methods in Molecular Medicine, Vol. 144, Microarrays in Clinical Diagnostics*
Edited by: T. Joos and P. Fortina © Humana Press Inc., Totowa, NJ

vitamin deficiencies and infertility in male patients owing to congenital bilateral absence of the vas deferens. The average life expectancy is 30 yrs, and the lifetime medical costs for the care of a CF child have been estimated at more than $1 million (1–3). CF can result from combinations of more than 1000 identified mutations in the cystic fibrosis transmembrane conductance regulator (CFTR) gene; in certain populations, as many as 1 in 25 persons is a carrier (heterozygous) for one of these mutations. Early screening can facilitate informed life decisions and provide the opportunity for early intervention, potentially reducing morbidity, mortality, and the substantial health care costs associated with the treatment of these patients. The American College of Medical Genetics (ACMG) and the American College of Obstetrics and Gynecology (ACOG) have recently recommended that DNA screening for CF be made available to all couples seeking preconception or prenatal care, not just those with a personal or family history, as was previously recommended (4). The ACMG Subcommittee on Cystic Fibrosis Screening has prepared a recommended standard mutation panel, which includes: (1) a core mutation panel consisting of 25 common CFTR mutations (accounting for approx 80% of cases in North America); and (2) a reflex panel consisting of six polymorphisms in the CFTR gene (Table 1) (4).

To date, screening methods have been either narrow, requiring multiple methodologies to be brought to bear for comprehensive coverage, or are cost prohibitive (3,5,6). This study describes the development of a direct hybridization assay using a microsphere-based suspension array technology, the Luminex® xMAP™ system, as a method for rapid, simultaneous detection of these mutations in the CFTR gene. Microsphere-based suspension array technologies offer a new platform for high-throughput genotyping and are being utilized with increasing frequency (7–18). Some advantages of these technologies include rapid data acquisition, excellent sensitivity, and multiplexed analysis capability (19–21).

The Luminex xMAP system incorporates 5.6-μm polystyrene microspheres that are internally dyed with two spectrally distinct fluorochromes. Using precise amounts of each of these fluorochromes, an array is created consisting of 100 different microsphere sets with specific spectral addresses. Each microsphere set can possess a different reactant on its surface. Because microsphere sets can be distinguished by their spectral addresses, they can be combined, allowing up to 100 different analytes to be measured simultaneously in a single reaction vessel. A third fluorochrome coupled to a reporter molecule quantifies the biomolecular interaction that has occurred at the microsphere surface. Microspheres are interrogated individually in a rapidly flowing fluid stream as they pass by two separate lasers in the Luminex® 100™ analyzer. A 635-nm 10-mW red diode laser excites the two fluo-

Table 1
Recommended Core Mutation Panel for General Population Cystic Fibrosis (CF) Carrier Screening

Standard mutation panel

ΔF508	ΔI507	G542X	G551D	W1282X	N1303K
R553X	621+1G→T	R117H	1717-1G→A	A455E	R560T
R1162X	G85E	R334W	R347P	711+1G→T	1898+1G→A
2184delA	1078delT	3849+10kbC→T	2789+5G→A	3659delC	1148T
3120+1G→A					

Reflex tests	I506V[a]	I507V[a]	F508C[a]	5T/7T/9T[b]

[a] Benign variants. This test distinguishes between a CF mutation and these benign variants. I506C, I507V, and F508C are performed only as reflex tests for unexpected homozygosity for ΔF508 and/or ΔI507.

[b] 5T in *cis* can modify the R117H phenotype or alone can contribute to congenital bilateral absence of vas deferens (CBAVD). 5T analysis is performed only as a reflex test for R117H positives *(4)*.

rochromes contained within the microspheres, and a 532-nm, 13-mW yttrium aluminum garnet (YAG) laser excites the reporter fluorochrome (R-phyco-erythrin or Alexa 532) bound to the microsphere surface. High-speed digital signal processing classifies the microsphere based on its spectral address and quantifies the reaction on the surface. Sample throughput is rapid, requiring only a few seconds per sample.

Direct hybridization is the simplest assay chemistry for single nucleotide discrimination and takes advantage of the fact that, for oligonucleotides approx- imately 15–20 nucleotides in length, the melting temperature for hybridization of a perfectly matched template compared with one with a sin-gle-base mismatch can differ by several degrees *(22,23)*. Single-nucleotide polymorphism (SNP) genotyping assays using a direct hybridization format on the *x*MAP platform have been described *(7,10–12,14,24)*. As in other assay chemistries utilizing a solid phase, the reaction kinetics can be adverse-ly affected by immobilization of a reactant on a solid surface. The effects are less severe for a microsphere in suspension than for a flat array, but the dif-fusion rate of the immobilized capture probe can be slower and the effective concentration is reduced compared with free DNA in solution *(23,25)*. However, taking these factors into consideration during probe/primer design and assay optimization usually circumvents any potential drawbacks of the direct assay format. Here we demonstrate the power of this technology for discrimination of SNPs and small deletions by using the CFTR gene as a model system.

2. Materials

1. Oligonucleotide capture probes with 5′ Amino Modifier C12 modification, solubilized in dH$_2$0 (*see* **Table 2** and **Note 1**).
2. Reverse complementary target oligonucleotides with 5′ biotin modification (**Table 3**).
3. Polymerase chain reaction (PCR) amplification primers with 5′ biotin modification for target strand only (**Table 4**).
4. Luminex xMAP carboxylated microspheres; light sensitive, store at 4°C.
5. Aluminum foil (for protecting microspheres from prolonged light exposure).
6. 1-Ethyl-3-(3dimethylaminopropyl) carbodiimide hydrochloride (EDC, Pierce, Rockford, IL); store dry desiccated at −20°C (*see* **Note 2**).
7. 1.5 mL Copolymer microcentrifuge tubes (USA Scientific, Ocala, FL, cat. no. 1415-2500) (*see* **Note 3**).
8. 100 m*M* 2-morpholinoethanesulfonic acid (MES), pH 4.5; filter-sterilize, store at 4°C (*see* **Note 4**).
9. 0.02% Tween-20; filter-sterilize, store at room temperature.
10. 0.1% Sodium dodecyl sulfate (SDS); filter sterilize, store at room temperature.
11. TE buffer: 10 m*M* Tris-HCl, 1 m*M* EDTA, pH 8.0; filter-sterilize, store at room temperature.
12. 1.5X TMAC hybridization solution: 4.5 *M* tetramethylammonium chloride, 75 m*M* Tris-HCl, pH 8.0, 6 m*M* EDTA, pH 8.0, 0.15% Sarkosyl [*N*-lauroylsarcosine sodium salt]; store at room temperature (*see* **Note 5**).
13. 1X TMAC hybridization solution: 3 *M* TMAC, 50 m*M* Tris-HCl, pH 8.0, 4 m*M* EDTA, pH 8.0, 0.1% Sarkosyl; store at room temperature (*see* **Note 5**).
14. Bath sonicator.
15. Vortex mixer.
16. Microcentrifuge.
17. Hemacytometer.
18. PCR amplification equipment, including thermal cycler and consumables.
19. HotStarTaq PCR Master Mix (Qiagen, Valencia, CA).
20. Costar® Thermowell™ model P 96-well polycarbonate plates (Corning, NY, cat. no. 6509).
21. Microseal™ 'A' sealing film (MJ Research, Waltham, MA) or equivalent.
22. Centrifuge capable of centrifuging 96-well plates.
23. Streptavidin-R-phycoerythrin reporter; light sensitive, store at 4°C.
24. Luminex 100 system.
25. 70% Ethanol or 70% isopropanol.
26. Distilled water.
27. Genomic DNA samples (**Table 5**).

3. Methods

The methods described below include: (1) design of oligonucleotide capture probes, (2) design of PCR amplification primers and multiplexed PCR reactions, (3) preparation of the probe-conjugated microsphere sets, (4) verification

Table 2
Oligonucleotide Capture Probes[a]

Probe	Target sequence	Modification[b]	Sequence 5′ → 3′	Microsphere set
Standard mutation panel				
1[c]	I507 & F508	5′-AmMC12	AACACCAA**AGA**T**GAT**ATTTT	006
2B	DI507	5′-AmMC12	ACACCAAAGATATTTTCTT	008
3B	DF508	5′-AmMC12	AAACACCAATGATATTTTC	015
4B	W1282	5′-AmMC12	CAACAGTG**G**AGGAAAGCC	012
5B	W1282X	5′-AmMC12	CAACAGTG**A**AGGAAAGCC	020
6	1717-1G	5′-AmMC12	TTGGTAATAG**G**ACATCTCCA	017
7	1717-1G→A	5′-AmMC12	TTGGTAATA**A**GACATCTCCA	019
8B	G542	5′-AmMC12	TATAGTTCTT**G**GAGAAGGTGGA	026
9B	G542X	5′-AmMC12	TATAGTTCT**T**TGAGAAGGTGGA	028
10C	G551 & R553	5′-AmMC12	AGTGGAGGTCAACGAGCAA	038
11B	G551D	5′-AmMC12	GTGGAGA**T**CAACGAGCAA	030
12C	R553X	5′-AmMC12	GTGGAGGTCAA**T**GAGCAA	032
13	R560	5′-AmMC12	CTTTAGCAA**G**GTGAATAACT	035
14	R560T	5′-AmMC12	CTTTAGCAA**C**GTGAATAACT	039
15	R117	5′-AmMC12	AGGAGGAAC**G**CTCTATCGCG	042
16	R117H	5′-AmMC12	AGGAGGAAC**A**CTCTATCGCG	025
17B	I148	5′-AmMC12	CTTCATCACA**T**TGGAATGCAGA	034
18B	I148T	5′-AmMC12	CTTCATCACA**C**TGGAATGCAGA	045
19C	621+1G	5′-AmMC12	TTTATAAGAAG**G**TAATACTTCCT	046
20E	621+1G→T	5′-AmMC12	ATTTATAAGAAG**T**TAATACTTCCTT	048
21	N1303	5′-AmMC12	GGGATCCAAG**T**TTTTTCTAA	051
22	N1303K	5′-AmMC12	GGGATCCAA**C**TTTTTTCTAA	052
23B	1078T	5′-AmMC12	CACCACAA**A**GAACCCTGA	054
24C	1078delT	5′-AmMC12	ACACCACAAGAACCCTGA	061
25	R334	5′-AmMC12	ATATTTTCC**G**GAGGATGATT	063
26	R334W	5′-AmMC12	ATATTTTCC**A**GAGGATGATT	064
27B	R347	5′-AmMC12	ACCGCCATG**C**GCAGAACAA	067
28B	R347P	5′-AmMC12	ACCGCCATG**G**GCAGAACAA	053
29C	711+1G	5′-AmMC12	ATTTGATGAA**G**TATGTACCTAT	059
30C	711+1G→T	5′-AmMC12	ATTTGATGAA**T**TATGTACCTAT	071
31	G85	5′-AmMC12	TGTTCTATG**G**AATCTTTTTA	066
32B	G85E	5′-AmMC12	ATGTTCTATGA**A**ATCTTTTTA	073
33	3849+10kbC	5′-AmMC12	GTCTTACTC**G**CCATTTAAT	077
34	3849+10kbC→T	5′-AmMC12	GTCTTACTC**A**CCATTTTAAT	075
35	A455	5′-AmMC12	CCAGCAACC**G**CCAACAACTG	011
36D	A455E	5′-AmMC12	TCCAGCAACCT**C**CAACAACTG	036
37	R1162	5′-AmMC12	TAAAGACTC**G**GCTCACAGAT	060
38	R1162X	5′-AmMC12	TAAAGACTC**A**GCTCACAGAT	068
39B	3659C	5′-AmMC12	TTGACTTG**G**TAGGTTTAC	022
40C	3659delC	5′-AmMC12	TTGACTTGTAGGTTTACC	079
41B	2789+5G	5′-AmMC12	TGGAAAGTGA**G**TATTCCATGTC	074
42D	2789+5G→A	5′-AmMC12	TTGGAAAGTGAA**T**ATTCCATGTC	014
43E	2184A	5′-AmMC12	GAAACAAA**A**AAACAATC	007
44E	2184delA	5′-AmMC12	AGAAACAAAAAACAATC	018
45B	1898+1G	5′-AmMC12	TATTTGAAAG**G**TATGTTCTTTG	013

(Continued)

Table 2
(Continued)

Probe	Target sequence	Modification[b]	Sequence 5′ → 3′	Microsphere set
46B	1898+1G→A	5′-AmMC12	TATTTGAAAGATATGTTCTTTG	027
47B	3120+1G	5′-AmMC12	CTTCATCCAGGTATGTAAAAAT	043
48B	3120+1G→A	5′-AmMC12	CTTCATCCAGATATGTAAAAAT	055
Reflex panel				
R2B	I506V	5′-AmMC12	CACCAAAGATGACATTTTC	009
R3B	I507V	5′-AmMC12	CACCAAAGACGATATTTC	021
R4B	F508C	5′-AmMC12	AACACCACAGATGATATTT	024
R5B	5T	5′-AmMC12	TGTGTGTTTTTAACAGGG	029
R6B	7T	5′-AmMC12	GTGTGTTTTTTTAACAGG	033
R7C	9T	5′-AmMC12	GTGTGTTTTTTTTTAACAG	037

[a] The position and sequence of the mutation or variation is indicated in bold type.
[b] 5′-Amino modifier C12.
[c] Probe 1 (I507 & F508) is also used in the reflex panel.

of microsphere coupling, (5) direct hybridization of biotinylated PCR amplification products to the multiplexed probe-coupled microsphere sets, and (6) results and data analysis.

3.1. Design of Oligonucleotide Capture Probes

Oligonucleotide capture probes specific for each normal and mutant sequence were designed and optimized for assay in a TMAC-containing hybridization buffer. TMAC stabilizes AT base pairs, minimizing the effect of base composition on hybridization *(26,27)*. For oligonucleotides up to 200 bp in length, hybridization efficiency in TMAC is a function of the length of the perfect match and is less dependent on base composition. Hybridization buffers incorporating 3 *M* or 4 *M* TMAC equalize the melting points of different probes and increase duplex yields, allowing probes with different characteristics to be used under identical hybridization conditions *(28,29)*. Initially, the capture probes were designed to be matched in length at 20 nucleotides, complementary in sequence to the biotinylated strand of the PCR product, with the mutation located at the center of the probe sequence. Mismatches in the center are known to have a more profound effect on the equilibrium state than mismatches near the 5′ or 3′ end *(30)*. Probes were modified with 5′ Amino Modifier C12 to provide a terminal amine and spacer for coupling to the carboxylated microspheres.

Probes were tested against a panel of patient DNA samples characterized for mutations in the CFTR gene (**Table 5**) by hybridization at 52°C for 15 min, and then optimized accordingly for sensitivity and specificity for the characterized samples. Nucleotides were added to the 5′ and 3′ ends of the probe sequences to improve hybridization efficiency of the probe to its perfect-match target, thus

Table 3
Reverse Complementary Oligonucleotide Targets[a]

Target	Target sequence	Modification	Sequence 5′ → 3′
Standard mutation panel			
C1[b]	I507 & F508	5′-Biotin	AAAAT**ATCATCT**TTGGTGTT
C2	ΔI507	5′-Biotin	AAAGAAAATATCTTTGGTGT
C3	ΔF508	5′-Biotin	AGAAAATATCATTGGTGTTT
C4	W1282	5′-Biotin	GGCTTTCCT**C**CACTGTTGC
C5	W1282X	5′-Biotin	GGCTTTCCT**T**CACTGTTGC
C6	1717-1G	5′-Biotin	TGGAGATGTC**C**TATTACCAA
C7	1717-1G→A	5′-Biotin	TGGAGATGTC**T**TATTACCAA
C8	G542	5′-Biotin	CCACCTTCTC**C**AAGAACTAT
C9	G542X	5′-Biotin	CCACCTTCTC**A**AAGAACTAT
C10	G551 & R553	5′-Biotin	CTTGCTC**G**TTGACCTCCACT
C11	G551D	5′-Biotin	CTTGCTCGTTGA**T**CTCCACT
C12	R553X	5′-Biotin	CTTGCTC**A**TTGACCTCCACT
C13	R560	5′-Biotin	AGTTATTCAC**C**TTGATAAAG
C14	R560T	5′-Biotin	AGTTATTCAC**G**TTGCTAAAG
C15	R117	5′-Biotin	CGCGATAGAG**C**GTTCCTCCT
C16	R117H	5′-Biotin	CGCGATAGAG**T**GTTCCTCCT
C17	I148	5′-Biotin	CTGCATTCCA**A**TGTGATGAA
C18	I148T	5′-Biotin	CTGCATTCCA**G**TGTGATGAA
C19	621+1G	5′-Biotin	GGAAGTATTA**C**CTTCTTATA
C20	621+1G→T	5′-Biotin	GGAAGTATTA**A**CTTCTTATA
C21	N1303	5′-Biotin	TTAGAAAAAA**C**TTGGATCCC
C22	N1303K	5′-Biotin	TTAGAAAAAA**G**TTGGATCCC
C23	1078T	5′-Biotin	CTCAGGGTTC**T**TTGTGGTGT
C24	1078delT	5′-Biotin	TCTCAGGGTTCTTGTGGTGT
C25	R334	5′-Biotin	AATCATCCTC**C**GGAAAATAT
C26	R334W	5′-Biotin	AATCATCCTC**T**GGAAAATAT
C27	R347	5′-Biotin	ATTGTTCTGC**G**CATGGCGGT
C28	R347P	5′-Biotin	ATTGTTCTGC**C**CATGGCGGT
C29	711+1G	5′-Biotin	TAGGTACATA**C**TTCATCAAA
C30	711+1G→T	5′-Biotin	TAGGTACATA**A**TTCATCAAA
C31	G85	5′-Biotin	TAAAAAGATT**C**CATAGAACA
C32	G85E	5′-Biotin	TAAAAAGATT**T**CATAGAACA
C33	3849+10kbC	5′-Biotin	ATTAAAATGG**C**GAGTAAGAC
C34	3849+10kbC→T	5′-Biotin	ATTAAAATGG**T**GAGTAAGAC
C35	A455	5′-Biotin	CAGTTGTTGG**C**GGTTGCTGG
C36	A455E	5′-Biotin	CAGTTGTTGG**A**GGTTGCTGG
C37	R1162	5′-Biotin	ATCTGTGAGC**C**GAGTCTTTA
C38	R1162X	5′-Biotin	ATCTGTGAGC**T**GAGTCTTTA

(Continued)

Table 3
(Continued)

Target	Target sequence	Modification	Sequence 5′ → 3′
C39	3659C	5′-Biotin	GGTAAACCTACCAAGTCAAC
C40	3659delC	5′-Biotin	AGGTAAACCTACAAGTCAAC
C41	2789+5G	5′-Biotin	ACATGGAATACTCACTTTCC
C42	2789+5G→A	5′-Biotin	ACATGGAATATTCACTTTCC
C43	2184A	5′-Biotin	AAGATTGTTTTTTTGTTTCT
C44	2184delA	5′-Biotin	AAGATTGTTTTTTGTTTCTG
C45	1898+1G	5′-Biotin	AAAGAACATACCTTTCAAAT
C46	1898+1G→A	5′-Biotin	AAAGAACATATCTTTCAAAT
C47	3120+1G	5′-Biotin	TTTTTACATACCTGGATGAA
C48	3120+1G→A	5′-Biotin	TTTTTACATATCTGGATGAA
Reflex panel			
CR2	I506V	5′-Biotin	GAAAATGTCATCTTTGGTGT
CR3	I507V	5′-Biotin	GAAAATATCGTCTTTGGTGT
CR4	F508C	5′-Biotin	AAAATATCATCTGTGGTGTT
CR5	5T	5′-Biotin	TCCCTGTTAAAAACACACAC
CR6	7T	5′-Biotin	CCCTGTTAAAAAAACACACA
CR7	9T	5′-Biotin	CCTGTTAAAAAAAAACACAC

[a]The position and sequence of the mutation or variation is indicated in bold type.
[b]Target C1 (I507 & F508) is also used in the reflex panel.

improving assay sensitivity. Nucleotides were removed from the 5′ and 3′ ends of the probe sequences to increase specificity of the probe for its perfect-match target, improving assay specificity. The position of the mutation within the probe sequence was adjusted when necessary to avoid potential formation of secondary structures, and we found that adequate specificity could be achieved when the mutation was between position 8 and position 14 of the 20-nucleotide probe. The final probe sequences used for this assay ranged from 17 to 26 nucleotides in length and are shown in **Table 2**.

3.2. Design of PCR Amplification Primers and Multiplexed PCR Reactions

PCR amplification primers were designed to amplify 19 regions of the CFTR gene containing the mutation sites, with the amplicons ranging from 89 to 212 bp in length (**Table 3**). One primer of each pair was biotinylated at the 5′ terminus for labeling the target strand of the amplicon. Using a small target DNA (approx 100–300 bp) minimizes the potential for steric hindrance to affect the

Table 4
PCR Amplification Primers

CFTR target	Mutation(s)	Primer	5′ Modification	Sequence 5′ → 3′	Size (bp)
Exon 10	ΔI507, ΔF508, I506V, I507V, F508C	BE10U	5′-Biotin	TTCTGTTCTCAGTTTTCCTGG	107
		E10D	None	TTGGCATGCTTTGATGACG	
Exon 20	W1282X	E20U	None	TTGAGACTACTGAACACTGAAGG	126
		BE20D	5′-Biotin	TTCTGGCTAAGTCCTTTTGC	
Intron 10	1717-1G→A	E11U	None	TCAGATTGAGCATACTAAAAGTGAC	89
		BE11D2	5′-Biotin	GAACTATATTGTCTTTCTCTGCAAAC	
Exon 11	G542X, G551D, R553X, R560T	E11U2	None	AAGTTTGCAGAGAAAGACAATATAG	135
		BE11D	5′-Biotin	GAATGACATTTACAGCAAATGC	
Exon 4	R117H	E4U	None	TTTGTAGGAAGTCACCAAAGC	145
		BE4D2	5′-Biotin	GAGCAGTGTCCTCACAATAAAGAG	
Exon 4/intron 4	I148T, 621+1G→T	E4U2	None	CTTCTCTTTATTGTGAGGACACTGC	169
		BE4D	5′-Biotin	ATGACATTAAAACATGTACGATACAG	
Exon 21	N1303K	BE21U	5′-Biotin	TGCTATAGAAAGTATTTATTTTTTCTGG	106
		E21D	None	AGCCTTACCTCATCTGCAAC	
Exon 7	1078delT, R334W, R347P	BE7U	5′-Biotin	GAACAGAACTGAAACTGACTCG	199
		E7D3	None	CAGGGAAATTGCCGAGTG	
Intron 5	711+1G→T	I5U	None	CAACTTGTTAGTCTCCTTTCC	99
		BI5D2	5′-Biotin	AGTTGTATAATTTATAACAATAGTGC	
Exon 3	G85E	E3U	None	CTGGCTTCAAAGAAAAATCC	117
		BE3D2	5′-Biotin	TGAATGTACAAATGAGATCCTTACC	
Chromosome 7	3849+10kbC→T	BC7U	5′-Biotin	GACTTGTCATCTTGATTTCTGG	148
		C7D	None	TTTGGTGCTAGCTGTAATTGC	
Exon 9	A455E	BE9U	5′-Biotin	TCACTTCTTGGTACTCCTGTCC	105
		E9D	None	CAAAAGAACTACCTTGCCTGC	
Exon 19-I	R1162X	BE19U	5′-Biotin	ATTGTGAAATTGTCTGCCATTC	167
		E19D[a]	None	CAATAATCATAACTTTCGAGAGTTG	
Exon 19-II	3659delC	BE19U2	5′-Biotin	TTTAAGTTCATTGACATGCCAAC	91
		E19D[a]	None	CAATAATCATAACTTTCGAGAGTTG	
Intron 14B	2789+5G→A	I14BU	None	GTGTCTTGTTCCATTCCAGG	147
		BI14BD	5′-Biotin	TGGATTACAATACATACAAACATAGTGG	
Exon 13	2184delA	E13U	None	AGATGCTCCTGTCTCCTGG	126
		BE13D	5′-Biotin	TGCACAATGGAAAATTTTCGTATAG	
Intron 12	1898+1G→A	I12U	None	TTAGACTCTCCTTTTGGATACC	110
		BI12D	5′-Biotin	GTCTTTCTTTTATTTTAGCATGAGC	
Intron 16	3120+1G→A	I16U	None	ATGACCTTCTGCCTCTTACC	118
		BI16D	5′-Biotin	ATGAAAACAAAATCACATTTGC	
Intron 8	5T/7T/9T	I8U	None	TAATGGATCATGGGCCATGTGC	212
		BI8D	5′-Biotin	ACTGAAGAAGAGGCTGTCATCACC	

CFTR, cystic fibrosis transmembrane conductance regulator gene.
[a] Primer E19D is used for amplification of both the exon 19-I and exon 19-II targets.

hybridization efficiency at the microsphere surface. In some cases, we and others have used larger targets (400–1200 bp) successfully, suggesting that hybridization efficiency is also dependent on the sequence and overall secondary structure of target *(31)*.

Table 5
Genomic DNA Samples

CFTR genotype	Source[a]
Normal/normal	Sigma, D6537
ΔF508/normal	Patient sample
ΔF508/ΔF508	Coriell Cell Repositories, NA04540
ΔI507/normal	Coriell Cell Repositories, NA11277
W1282/normal	Coriell Cell Repositories, NA11723
1717-1G→A/normal	Coriell Cell Repositories, NA12444
G542X/G542X	Coriell Cell Repositories, NA11496B
G542X/normal	Coriell Cell Repositories, NA11497B
ΔF508/G551D	Coriell Cell Repositories, NA11274
ΔF508/R553X	Coriell Cell Repositories, NA07469
G551D/R553X	Coriell Cell Repositories, NA11761
ΔF508/R560T	Coriell Cell Repositories, NA11284
ΔF508/R117H	Coriell Cell Repositories, NA13591
I148T/normal	Patient sample
ΔF508/621+1G→T	Coriell Cell Repositories, NA11281
N1303K/G1349D	Coriell Cell Repositories, NA11472A
ΔF508/1078delT	Patient sample
R334W/?	Coriell Cell Repositories, NA12960
ΔI507/R347P	Patient sample
G551D/R347P	Coriell Cell Repositories, NA12785
621+1G→T/711+1G→T	Coriell Cell Repositories, NA11280
621+1G→T/G85E	Coriell Cell Repositories, NA11282
3849+10kbC→T/3849+10kbC→T	Coriell Cell Repositories, NA11860
A455E/normal	Patient sample
621+1G→T/A455E	Coriell Cell Repositories, NA11290
R1162X/normal	Coriell Cell Repositories, NA12585
ΔF508/3659delC	Coriell Cell Repositories, NA11275
2789+5G→A/2789+5G→A	Coriell Cell Repositories, NA11859
2184delA/normal	Patient sample
1898+1G→A/normal	Patient sample
621+1G→T/3120+1G→A	Coriell Cell Repositories, NA07441
3120+1G→A/3120+1G→A	Patient sample
F508C/normal	Coriell Cell Repositories, NA13033
I506V/normal	Coriell Cell Repositories, NA13032
R347H/normal	Patient sample
ΔF508/3120G→A	Patient sample
S549N/normal	Patient sample
S549R/normal	Patient sample

CFTR, cystic fibrosis transmembrane conductance regulator gene.

[a] Samples used in this study were purchased either purchased from Sigma or Coriell Cell Repositories (Camden, NJ) or were provided by other laboratories. The catalog number is indicated for purchased samples.

1. Targets were amplified by multiplexed PCR consisting of three six-member reactions (M1, M2, and M3) for the standard mutation panel. The M1 reaction was also used for detection of the I506V, I507V, and F508C polymorphisms in the reflex panel.
2. A single target (intron 8) was amplified separately for detection of the intron 8 5T/7T/9T polymorphism in the reflex panel. Organization of the multiplexed PCR reactions and primer concentrations are shown in **Table 6**.
3. The primer concentration for the intron 8 reaction was 0.2 μ*M*.
4. The 50 μL PCR reactions contained 1X PCR buffer (Qiagen), 200 μ*M* of each dNTP, primers, 1.5 m*M* MgCl$_2$, 2.5 U HotStarTaq DNA polymerase (Qiagen), and 100 ng template DNA.
5. PCR reactions were amplified in a DNA Engine™ thermocycler (MJ Research) by incubation at 95°C for 15 min to activate the enzyme, followed by 35 cycles of denaturation at 94°C for 30 s, annealing at 50°C for 1 min, and extension at 72°C for 1 min. Final extension was done at 72°C for 7 min, and reactions were held at 4°C.
6. Production of each target was verified by gel electrophoresis of 5 μL of each reaction on 4% agarose gels **(Fig. 1)**.
7. For the standard mutation panel, 5 μL of M1 were combined with 1 μL of M2 and 5 μL of M3, and 5 μL of the pooled reactions were used for hybridization (*see* **Note 6**).
8. For the reflex panel, 5 μL of M1 were used for detection of the exon 10 polymorphisms, and 5 μL of intron 8 were used for detection of the intron 8 5T/7T/9T variant.

3.3. Coupling of Amino-Modified Oligonucleotide Capture Probes to Carboxylated Microspheres

Microspheres should be protected from prolonged exposure to light throughout this procedure.

1. Bring a fresh aliquot of –20°C, desiccated Pierce EDC powder to room temperature (*see* **Note 2**).
2. Resuspend the amine-substituted oligonucleotide capture probe to 1 m*M* in dH$_2$O (*see* **Note 1**)
3. Resuspend the stock microspheres by vortex and sonication for approx 20 s.
4. Transfer 5.0×10^6 of the stock microspheres to a USA Scientific microcentrifuge tube (*see* **Note 3**).
5. Pellet the stock microspheres by microcentrifugation at ≥8000 *g* for 1–2 min.
6. Remove the supernatant and resuspend the pelleted microspheres in 50 μL of 0.1 *M* MES, pH 4.5, by vortex and sonication for approx 20 s (*see* **Note 4**).
7. Prepare a 1:10 dilution of the 1 m*M* oligonucleotide capture probe in dH$_2$O.
8. Add 2 μL of the 1:10 diluted capture probe to the resuspended microspheres and mix by vortex (*see* **Note 7**).
9. Prepare a fresh solution of 10 mg/mL EDC in dH$_2$O. Return the EDC powder to desiccant to re-use for the second EDC addition.
10. Add 2.5 μL of fresh 10 mg/mL EDC to the microspheres and mix by vortex.

Table 6
Multiplexed PCR Reactions and Primer Concentrations for the Standard Mutation Panel

Target	Primer concentration μM
Multiplex PCR reaction 1 (M1)	
Exon 10	0.2
Exon 20	0.2
Exon 7	0.2
Intron 16	0.4
Exon 4	0.2
Exon 19-II	0.2
Multiplex PCR reaction 2 (M2)	
Exon 21	0.2
Exon 11	0.4
Exon4/intron 4	0.2
Exon 3	0.2
Chromosome 7	0.2
Intron 5	0.8
Multiplex PCR reaction 3 (M3)	
Exon 9	0.4
Exon 13	0.4
Exon 19-I	0.8
Intron 12	0.4
Intron 14B	0.4
Intron 10	0.8

11. Incubate for 30 min at room temperature in the dark.
12. Prepare a second fresh solution of 10 mg/mL EDC in dH_2O.
13. Add 2.5 μL of fresh 10 mg/mL EDC to the microspheres and mix by vortex.
14. Incubate for 30 min at room temperature in the dark.
15. Add 1.0 mL of 0.02% Tween-20 to the coupled microspheres.
16. Pellet the coupled microspheres by microcentrifugation at ≥8000 *g* for 1–2 min.
17. Remove the supernatant and resuspend the coupled microspheres in 1.0 mL of 0.1% SDS by vortex.
18. Pellet the coupled microspheres by microcentrifugation at ≥8000 *g* for 1–2 min.
19. Remove the supernatant and resuspend the coupled microspheres in 100 μL of TE, pH 8.0, by vortex and sonication for approx 20 s.
20. Enumerate the coupled microspheres by hemacytometer:
    a. Dilute the resuspended, coupled microspheres 1:100 in dH_2O.
    b. Mix thoroughly by vortex.
    c. Transfer 10 μL to the hemacytometer.

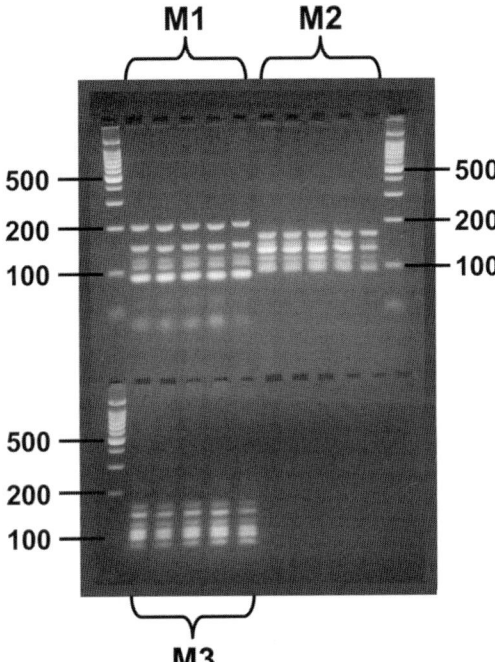

Fig. 1. Gel electrophoresis of multiplexed PCR reactions: 5 μL of five M1, M2, and M3 reactions were resolved in a 4% agarose gel and stained with ethidium bromide. Positions of the 500-, 200-, and 100-bp markers are indicated. The 118-bp intron 16 target in M1 could not be visualized in this electrophoresis system.

    d. Count the microspheres within the four large corners of the hemacytometer grid.

    e. Microspheres/μL = (sum of microspheres in four large corners) × 2.5 × 100 (dilution factor). Maximum is 50,000 microspheres/μL.

21. Store the coupled microspheres refrigerated at 4°C in the dark.

3.4. Verification of Microsphere Coupling by Hybridization to Biotinylated Reverse Complementary Oligonucleotide Targets

Microspheres should be protected from prolonged exposure to light throughout this procedure.

1. Select appropriate oligonucleotide-coupled microsphere sets.
2. Resuspend microspheres by vortex and sonication for approx 20 s.
3. Prepare a working microsphere mixture by diluting coupled microsphere stocks to 150 microspheres of each set/μL in 1.5X TMAC hybridization solution (*see* **Note 5**). In all, 33 μL of working microsphere mixture are required for each reaction.

4. Mix the working microsphere mixture by vortex and sonication for approx 20 s.
5. To each sample or background well, add 33 µl of working microsphere mixture.
6. To each background well, add 17 µL TE, pH 8.0.
7. To each sample well, add 5–200 fmol of biotinylated reverse complementary oligonucleotide and TE, pH 8.0, to a total volume of 17 µL.
8. Mix reaction wells gently by pipetting up and down several times with a multichannel pipetor.
9. Cover the reaction plate to prevent evaporation and incubate at 95°C for 1–3 min to denature any secondary structure in the sample oligonucleotides (*see* **Note 8**).
10. Incubate the reaction plate at 52°C (hybridization temperature for this assay) for 15 min (*see* **Note 8**).
11. Prepare fresh reporter mix by diluting streptavidin-R-phycoerythrin to 10 µg/mL in 1X TMAC hybridization solution (*see* **Note 5**). In all, 12–25 µL of reporter mix should be used for each reaction well.
12. Add 12–25 µL of reporter mix to each well and mix gently by pipeting up and down several times with a multichannel pipetor.
13. Incubate the reaction plate at 52°C for 5 min.
14. Analyze 50 µL at 52°C on the Luminex 100 analyzer according to the system manual.

In a typical assay, the Luminex 100 analyzer is instructed to measure the reporter fluorescence for a minimum of 100 microspheres of each microsphere set present in the reaction (i.e., a minimum of 100 events is collected for each microsphere set). The results are reported as the median of the fluorescent intensity (MFI) measured for each microsphere set. The MFI values are used to generate standard curves for each oligonucleotide target to assess the relative coupling efficiency of each capture probe to its respective microsphere set and to check for any cross-hybridization between targets. A typical standard curve is shown in **Fig. 2**.

3.5. Direct Hybridization of Biotinylated PCR Amplification Products to the Multiplexed Probe-Coupled Microsphere Sets

Microspheres should be protected from prolonged exposure to light throughout this procedure.

1. Select the appropriate oligonucleotide-coupled microsphere sets.
2. Resuspend the microspheres by vortex and sonication for approx 20 s.
3. Prepare a working microsphere mixture by diluting coupled microsphere stocks to 150 microspheres of each set/µL in 1.5X TMAC hybridization solution (*see* **Note 5**). In all, 33 µL of working microsphere mixture are required for each reaction.
4. Mix the working microsphere mixture by vortex and sonication for approx 20 s.
5. To each sample or background well, add 33 µL of working microsphere mixture.
6. To each background well, add 17 µL TE, pH 8.
7. To each sample well add 5 µL of the appropriate PCR reaction(s) and TE, pH 8.0, to a total volume of 17 µL (*see* **Note 6**).

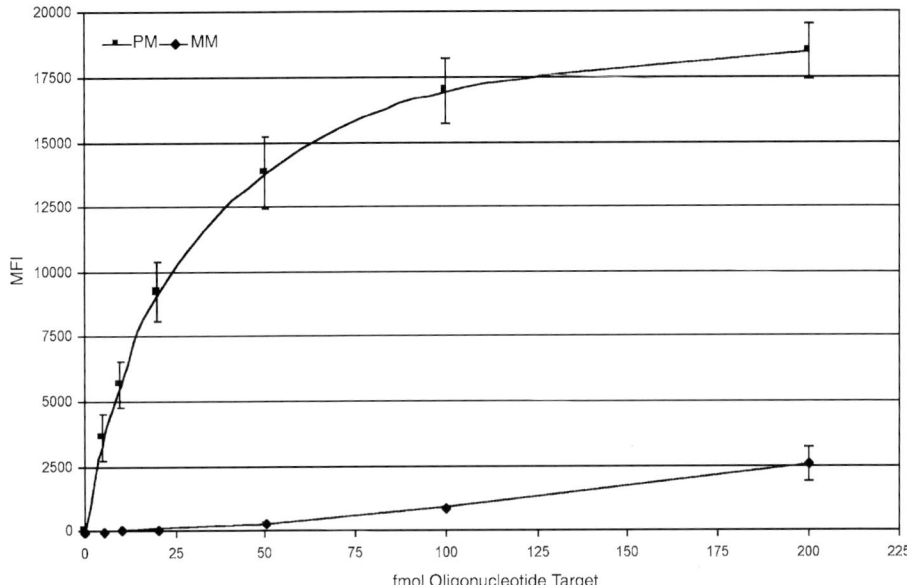

Fig. 2. Oligonucleotide target hybridization. Typical hybridization curves for perfect-match (PM) and single mismatch (MM) oligonucleotide targets are shown. Median fluorescent intensity (MFI) values are the average and standard deviation of four independent experiments for PM and two for MM. Cross-hybridization to the mismatched target can be observed at high (saturating) target concentrations.

8. Mix reaction wells gently by pipeting up and down several times with a multichannel pipetor.
9. Cover the reaction plate to prevent evaporation and incubate 95°C for 5 min to denature the amplified biotinylated DNA (*see* **Note 8**).
10. Incubate the reaction plate at 52°C (hybridization temperature for this assay) for 15 min (*see* **Note 8**).
11. Centrifuge the reaction plate at ≥2250 *g* for 3 min to pellet the microspheres (*see* **Note 9**).
12. During centrifugation, prepare fresh reporter mix by diluting streptavidin-R-phycoerythrin to 4 µg/mL in 1X TMAC hybridization buffer (*see* **Note 5**); 75 µL of reporter mix are required for each reaction well.
13. After centrifugation, carefully remove the supernatant using an eight-channel pipetor to extract the supernatant simultaneously from each column of wells. Be careful not to disturb the pelleted microspheres.
14. Return the reaction plate to 52°C.
15. Add 75 µL of reporter mix to each well, and mix gently by pipeting up and down several times with a multichannel pipetor.
16. Incubate the reaction plate at 52°C for 5 min.

17. Analyze 50 µL at 52°C on the Luminex 100 analyzer according to the system manual.

3.6. Results and Data Analysis

For each well, a minimum of 100 events is collected for each microsphere set, and the results are reported as the MFI of each microsphere set. The data are written to a .csv file, which can be imported to Excel or other analysis software for data reduction. A background well, consisting of all reaction components except a DNA sample, is used to determine the background reporter fluorescence associated with each microsphere set. The background MFI values are subtracted from the sample MFI values to determine the net MFI. Net MFI values are then normalized to an allelic ratio for each target by dividing the net MFI of the allele by the sum of the net MFI for all of the alleles of the target. An example for a background and normal DNA sample in the standard mutation panel is shown in **Fig. 3**. Generally, the allelic ratio is ≥0.75 for a homozygous positive allele, between 0.25 and 0.75 for a heterozygous positive allele, and ≤0.25 for a negative allele. However, testing a panel of known DNA samples is recommended to establish an absolute cutoff value for each allele. A total of 38 characterized DNA samples were tested for the standard mutation panel, and a total of 16 samples were tested for the reflex panel. Allelic ratio data for all samples tested are shown in **Tables 7 and 8**.

4. Notes

1. Amine-substituted oligonucleotide probes should be resuspended and diluted in dH_2O. Tris, azide or other amine-containing buffers must not be present during the coupling procedure. If oligonucleotides were previously solubilized in an amine-containing buffer, desalting by column or precipitation and resuspension into dH_2O is required.

2. We recommend using EDC from Pierce for best results. EDC is labile in the presence of water. The active species is hydrolyzed in aqueous solutions at a rate constant of just a few seconds, so care should be taken to minimize exposure to air and moisture *(32)*. EDC should be stored desiccated at −20°C in dry, single-use aliquots with secure closures. A fresh aliquot of EDC powder should be used for each coupling episode. Allow the dry aliquot to warm to room temperature before opening. Prepare a fresh 10 mg/mL EDC solution immediately before each of the two additions, close the dry aliquot tightly, and return to desiccant between preparations. The dry aliquot should be discarded after the second addition.

3. Uncoupled microspheres tend to be somewhat sticky and will adhere to the walls of most microcentrifuge tubes, resulting in poor postcoupling microsphere recovery. We have found that copolymer microcentrifuge tubes from USA Scientific (cat. no. 1415-2500) perform best for coupling and yield the highest microsphere recoveries post coupling.

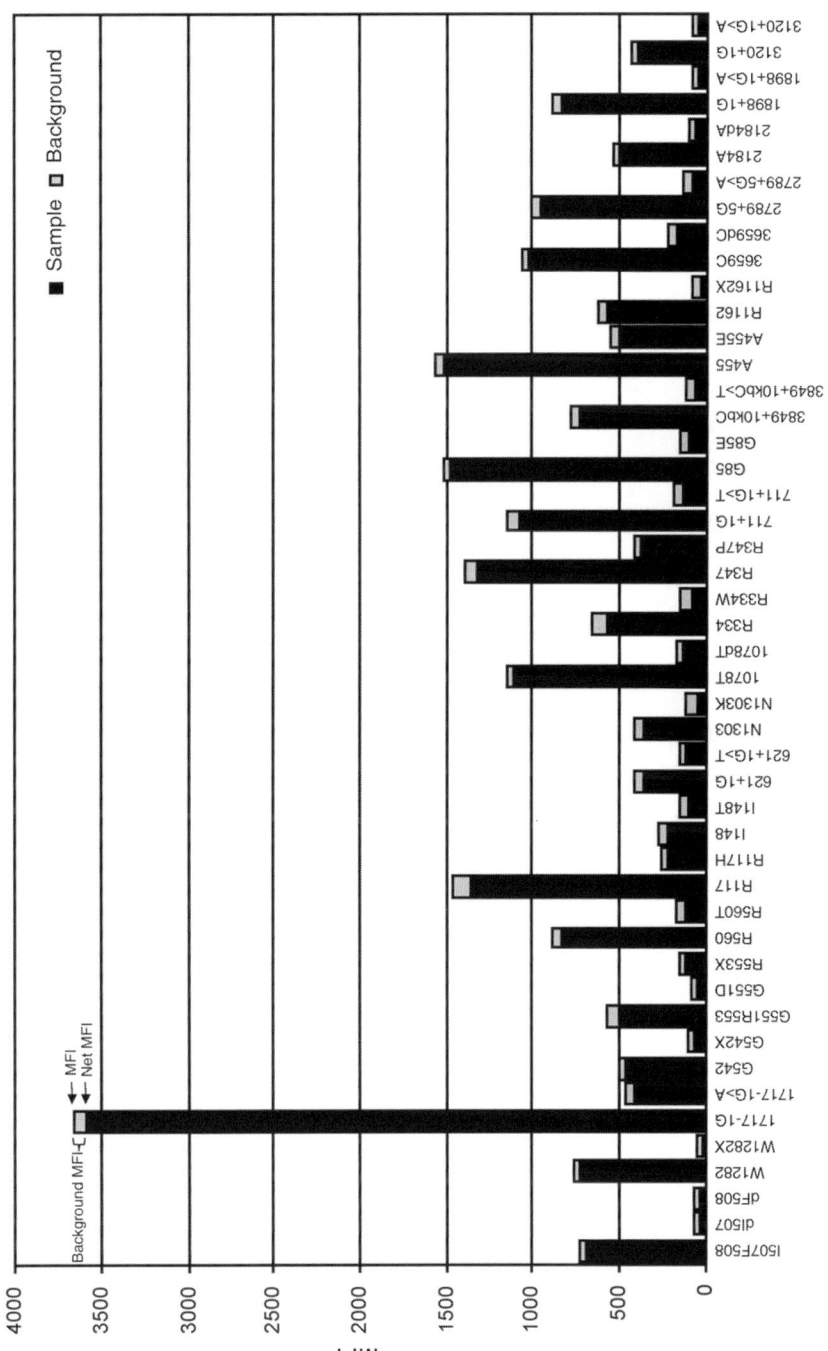

Fig. 3. (A) Graphical presentation of median fluorescent intensity (MFI) data for a homozygous normal sample showing the total, background, and net MFI for each allele. **(B)** Graphical presentation of the normalized allelic ratios for the homozygous normal sample shown in (A).

163

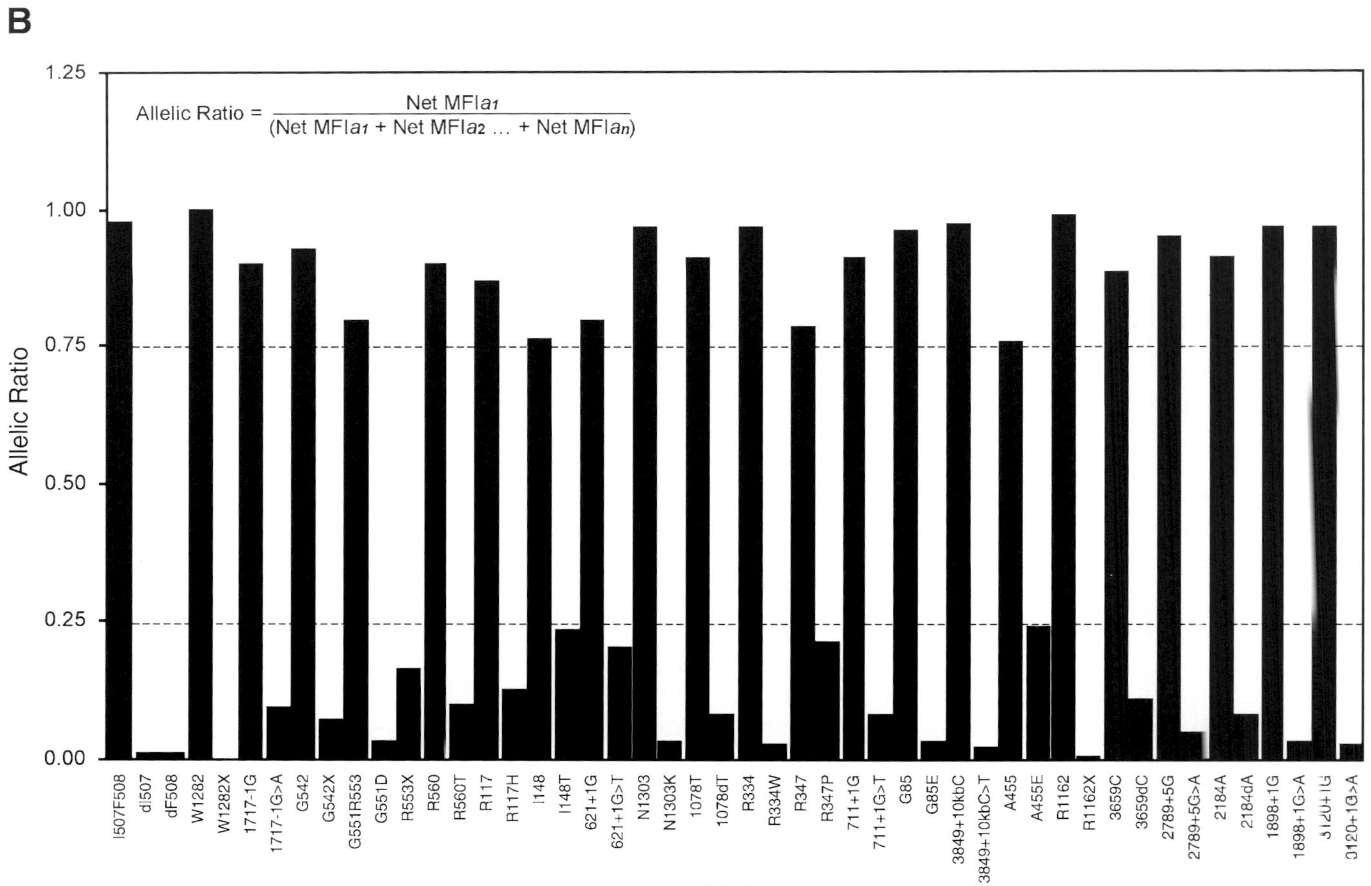

B

Allelic Ratio = $\dfrac{\text{Net MFI}a_1}{(\text{Net MFI}a_1 + \text{Net MFI}a_2 \ldots + \text{Net MFI}a_n)}$

Allele

I507F508
dI507
dF508
W1282
W1282X
1717-1G
1717-1G>A
G542
G542X
G551R553
G551D
R553X
R560
R560T
R117
R117H
I148
I148T
621+1G
621+1G>T
N1303
N1303K
1078T
1078dT
R334
R334W
R347
R347P
711+1G
711+1G>T
G85
G85E
3849+10kbC
3849+10kbC>T
A455
A455E
R1162
R1162X
3659C
3659dC
2789+5G
2789+5G>A
2184A
2184dA
1898+1G
1898+1G>A
3120+1G
3120+1G>A

Allelic Ratio

1.25
1.00
0.75
0.50
0.25
0.00

Fig. 3. (B) *(Continued.)*

164

Table 7
Allelic Ratio Data for the Standard Mutation Panel[a]

Genotype	I507 & F508	ΔI507	ΔF508	W1282	W1282X	1717-1G	1717-1G→A	G542	G542X	G551 & R553	G551D	R553X	R560	R560T	R117	R117H
Normal/Normal	0.98	0.01	0.01	1.00	0.00	0.90	0.10	0.93	0.07	0.80	0.03	0.17	0.90	0.10	0.87	0.13
ΔF508/Normal	0.38	0.00	0.62	1.01	-0.01	0.90	0.10	1.00	0.00	1.01	-0.05	0.04	1.01	-0.01	0.96	0.04
ΔF508/ΔF508	-0.01	0.00	1.01	1.02	-0.02	0.90	0.10	1.02	-0.02	1.00	-0.03	0.04	1.02	-0.02	0.96	0.04
ΔI507/Normal	0.47	0.49	0.04	0.99	0.01	0.89	0.11	0.93	0.07	0.83	0.04	0.13	0.91	0.09	0.86	0.14
W1282/Normal	0.98	0.03	0.00	0.50	0.50	0.89	0.11	0.92	0.08	0.77	0.02	0.21	0.93	0.07	0.85	0.15
1717-1G→A/Normal	0.97	0.03	0.00	0.99	0.01	0.50	0.50	0.94	0.06	0.87	0.03	0.10	0.93	0.07	0.87	0.13
G542X/G542X	0.99	0.02	-0.01	0.99	0.01	0.89	0.11	0.19	0.81	0.89	0.02	0.10	0.93	0.07	0.83	0.17
G542X/Normal	0.98	0.02	0.01	1.02	-0.02	0.89	0.11	0.61	0.39	0.84	0.03	0.14	0.91	0.09	0.86	0.14
ΔF508/G551D	0.46	0.01	0.53	1.00	0.00	0.91	0.09	0.92	0.08	0.47	0.40	0.13	0.91	0.09	0.84	0.16
ΔF508/R553X	0.43	0.01	0.56	1.01	-0.01	0.88	0.12	0.92	0.08	0.46	0.03	0.51	0.90	0.10	0.85	0.15
G551D/R553X	0.97	0.02	0.01	1.02	-0.02	0.87	0.13	0.97	0.03	0.04	0.38	0.58	0.90	0.10	0.87	0.13
ΔF508/R560T	0.45	0.02	0.54	1.00	0.00	0.90	0.10	0.93	0.07	0.83	0.04	0.13	0.47	0.53	0.86	0.14
ΔF508/R117H	0.46	0.01	0.53	1.01	-0.01	0.90	0.10	0.94	0.06	0.86	0.04	0.10	0.91	0.09	0.49	0.51
I48T/Normal	0.97	-0.02	0.05	1.00	0.00	0.90	0.10	0.97	0.03	0.86	0.03	0.03	0.87	0.13	0.83	0.17
ΔF508/621+1G→T	0.44	0.01	0.56	1.01	-0.01	0.90	0.10	0.96	0.04	0.78	0.05	0.17	0.92	0.08	0.85	0.15
N1303K/G1349D	1.00	0.02	-0.02	1.01	-0.01	0.92	0.08	0.93	0.07	0.94	0.02	0.04	0.88	0.12	0.86	0.14
ΔF508/1078delT	0.38	0.02	0.60	1.00	0.00	0.91	0.09	0.95	0.05	0.85	0.09	0.05	0.88	0.12	0.86	0.14
R334W/?	0.98	0.02	0.00	1.00	0.00	0.90	0.10	0.93	0.07	0.78	0.06	0.16	0.91	0.09	0.85	0.15
ΔI507/R347P	0.43	0.52	0.05	1.01	-0.01	0.90	0.10	0.98	0.02	0.92	0.02	0.06	0.88	0.12	0.84	0.16
G551D/R347P	0.98	0.01	0.01	1.00	0.00	0.89	0.11	0.92	0.08	0.45	0.41	0.14	0.91	0.09	0.85	0.15
621+1G→T/711+1G→T	0.98	0.02	0.00	0.99	0.01	0.89	0.11	0.94	0.06	0.85	0.03	0.12	0.90	0.10	0.87	0.13
621+1G→T/G85E	0.98	0.01	0.01	1.00	0.00	0.91	0.09	0.92	0.08	0.86	0.04	0.10	0.91	0.09	0.85	0.15
3849+10kbC→T/3849+10kbC→T	0.98	0.03	0.00	0.99	0.01	0.88	0.12	0.96	0.04	0.86	0.03	0.11	0.90	0.10	0.85	0.15
A455E/Normal	0.99	0.01	0.00	1.03	-0.03	0.91	0.09	0.94	0.06	0.93	0.04	0.03	0.88	0.12	0.84	0.16
621+1G→T/A455E	1.00	0.01	-0.01	0.99	0.01	0.92	0.08	0.93	0.07	0.86	0.02	0.12	0.92	0.08	0.86	0.14
R1162X/Normal	0.97	0.02	0.01	0.99	0.01	0.90	0.10	0.92	0.08	0.85	0.02	0.13	0.92	0.08	0.87	0.13
ΔF508/3659delC	0.35	0.01	0.64	1.01	-0.01	0.92	0.08	0.98	0.02	0.92	0.01	0.08	1.00	0.00	0.92	0.08
2789+5G→A/2789+5G→A	0.93	-0.01	0.09	1.11	-0.11	0.88	0.12	1.01	-0.01	1.00	-0.01	0.01	0.92	0.08	0.90	0.10
2184delA/Normal	0.99	0.01	0.00	1.00	0.00	0.91	0.09	0.95	0.05	0.93	0.04	0.04	0.88	0.12	0.84	0.16
1898+1G→A/Normal	1.01	0.01	-0.03	0.98	0.02	0.93	0.07	0.92	0.08	0.85	0.03	0.12	0.96	0.04	0.90	0.10
621+1G→T/3120+1G→A	0.97	0.02	0.01	1.01	-0.01	0.86	0.14	0.98	0.02	0.97	0.01	0.02	0.89	0.11	0.87	0.13
3120+1G→A/3120+1G→A	1.00	0.05	-0.01	0.98	0.02	0.93	0.07	0.94	0.06	0.92	0.02	0.06	0.88	0.12	0.84	0.16
F508C/Normal	0.94	0.02	0.02	1.02	-0.02	0.90	0.10	0.93	0.07	0.80	0.03	0.17	0.93	0.07	0.86	0.14
I506V/Normal	0.98	0.01	-0.01	1.01	-0.01	0.90	0.10	0.98	0.02	0.80	0.01	0.19	0.90	0.10	0.86	0.14
R347H/Normal	0.97	0.01	0.02	1.01	-0.01	0.90	0.10	0.93	0.07	0.93	0.03	0.04	0.89	0.11	0.86	0.14
ΔF508/3120G→A	0.41	0.01	0.59	0.99	0.01	0.90	0.10	0.94	0.06	0.91	0.03	0.05	0.85	0.15	0.85	0.15
S549N/Normal	1.02	0.00	-0.02	1.00	0.00	0.93	0.07	0.94	0.06	0.93	0.03	0.05	0.86	0.14	0.85	0.15
S549R/Normal	0.99	0.01	0.00	0.99	0.01	0.87	0.13	0.94	0.06	0.90	0.03	0.06	0.86	0.14	0.85	0.15

[a]Positive alleles are indicated in bold type. An allelic ratio of ≥0.70 was used as the cutoff for homozygous positive A455 alleles.

Table 7 (continued)

Genotype	I148	I148T	621+1G	621+1G→T	N1303	N1303K	1078T	1078delT	R334	R334W	R347	R347P	711+1G	711+1G→T	G85	G85E
Normal/Normal	0.76	0.24	0.80	0.20	0.97	0.03	0.92	0.08	0.97	0.03	0.79	0.21	0.91	0.09	0.96	0.05
ΔF508/Normal	0.76	0.24	0.88	0.12	1.07	-0.07	0.98	0.02	1.01	-0.01	0.98	0.02	0.89	0.11	1.00	0.00
ΔF508/ΔF508	0.76	0.24	0.89	0.11	1.07	-0.07	0.98	0.02	1.02	-0.02	0.98	0.02	0.90	0.10	1.00	0.00
ΔI507/Normal	0.82	0.18	0.80	0.20	0.97	0.03	0.91	0.09	0.97	0.03	0.79	0.21	0.89	0.11	0.95	0.05
W1282/Normal	0.77	0.23	0.81	0.19	0.98	0.02	0.91	0.09	0.94	0.06	0.80	0.20	0.90	0.13	0.93	0.07
1717-1G→A/Normal	0.80	0.20	0.81	0.19	0.97	0.03	0.92	0.08	1.00	0.00	0.81	0.19	0.92	0.03	0.97	0.03
G542X/G542X	0.80	0.20	0.79	0.21	1.02	-0.02	0.92	0.08	0.99	0.01	0.78	0.22	0.89	0.11	0.95	0.05
G542X/Normal	0.77	0.23	0.80	0.20	0.98	0.02	0.91	0.09	0.99	0.01	0.80	0.20	0.89	0.11	0.95	0.05
ΔF508/G551D	0.79	0.21	0.83	0.17	0.98	0.02	0.92	0.08	0.99	0.01	0.80	0.20	0.90	0.10	0.95	0.05
ΔF508/R553X	0.80	0.20	0.81	0.19	1.04	-0.04	0.91	0.09	0.98	0.02	0.80	0.20	0.89	0.1	0.95	0.05
G551D/R553X	0.80	0.20	0.80	0.20	1.00	0.00	0.93	0.07	0.99	0.01	0.81	0.19	0.92	0.08	0.93	0.07
ΔF508/R560T	0.82	0.18	0.82	0.18	1.00	0.00	0.91	0.09	0.96	0.04	0.80	0.20	0.89	0.11	0.96	0.04
ΔF508/R117H	0.81	0.19	0.83	0.17	0.99	0.01	0.91	0.09	0.97	0.03	0.79	0.21	0.90	0.10	0.95	0.05
I148T/Normal	0.44	0.56	0.79	0.21	0.94	0.06	0.93	0.07	1.01	-0.01	0.78	0.22	0.96	0.04	0.95	0.05
ΔF508/621+1G→T	0.78	0.22	0.55	0.45	0.98	0.02	0.90	0.10	0.97	0.03	0.80	0.20	0.92	0.08	0.95	0.05
N1303K/G1349D	0.80	0.20	0.81	0.19	0.48	0.52	0.95	0.05	1.00	0.00	0.82	0.18	0.95	0.05	0.97	0.03
ΔF508/1078delT	0.80	0.20	0.82	0.18	0.98	0.02	0.57	0.43	1.01	-0.01	0.80	0.20	0.96	0.04	0.97	0.03
R334W/?	0.76	0.24	0.81	0.19	0.98	0.02	0.91	0.09	0.36	0.64	0.78	0.22	0.89	0.11	0.95	0.05
ΔI507/R347P	0.80	0.20	0.80	0.20	0.99	0.01	0.93	0.07	1.00	0.00	0.48	0.52	0.95	0.05	0.95	0.05
G551D/R347P	0.81	0.19	0.80	0.20	1.02	-0.02	0.92	0.08	1.00	0.00	0.46	0.54	0.91	0.09	0.95	0.05
621+1G→T/711+1G→T	0.86	0.14	0.53	0.47	1.00	0.00	0.93	0.07	0.98	0.02	0.81	0.19	0.53	0.47	0.94	0.06
621+1G→T/G85E	0.79	0.21	0.55	0.45	1.00	0.00	0.93	0.07	0.99	0.01	0.79	0.21	0.92	0.08	0.59	0.41
3849+10kbC→T/3849+10kbC→T	0.78	0.22	0.78	0.22	0.94	0.06	0.92	0.08	0.97	0.03	0.80	0.20	0.90	0.10	0.95	0.05
A455E/Normal	0.78	0.22	0.76	0.24	1.05	-0.05	0.95	0.05	1.02	-0.02	0.81	0.19	0.94	0.06	0.95	0.05
621+1G→T/A455E	0.80	0.20	0.54	0.46	1.00	0.00	0.93	0.07	0.99	0.01	0.79	0.21	0.89	0.11	0.96	0.04
R1162X/Normal	0.78	0.22	0.80	0.20	1.00	0.00	0.93	0.07	0.98	0.02	0.80	0.20	0.89	0.11	0.94	0.06
ΔF508/3659delC	0.83	0.17	0.84	0.16	1.01	-0.01	0.97	0.03	0.98	0.02	0.92	0.08	0.97	0.03	1.00	0.00
2789+5G→A/2789+5G→A	0.76	0.24	0.91	0.09	1.07	-0.07	0.97	0.03	1.05	-0.05	0.80	0.20	0.87	0.13	0.98	0.02
2184delA/Normal	0.78	0.22	0.79	0.21	0.95	0.05	0.96	0.04	0.99	0.01	0.81	0.19	0.96	0.04	0.96	0.04
1898+1G→A/Normal	0.86	0.14	0.81	0.19	0.99	0.01	0.99	0.01	0.89	0.11	0.85	0.15	0.77	0.23	0.93	0.07
621+1G→T/3120+1G→A	0.79	0.21	0.54	0.46	1.01	-0.01	0.92	0.08	1.00	0.00	0.78	0.22	0.87	0.13	0.96	0.04
3120+1G→A/3120+1G→A	0.78	0.22	0.82	0.18	1.00	0.00	0.93	0.07	1.00	0.00	0.80	0.20	0.96	0.04	0.96	0.04
F508C/Normal	0.75	0.25	0.82	0.18	0.99	0.01	0.90	0.10	0.97	0.03	0.81	0.19	0.90	0.10	0.95	0.05
I506V/Normal	0.81	0.19	0.82	0.18	1.00	0.00	0.90	0.10	0.98	0.02	0.80	0.20	0.90	0.10	0.96	0.04
R347H/Normal	0.82	0.18	0.81	0.19	0.98	0.02	0.94	0.06	1.00	0.00	0.79	0.21	0.96	0.04	0.95	0.05
ΔF508/3120G→A	0.79	0.21	0.81	0.19	1.00	0.00	0.92	0.08	1.00	0.00	0.80	0.20	0.97	0.03	0.96	0.04
S549N/Normal	0.78	0.22	0.82	0.18	0.99	0.01	0.96	0.04	1.00	0.00	0.80	0.20	0.98	0.02	0.96	0.04
S549R/Normal	0.78	0.22	0.79	0.21	1.01	-0.01	0.94	0.06	0.96	0.04	0.81	0.19	0.96	0.04	0.94	0.06

Genotype	3849+10kbC	3849+10kbC→T	A455*	A455E	R1162	R1162X	3659C	3659delC	2789+5G	2789+5G→A	2184A	2184delA	1898+1G	1898+1G→T	3120+1G	3120+1G→A
Normal/Normal	0.98	0.03	0.76	0.24	0.99	0.01	0.89	0.11	0.95	0.05	0.91	0.09	0.97	0.03	0.97	0.03
ΔF508/Normal	1.00	0.00	0.72	0.28	1.02	-0.02	0.80	0.20	0.99	0.01	0.94	0.06	1.01	-0.01	1.06	-0.06
ΔF508/ΔF508	0.99	0.01	0.75	0.25	1.01	-0.01	0.79	0.21	0.99	0.01	0.94	0.06	1.00	0.00	1.02	-0.02
ΔI507/Normal	0.97	0.03	0.75	0.25	1.00	0.00	0.88	0.12	0.94	0.06	0.93	0.07	0.98	0.02	0.92	0.08
W1282/Normal	0.99	0.01	0.77	0.23	1.00	0.00	0.87	0.13	0.96	0.04	0.93	0.07	0.99	0.01	0.98	0.02
1717-1G→A/Normal	0.98	0.02	0.75	0.25	0.97	0.03	0.89	0.11	0.92	0.08	0.92	0.08	0.98	0.02	0.95	0.05
G542X/G542X	1.00	0.00	0.74	0.26	0.98	0.02	0.87	0.13	0.94	0.06	0.95	0.05	0.98	0.02	0.91	0.09
G542X/Normal	0.99	0.01	0.74	0.26	0.99	0.01	0.89	0.11	0.96	0.04	0.93	0.07	0.98	0.02	0.97	0.03
ΔF508/G551D	0.98	0.02	0.75	0.25	0.98	0.02	0.87	0.13	0.95	0.05	0.95	0.05	0.97	0.03	0.97	0.03
ΔF508/R553X	1.01	-0.01	0.75	0.25	0.99	0.01	0.89	0.11	0.94	0.06	0.97	0.03	0.98	0.02	0.96	0.04
G551D/R553X	0.99	0.01	0.76	0.24	0.98	0.02	0.86	0.14	0.93	0.07	0.95	0.05	0.97	0.03	0.92	0.08
ΔF508/R560T	0.98	0.02	0.76	0.24	1.00	0.00	0.88	0.12	0.94	0.06	0.95	0.05	0.98	0.02	0.93	0.07
ΔF508/R117H	0.98	0.02	0.76	0.24	0.99	0.01	0.88	0.12	0.93	0.07	0.90	0.10	0.98	0.02	0.96	0.04
I148T/Normal	0.98	0.02	0.73	0.27	0.87	0.13	0.84	0.16	0.89	0.11	0.91	0.09	0.99	0.01	0.88	0.12
ΔF508/621+1G→T	1.01	-0.01	0.76	0.24	1.00	0.00	0.88	0.12	0.94	0.06	0.94	0.06	0.97	0.03	0.97	0.03
N1303K/G1349D	0.98	0.02	0.72	0.28	0.87	0.13	0.88	0.12	0.91	0.09	0.91	0.09	0.97	0.03	0.90	0.10
ΔF508/1078delT	0.99	0.01	0.73	0.27	0.88	0.12	0.89	0.11	0.89	0.11	0.91	0.09	0.98	0.02	0.90	0.10
R334W/?	0.97	0.03	0.77	0.23	0.99	0.01	0.86	0.14	0.95	0.05	0.94	0.06	0.98	0.02	0.97	0.03
ΔI507/R347P	0.98	0.02	0.75	0.25	0.85	0.15	0.89	0.11	0.89	0.11	0.89	0.11	0.98	0.02	0.89	0.11
G551D/R347P	0.99	0.01	0.75	0.25	1.00	0.00	0.88	0.12	0.94	0.06	0.93	0.07	0.98	0.02	0.93	0.07
621+1G→T/711+1G→T	0.97	0.03	0.77	0.23	0.98	0.02	0.89	0.11	0.92	0.08	0.92	0.08	0.98	0.02	0.93	0.07
621+1G→T/G85E	0.99	0.01	0.75	0.25	0.99	0.01	0.89	0.11	0.93	0.07	0.92	0.08	0.99	0.01	0.94	0.06
3849+10kbC→T/3849+10kbC→T	0.12	0.88	0.76	0.24	0.99	0.01	0.87	0.13	0.93	0.07	0.93	0.07	0.98	0.02	0.90	0.10
A455E/Normal	0.98	0.02	0.64	0.36	0.86	0.14	0.85	0.15	0.90	0.10	0.89	0.11	0.98	0.02	0.87	0.13
621+1G→T/A455E	0.98	0.02	0.68	0.32	0.99	0.01	0.88	0.12	0.93	0.07	0.93	0.07	0.98	0.02	0.92	0.08
R1162X/Normal	0.99	0.01	0.76	0.24	0.57	0.43	0.87	0.13	0.93	0.07	0.93	0.07	0.98	0.02	0.95	0.05
ΔF508/3659delC	0.99	0.01	0.84	0.16	0.96	0.04	0.36	0.64	0.94	0.06	0.96	0.04	1.01	-0.01	0.96	0.04
2789+5G→A/2789+5G→A	1.01	-0.01	0.81	0.19	1.05	-0.05	0.96	0.04	0.23	0.77	1.05	-0.05	1.01	-0.01	1.07	-0.07
2184delA/Normal	0.99	0.01	0.73	0.27	0.88	0.12	0.84	0.16	0.87	0.13	0.51	0.49	0.99	0.01	0.90	0.10
1898+1G→A/Normal	1.01	-0.01	0.85	0.15	0.98	0.02	0.83	0.17	0.96	0.04	0.97	0.03	0.51	0.49	0.94	0.06
621+1G→T/3120+1G→A	0.96	0.04	0.78	0.22	0.98	0.02	0.90	0.10	0.91	0.09	1.00	0.00	0.97	0.03	0.54	0.46
3120+1G→A/3120+1G→A	0.99	0.01	0.73	0.27	0.89	0.11	0.87	0.13	0.87	0.13	0.90	0.10	0.99	0.01	0.25	0.75
F508C/Normal	0.98	0.02	0.77	0.23	1.00	0.00	0.88	0.12	0.96	0.04	0.93	0.07	0.98	0.02	0.93	0.07
I506V/Normal	0.98	0.02	0.77	0.23	1.00	0.00	0.88	0.12	0.96	0.04	0.91	0.09	0.96	0.04	0.95	0.05
R347H/Normal	0.99	0.01	0.72	0.28	0.88	0.12	0.81	0.19	0.88	0.12	0.91	0.09	0.99	0.01	0.92	0.08
ΔF508/3120G→A	0.99	0.01	0.73	0.27	0.88	0.12	0.86	0.14	0.88	0.12	0.91	0.09	0.98	0.02	0.89	0.11
S549N/Normal	1.01	-0.01	0.72	0.28	0.89	0.11	0.87	0.13	0.89	0.11	0.91	0.09	1.00	0.00	0.90	0.10
S549R/Normal	0.98	0.02	0.75	0.25	0.86	0.14	0.88	0.12	0.88	0.12	0.89	0.11	0.98	0.02	0.90	0.10

Table 8
Allelic Ratio Data for the Reflex Panel[a]

Genotype	I507 & F508	I506V	I507V[b]	F508C
Exon 10 variants				
ΔF508/ΔF508[c]	—	—	—	—
ΔF508/Normal	**0.93**	0.03	0.02	0.02
Normal/Normal	**0.94**	0.03	0.01	0.01
ΔI507/Normal	**0.97**	0.04	−0.03	0.01
I506V/Normal	**0.45**	0.01	−0.01	**0.54**
F508C/Normal	**0.40**	**0.58**	0.00	0.01

Intron 8 variants			
Genotype	**5T**	**7T**	**9T**
7T/7T	−0.06	**1.06**	0.01
7T/7T	−0.01	**1.00**	0.01
7T/7T	−0.01	**1.01**	0.00
9T/9T	0.05	0.05	**0.90**
9T/9T	0.07	0.05	**0.87**
7T/9T	0.04	**0.45**	**0.51**
7T/9T	0.03	**0.40**	**0.56**
5T/7T	**0.42**	**0.60**	−0.01
5T/7T	**0.45**	**0.59**	−0.04
5T/9T	**0.36**	0.00	**0.64**

[a] Positive alleles are indicated in bold type.
[b] Samples positive for the I507V allele were not available.
[c] ΔF508/ΔF508 is negative for these alleles. Net MFI values were ≤9.

4. The 100 mM MES, pH 4.5, should be filter-sterilized and either prepared fresh or stored at 4°C between uses. Do not store at room temperature. The pH must be in the 4.5–4.7 range for optimal coupling efficiency.

5. We use 5 M tetramethylammonium chloride (TMAC) solution from Sigma (T-3411) for preparation of 1.5X and 1X TMAC hybridization solutions. We find that this TMAC formulation does not have a strong "ammonia" odor. TMAC hybridization solutions should be stored at room temperature to prevent precipitation of the Sarkosyl. TMAC hybridization solutions can be warmed to hybridization temperature to resolubilize precipitated Sarkosyl.

6. The hybridization kinetics and thermodynamic affinities of matched and mismatched sequences can be driven in a concentration-dependent manner *(33)*. At concentrations beyond the saturation level, the hybridization efficiency can decrease, presumably owing to competition of the complementary strand and renaturation of the PCR product *(7,20)*. Therefore, it is important to determine the range of target concentrations that yield efficient hybridization without sacrificing

discrimination. Volumes of 0.5–5 μL were tested for each PCR reaction, and 5 μL of a 5:1:5 pool of multiplex PCR reactions M1, M2, and M3 was found to be optimal for the standard mutation panel. At higher concentrations of M2, the hybridization efficiency of the exon 11 target was decreased, with a concomitant drop in reporter signal on the G542X-, G551D-, and R553X-specific microsphere sets.

7. The optimal amount of a particular oligonucleotide capture probe for coupling to carboxylated microspheres is determined by coupling various amounts in the range of 0.04–5 nmol per 5×10^6 microspheres. For this assay, we found 0.2 nmol per 5×10^6 microspheres in a 50 μL reaction to be optimal. The coupling procedure can be scaled up or down. Above 5×10^6 microspheres, use the minimum volume required to resuspend the microspheres. Below 5×10^6 microspheres, maintain the microsphere concentration and scale down the volume accordingly.

8. Denaturation and hybridization can be performed in a thermal cycler. Use a heated lid and a spacer (if necessary) to prevent evaporation. Maintain hybridization temperature throughout the labeling and analysis steps.

9. Whether it is necessary to remove the hybridization supernatant before the labeling step will depend on the quantity of biotinylated PCR primers and unhybridized biotinylated PCR products that is present and available to compete with the hybridized biotinylated PCR product for binding to the streptavidin-R-phycoerythrin reporter. We found it necessary to remove the hybridization supernatant prior to labeling, but this should be determined for each individual assay.

References

1. Grody, W. W. (1999) Cystic fibrosis: molecular diagnosis, population screening, and public policy. *Arch. Pathol. Lab. Med.* **123,** 1041–1046.
2. Balinsky, W. and Zhu, C. W. (2004) Pediatric cystic fibrosis: evaluating costs and genetic testing. *J. Pediatr. Health Care* **18,** 30–34.
3. Rowley P. T., Loader S., and Kaplan, R. M. (1998) Prenatal screening for cystic fibrosis carriers: an economic evaluation. *Am. J. Hum. Genet.* **63,** 1160–74.
4. Grody, W. W., Cutting, G. R., Klinger, K. W., Richards, C. S., Watson, M. S., and Desnick, R. J. (2001) Laboratory standards and guidelines for population-based cystic fibrosis carrier screening. *Genet. Med.* **3,** 149–154.
5. Schwarz, M. and Malone, G. (1996) Methods for screening in cystic fibrosis, in *Methods in Molecular Medicine: Molecular Diagnosis of Genetic Diseases* (Elles, R., ed.), Humana, Totowa, NJ, pp. 99–119.
6. Richards, C. S. and Grody, W. W. (2004) Prenatal screening for cystic fibrosis: past, present and future. *Expert. Rev. Mol. Diagn.* **4,** 49–62.
7. Armstrong, B., Stewart, M., and Mazumder, A. (2000) Suspension arrays for high throughput, multiplexed single nucleotide polymorphism genotyping. *Cytometry* **40,** 102–108.
8. Cai, H., White, P. S., Torney, D., et al. (2000) Flow cytometry-based minisequencing: a new platform for high-throughput single-nucleotide polymorphism scoring. *Genomics* **66,** 135–143.

9. Chen, J., Iannone, M. A., Li, M.-S., et al. (2000) A microsphere-based assay for multiplexed single nucleotide polymorphism analysis using single base chain extension. *Genome Res.* **10,** 549–557.

10. Colinas, R. J., Bellisario, R., and Pass, K. A. (2000) Multiplexed genotyping of beta-globin variants from PCR-amplified newborn blood spot DNA by hybridization with allele-specific oligodeoxynucleotides coupled to an array of fluorescent microspheres. *Clin. Chem.* **46,** 996–998.

11. Dunbar, S. A. and Jacobson, J. W. (2000) Application of the Luminex LabMAP in rapid screening for mutations in the cystic fibrosis transmembrane conductance regulator gene: a pilot study. *Clin. Chem.* **46,** 1498–1500.

12. Fulton, R. J., McDade, R. L., Smith, P. L., Kienker, L. J., and Kettman, J. R. (1997) Advanced multiplexed analysis with the FlowMetrix™ system. *Clin. Chem.* **43,** 1749–1756.

13. Iannone, M. A., Taylor, J. D., Chen, J., et al. (2000) Multiplexed single nucleotide polymorphism genotyping by oligonucleotide ligation and flow cytometry. *Cytometry* **39,** 131–140.

14. Iannone, M. A., Taylor, J. D., Chen, J., M.-S., Ye, F., and Weiner, M. P. (2003) Microsphere-based single nucleotide polymorphism genotyping. *Methods Mol. Biol.* **226,** 123–134.

15. Kaderali, L., Deshpande, A., Nolan, J. P., and White, P. S. (2003) Primer-design for multiplexed genotyping. *Nucleic Acids Res.* **31,** 1796–1802.

16. Musher, D., Goldsmith, E., Dunbar, S., et al. (2002) The association between hypercoagulable states or increased platelet adhesion/aggregation and bacterial colonization of intravenous catheters. *J. Infect. Dis.* **186,** 769–773.

17. Taylor, J. D., Briley, D., Nguyen, Q., et al. (2001) Flow cytometric platform for high-throughput single nucleotide polymorphism analysis. *Biotechniques* **30,** 661–669.

18. Ye, F., Li, M.-S., Taylor, J. D., et al. (2001) Fluorescent microsphere-based read-out technology for multiplexed human single nucleotide polymorphism analysis and bacterial identification. *Hum. Mutat.* **17,** 305–316.

19. Kellar, K. L. and Iannone, M. A. (2002) Multipexed microsphere-based flow cytometric assays. *Exp. Hematol.* **30,** 1227–1237.

20. Nolan, J. P. and Mandy, F. F. (2001) Suspension array technology: new tools for gene and protein analysis. *Cell Mol. Biol.* **47,** 1241–1256.

21. Nolan, J. P. and Sklar, L. A. (2002) Suspension array technology: evolution of the flat-array paradigm. *Trends Biotechnol.* **20,** 9–12.

22. Ikuta, S., Takagi, K., Wallace, R. B., and Itakura, K. (1987) Dissociation kinetics of 19 base paired oligonucleotide-DNA duplexes containing different single mismatched base pairs. *Nucleic Acids Res.* **15,** 797–811.

23. Livshits, M. A. and Mirzabekov, A. D. (1996) Theoretical analysis of the kinetics of DNA hybridization with gel-immobilized oligonucleotides. *Biophys. J.* **71,** 2795–2801.

24. Dunbar, S. A., Vander Zee, C. A., Oliver, K. G., Karem, K. L., and Jacobson, J. W. (2003) Quantitative, multiplexed detection of bacterial pathogens: DNA and

protein applications of the Luminex LabMAP™ system. *J. Microbiol. Methods* **53,** 245–252.

25. Peterson, A. W., Wolf, L. K., and Georgiadis, R. M. (2002) Hybridization of mismatched or partially matched DNA at surfaces. *J. Am. Chem. Soc.* **124,** 14601–14607.

26. Jacobs, K.A., Rudersdorf, R., Neill, S. D., Dougherty, J. P., Brown, E. L., and Fritsch, E. F. (1988) The thermal stability of oligonucleotide duplexes is sequence independent in tetraalkylammonium salt solutions: application to identifying recombinant DNA clones. *Nucleic Acids Res.* **16,** 4637–4650.

27. Wood, W. I., Gitschier, J., Lasky, L. A., and Lawn, R. M. (1985) Base-composition-independent hybridization in tetramethylammonium chloride: a method for oligonucleotide screening of highly complex gene libraries. *Proc. Natl. Acad. Sci. USA* **82,** 1585–1588.

28. Maskos, U. and Southern, E. M. (1992) Parallel analysis of oligodeoxyribonucleotide (oligonucleotide) interactions. I. Analysis of factors influencing duplex formation. *Nucleic Acids Res.* **20,** 1675–1678.

29. Maskos, U. and Southern, E. M. (1993) A study of oligonucleotide reassociation using large arrays of oligonucleotides synthesized on a large support. *Nucleic Acids Res.* **21,** 4663–4669.

30. Gotoh, M., Hasegawa, Y., Shinohara, Y., Schimizu, M., and Tosu, M. (1995) A new approach to determine the effect of mismatches on kinetic parameters in DNA hybridization using an optical biosensor. *DNA Res.* **2,** 285–293.

31. Diaz, M. R. and Fell, J. W. (2004) Highthrough-put detection of pathogenic yeasts in the genus of *Trichosporon*. *J. Clin. Microbiol.* **42:** 3696–3706

32. Hermanson, G. T. (1996) Zero-length cross-linkers, in *Bioconjugate Techniques,* Academic Press, San Diego, CA, pp. 169–186.

33. Wetmur, J. G. (1991) DNA probes: applications of the principles of nucleic acid hybridization. *Crit. Rev. Biochem. Mol. Biol.* **26,** 227–259.

9

Protein Microarray-Based Screening of Antibody Specificity

Rhonda Bangham, Gregory A. Michaud, Barry Schweitzer, and Paul F. Predki

Summary

The increased use of antibodies as therapeutics, as well as the growing demand for large numbers of antibodies for high-throughput protein analyses, has been accompanied by a need for more specific antibodies. An array containing every protein for the relevant organism represents the ideal format for an assay to test antibody specificity since it allows the simultaneous screening of thousands of proteins in relatively normalized quantities. Indeed, the use of a yeast proteome array to profile the specificity of several antibodies directed against yeast proteins has recently been described. In this chapter, we present a detailed description of the methods used to probe protein arrays with antibodies as well as the technical issues to consider when carrying out such experiments.

Key Words: Microarrays; antibody specificity; crossreactivity.

1. Introduction

DNA microarrays have become powerful tools for characterizing the relationships between large groups of genes. Although this type of microarray has become an established tool for expression profiling technology in biological and medical research, many aspects of cellular dynamics (such as tissue-specific protein expression, posttranslational modifications, protein activity, and protein–protein interactions), cannot be studied at the DNA level. Consequently, attention has turned more recently to proteomics—the study of all the proteins expressed by a given genome.

It is now possible to study entire proteomes with the goals of elucidating protein expression, subcellular localization, biochemical activities, and protein pathways. A variety of approaches exist to carry out such studies, including 2D gel electrophoresis, mass spectroscopy, high-throughput immunofluorescence studies with cell mounts, and yeast two-hybrid methods (reviewed in **ref. 1**).

From: *Methods in Molecular Medicine, Vol. 144, Microarrays in Clinical Diagnostics*
Edited by: T. Joos and P. Fortina © Humana Press Inc., Totowa, NJ

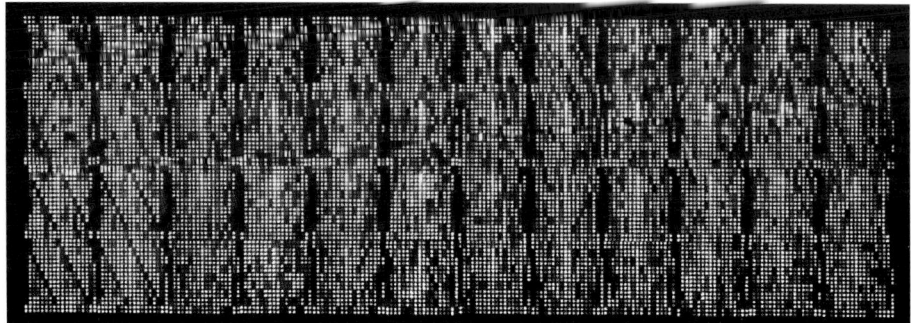

Fig. 1. Fluorescent image of the yeast ProtoArray™ (*see* Color Plate 16 following p.178). Microscope slides were spotted with over 4000 different proteins that were cloned, expressed and purified from yeast. Each protein on the array was detected using a Cy5-labeled antibody directed against an attached epitope tag. Slides were scanned using an Axon 4000B microarray laser scanner. A protocol for antibody probing of protein microarrays is given in **Subheading 3.** (Copyright 2003 Protometrix Inc.)

More recently, it has also become possible to analyze the activities of hundreds to thousands of different proteins using protein microarrays *(2,3)*. Cloning, expression, and purification strategies for high-throughput protein generation that do not discriminate among the diverse characteristics of individual proteins have been developed *(2–4)*. In addition, many new surface chemistries are available for protein immobilization that permit high detection sensitivity while preserving the functionality of the printed proteins. These include, but are not limited to, modified glass slides *(2,5)*, nitrocellulose-coated slides *(6)*, oriented immobilization via affinity tags *(3,7)* and 3D polyacrylamide gel pads *(8)*. These advances have fostered the development of high-content protein arrays such as a recently reported functional array consisting of nearly the entire proteome of the yeast *Saccharomyces cerevisiae (3)* (*see* also **Fig. 1**).

Protein microarrays have the potential to make significant contributions to both basic and applied research *(9,10)*. The breadth of this potential was first demonstrated in the work of MacBeath and Schreiber, who used relatively simple protein arrays to demonstrate a variety of applications such as protein–protein interactions, protein–small molecule interactions, and protein kinase substrate phosphorylation *(2)*. The first reported use of functional proteome-scale microarrays for biological discovery was reported by Snyder and coworkers *(3)*. In this study, arrays of thousands of yeast proteins were manufactured and used for protein–protein interaction and lipid binding screens. Since this publication, the number of discovery-based papers using protein microarrays has been increasing steadily *(9)*.

One application in which protein microarrays have enormous potential is the profiling of antibodies in clinical specimens for the diagnosis of disease, monitoring of disease progression, assessing response to treatment, and developing therapies to diseases such as cancer and autoimmune disorders. In fact, several studies have already shown sensitive and specific detection of antibodies for certain autoimmune diseases *(11,12)* Using an indirect fluorescence assay, specialized protein arrays consisting of a subset of proteins and peptides related to connective tissue diseases have demonstrated strong autoantibody reactivity with a four-to eightfold increase in sensitivity over conventional enzyme-linked immunosorbent assays (ELISAs).

A related area in which protein microarrays may have great utility is in profiling antibody specificity. New applications such as antibody arrays *(15,16)* and monoclonal antibody therapeutics *(17)* have increased the demand for more target-specific antibodies that exhibit minimal crossreactivity with disparate biomolecules. The ability to isolate and screen antibodies against a large number of different proteins is critical for providing sufficiently specific antibodies. A variety of approaches currently exist for antibody profiling including immunoblotting, immunoprecipitation, ELISAs, and radioimmunoassays (RIAs). These techniques are labor intensive and costly and consume large volumes of sample. Applying protein microarray technology to profiling antibodies has many advantages over these more conventional methods, including the ability to screen thousands of proteins simultaneously for crossreactivity in a quick and simple procedure. The identity of all represented proteins is known, including sequence information, minimizing the need for extensive analyses aimed at identifying the crossreactive species and protein concentration. In addition, the sample size needed to perform these experiments is very small, which is critical when the sample to be screened (i.e., patient sera) is precious.

The first demonstration of the use of positionally addressable protein microarrays for antibody specificity profiling was recently reported by Michaud et al. *(18)*. In this study, polyclonal and monoclonal antibodies generated against yeast proteins were probed against an array of more than 4000 purified yeast proteins. Not surprisingly, monoclonal antibodies showed greater specificity than polyclonal antibodies. Even monoclonal antibodies, however, exhibited some degree of crossreactivity. Further analyses indicated that the nature of the crossreactivity could not be predicted strictly on the basis of sequence analysis, indicating that this type of experimental approach to profiling antibody specificity will be an important tool for developing or using antibodies for research or medical purposes. This chapter describse a method that utilizes protein microarrays as a screening tool for the characterization of antibody binding, specificity, and crossreactivity.

2. Materials

1. Blocking buffer. 1X phosphate-buffered saline (PBS), 0.1% Tween-20, 1% protease-free bovine serum albumin (BSA). Prepare fresh as required.
2. 30% Protease-free BSA solution (Sigma, St. Louis, MO).
3. PBS probe buffer: 1X PBS, 5 mM MgCl$_2$, 5% glycerol, 0.05% Triton X-100, 1% protease-free BSA. Can be stored at 4°C indefinitely. (We recommend adding BSA just prior to use).
4. Primary antibody.
5. Fluorescent-labeled secondary antibody (reactive with primary antibody).
6. Nalgene trays or equivalent chambers for antibody incubations with slides.
7. Platform shaker.
8. Slide rack.
9. Microarray fluorescent scanner.

3. Methods

The method described in this chapter is a protein-antibody sandwich assay where by proteins have been spotted onto the surface, and two antibodies, one to recognize a specific target and a second to detect the primary antibody, are used as probes.

1. An unlabeled primary antibody is incubated with the protein microarray, and the array is then washed and incubated with a secondary antibody labeled with a fluorescent dye.
2. All steps can be completed in 1 d, including data processing. All steps should be performed at 4–6°C to maximize detection sensitivity and to preserve the integrity of the proteins on the array. All reagents should also be at this temperature when you are ready to begin the experiment.
3. It is very important to consider environmental conditions such as humidity and air quality before working with microarrays. Small changes in humidity may result in condensation on the surface of the array, leading to irregular spot morphology and subsequent difficulties in data analysis. Particulate matter in the air can also adversely affect data quality. Dust particles not visible to the eye can autofluoresce, leaving artifacts on the final scanned image that can mask spots and affect data extraction. Therefore, the experimenter should make every attempt to perform protein array experiments under clean conditions.

3.1. Probing Protein Arrays With Antibodies

3.1.1. Preparation of Protein Arrays

Preparation of the yeast proteome microarrays used in our examples is only described here briefly.

1. More than 4000 different yeast proteins were purified using affinity chromatography and printed on FAST (nitrocellulose pad size, 20×60 mm) slides (Schleicher & Schuell) with a Genemachines Omnigrid arrayer. All features were spotted in duplicate.

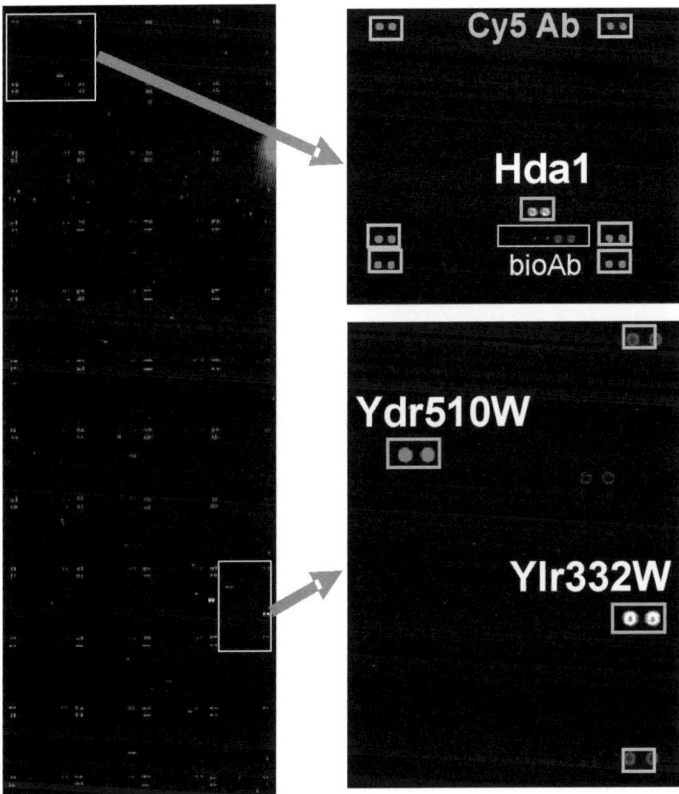

Fig. 2. Fluorescent images of antibody probing of the yeast proteome microarray (*see* Color Plate 17 following p.178). The expanded subarray on the top shows the anti-Hda1 antibody reacting with its cognate protein. The blue boxes are drawn around spots of Cy5-labeled antibody used as fiduciary markers. The yellow box is drawn around the spots representing biotinylated antibody used as a positive control. The expanded subarray on the bottom shows the anti-Hda1 antibody crossreacting with two other proteins.

2. Fluorescently-labeled antibody was printed in every subarray to facilitate gridding during data acquisition (**Fig. 2**; *see* Color Plate 17 following p.178). Each array contains 48 subarrays with 18×18 geometry with 250-μm center-to-center spacing. A gradient of BSA has been included as a negative control for assessing nonspecific protein interactions.
3. To calculate the amount of protein represented in each spot, a gradient of pure glutathione-*S*-transferase (GST) was printed in every subarray to generate a standard curve. Probing slides with an antibody that recognizes GST allows for the calculation of the equivalent solution protein concentrations for every protein printed on the array by extrapolating the concentrations from this standard curve.
4. Slides are stored at $-20°C$ in a sealed bag until ready to use.

3.1.2. Blocking the Protein Microarrays

The first step in probing a protein array with antibodies is to block sites on the microarray surface with a nonspecific blocking agent such as BSA. This will prevent nonspecific binding of proteins to the microarray substrate, effectively increasing the specificity of protein–protein interactions and expanding the dynamic range of the data.

1. Prepare blocking buffer fresh (*see* **Note 1**). Buffer must be cold (4–6°C) prior to use.
2. Transfer enough blocking bluffer to a tray or slide rack to ensure complete coverage of the slides. Slides can be blocked lying flat or upright depending on choice of tray.
3. Place the slides in the blocking buffer (*see* **Note 2**).
4. Incubate slides for 1 h with gentle agitation on a platform shaker.
5. Dispose of the blocking buffer, and immediately incubate with the primary antibody (*see* **Note 3**).

3.1.3. Primary and Secondary Antibody Binding to Array

Conditions should be optimized for probing the arrays with any antibody. Titering the antibody will determine the optimal concentration needed to achieve maximal sensitivity with minimal background. The most favorable dilution should be determined empirically from the manufacturer's recommendation when using a purchased antibody. Three 10-fold dilutions are recommended. If the signal for the cognate protein is weak, or no protein signals are detectable on the array, consider using a higher concentration of antibody. Consequently, if background intensity is masking recognition of spots on the array or there are large numbers of nonspecific interactions, consider using a more dilute antibody concentration.

1. Add 20 mL probe buffer with 1% BSA and primary antibody to slides in the trays. For the negative control slide, add buffer with 1% BSA only (*see* **Note 4**).
2. Incubate while shaking for 2 h.
3. Discard buffer.
4. Wash the slides with 20 mL of probe buffer. Mix gently for 5 min on a platform shaker. Discard the buffer and repeat two more times.
5. Centrifuge the secondary antibody (16,000g, 5 min) to sediment any dye-labeled protein aggregates.
6. Add 20 mL of probe buffer with 1% BSA and a secondary antibody (directed against the primary antibody) to the tray. Also add secondary antibody only to a slide that will serve as the negative control slide. A titer of the secondary antibody should be performed to determine the concentration required for minimizing background and optimizing sensitivity.
7. Incubate while shaking on a platform shaker for 2 h.
8. Discard buffer.

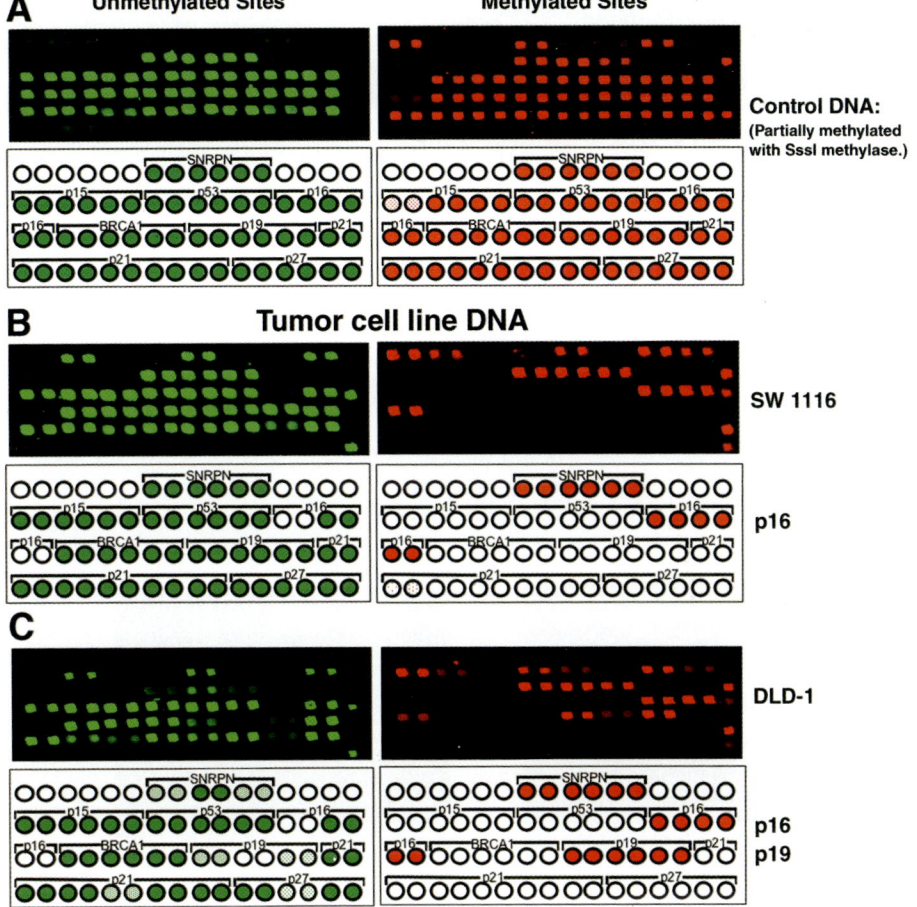

Color Plate 15. Universal array images of methylation profiles of selected promoter regions in normal and colorectal tumor cell line genomic DNAs (Chapter 2, Fig. 8, *see* pp. 48, 49).

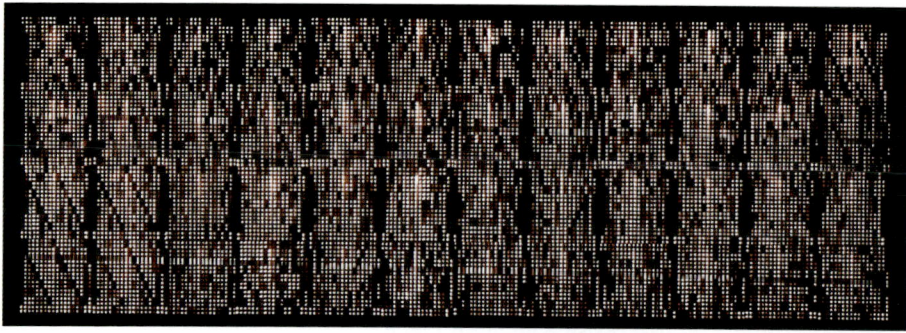

Color Plate 16. Fluorescent image of the yeast ProtoArray™ (Chapter 9, Fig. 1, *see* p. 174).

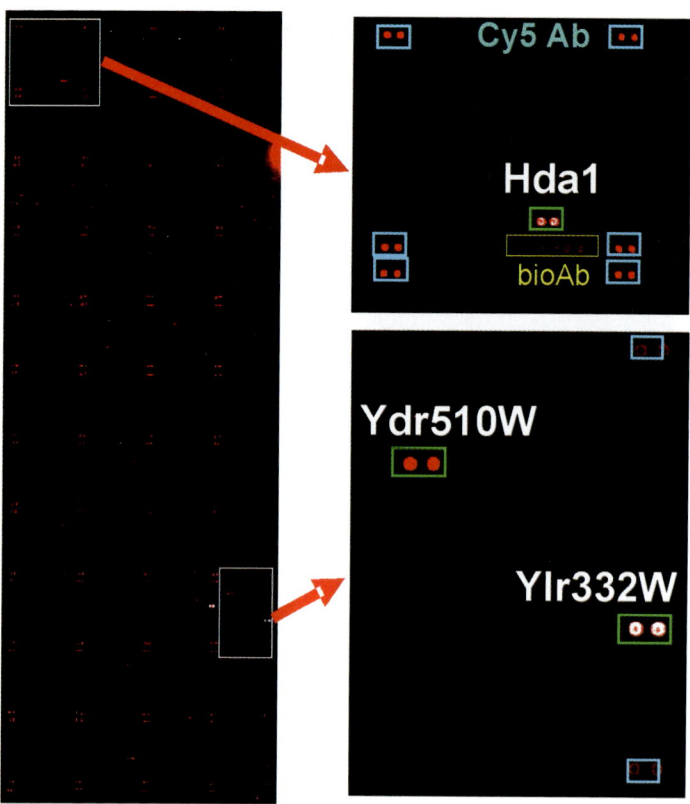

Color Plate 17. Fluorescent images of antibody probing of the yeast proteome microarray (Chapter 9, Fig. 2, *see* p. 177).

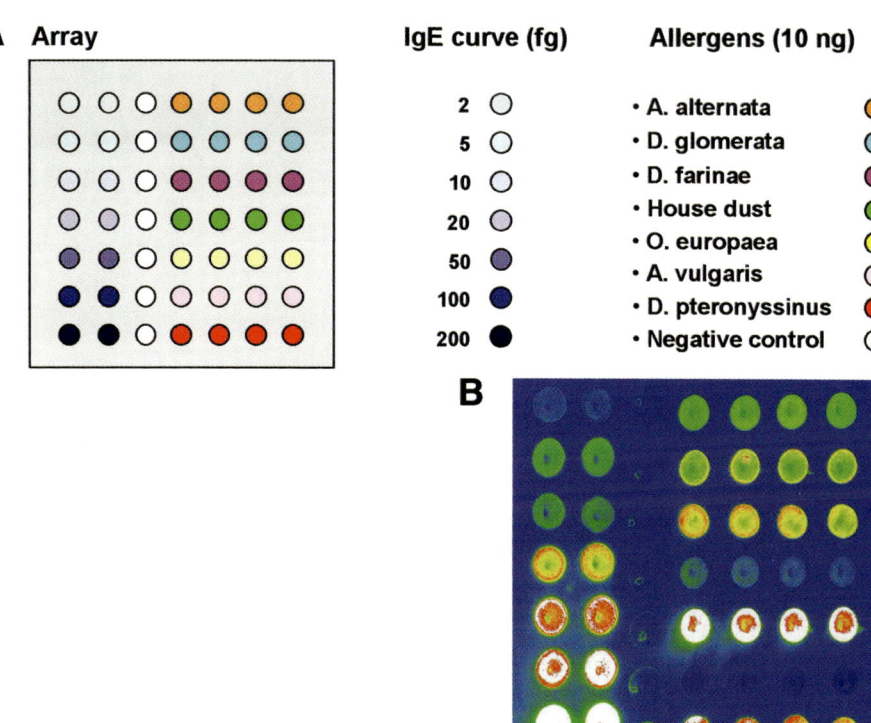

Color Plate 18. Schematic array format and fluorescent scan for allergen investigation (Chapter 11, Fig. 1, *see* pp. 201, 202).

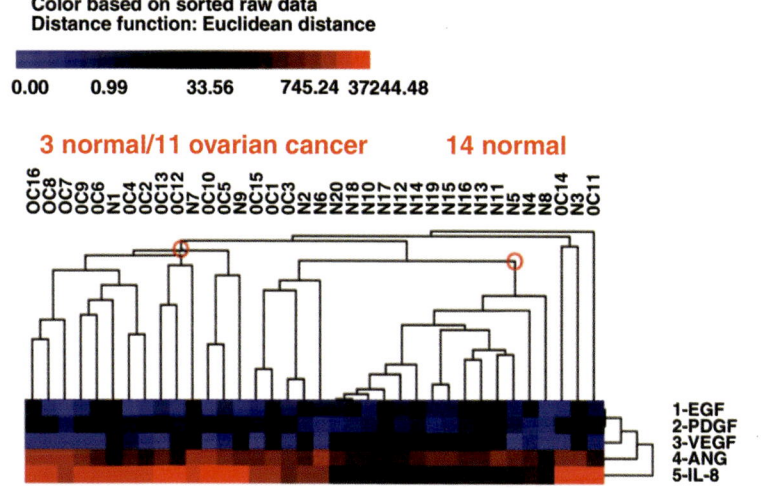

Color Plate 19. Clustering analysis of angiogenic factors measured by enhanced protein profiling arrays between normal subjects and patients with gynecological diseases (Chapter 12, Fig. 3, *see* p. 219).

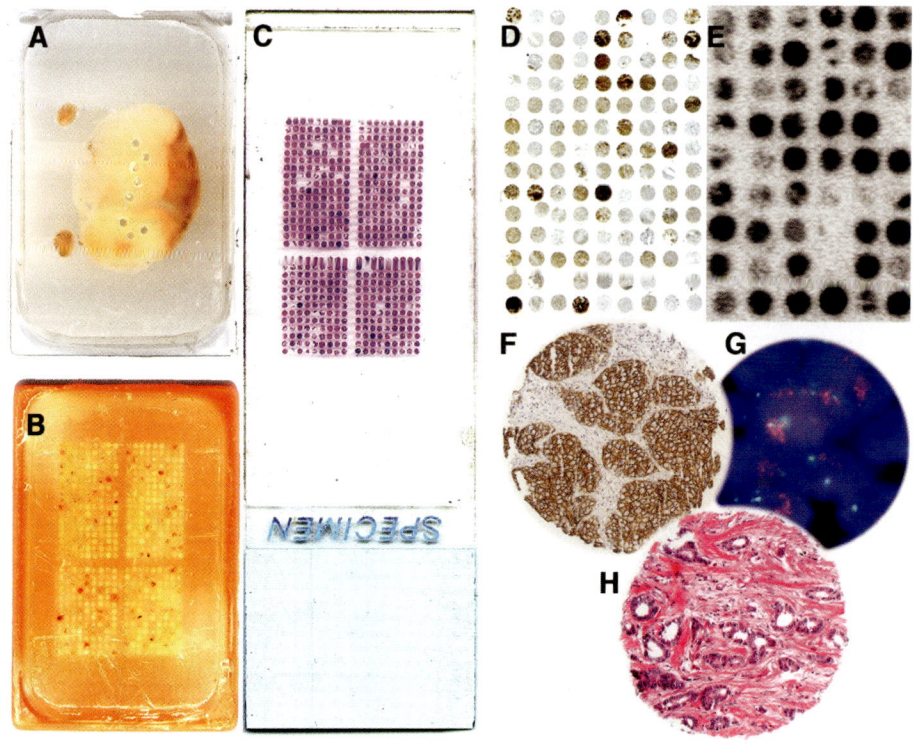

Color Plate 20. Tissue microarray manufacturing and applications (Chapter 15, Fig. 2, *see* p. 264, 265).

9. Wash the slides with 20 mL of probe buffer for 10 min with gentle shaking on a platform shaker. Discard the buffer and repeat two more times.
10. After the final wash, remove slides from the tray by grasping the slide edges (do not touch center of array with any object), and remove excess liquid by tapping lightly onto a dry cloth. Place slides into a slide box.
11. Centrifuge at (<1800g) for 1 min in a table-top centrifuge equipped with a rotor for handling slide boxes for 1 min. Be sure to provide a balance.
12. Place slides in a dark place until ready to be imaged. (We recommend scanning slides within 12 h after completion of experiment.)

3.2. Array Imaging

The following protocol utilizes the Axon 4000B scanner for acquiring images. Steps can be modified for other microarray laser scanners. Cy5 labels are scanned with a red excitation laser at 635 nm and Cy3 with a green laser at 532 nm (*see* **Note 5**) Images are analyzed using Genepix Pro 4.0 software (Axon Instruments).

Slides must be completely dry before scanning.

3.2.1. Acquiring the Image

1. Turn on the Axon 4000B scanner and then open Genepix software from the desktop.
2. Slide the door of the scanner open and lift the slide holder up. Place the slide face down with the barcode nearest the front of the machine. Close the slide holder and door. Slide is ready to be scanned.
3. Select the wavelength appropriate to the fluorescence excitation/emission spectrum of the dye.
4. Set the photomultiplier tube (PMT) gain. Recommended starting PMT is 600.
5. Set laser power to 100%, pixel size to 10 μm (or highest resolution), and focus position to 0 μm (*see* **Note 6**).
6. Perform a preview scan (*see* **Note 7**). While the scan is being performed, the PMT can be adjusted. If the cognate protein is present on the array and is saturated (spots will be white), the PMT should be adjusted so that these spots are just below saturation (65,535). Be sure the scan area encompasses all the spots on the array.
7. Scan the image (*see* **Notes 8** and **9**).

3.2.2. Extracting the Data With Genepix Pro Software

1. Once the image has been acquired and saved, load an array list file (.gal file). This file can be made with the Genepix Pro software. Basically, it will transform a text file with the identities of the spots into a file (.gal) that can be overlaid on the image as a grid that assigns each spot on the chip its identity based on the printing parameters. Refer to the Help menu for details on how to make this array list file.
2. Overlay the grid to roughly cover the spots on the array (*see* **Note 10**).

3. Go to "Align Blocks" and select "Align features in all blocks." Accuracy of the alignment should be ensured and manually reconfigured if necessary.
4. Click on "Analyze Data," and Genepix Pro will generate a report for all feature and background pixel intensites for the array.
5. Save results as an Excel (.xls) file.

3.3. Analyzing the Data

The following method uses Microsoft Excel and Access software to process the data. If a number of slides are going to be analyzed, it is best to write a macro to process the data more quickly.

1. Calculate median F-B for each spot. Use the values appropriate for the scanning wavelength used (i.e., 532 for Cy3 or 635 for Cy5). This is the signal–background ratio for each spot.
2. Average the F-B values for duplicate spots of the same protein.
3. Calculate the standard deviation and coefficient of variance (CV) for the duplicate spots.
4. Filter out the control spots from the protein spots of interest.
5. Calculate the average signal–background for the slide (minus control spots) by averaging the signal–background ratios for all the yeast proteins.
6. Calculate the standard deviation for the yeast proteins on the slide by measuring the standard deviation of the signal–background ratios for all the yeast proteins.
7. Determine the statistical significance of the features by setting an F-B cutoff value. In general, we have assigned hits as significant if the F-B value is greater than or equal to the average signal–background ratio of the slide plus three standard deviations (3*SD [step 7] + Average [step 6]). However, the user should consider the signal distribution for each slide when determining statistical significance.

Once features have been assigned some statistical significance (*see* **Note 11**), it is important to consider the CV values that were calculated for every feature when using an automated method of data analysis. CVs greater than 30% generally indicate artifactual problems on the array. It is best to refer back to the image to see whether local artifacts such as dust particles or anomalous local variance have influenced the data. Import the Excel files into Access.

1. Design queries to align (based on array location) the features for each protein. The output of these queries should contain the name of each protein, array location, F-B background value, and standard deviation of F-B for every feature from each slide.
2. Export results from queries files into Excel.
3. Proteins having signals (F-B) that are greater than two fold over the signal generated on the negative slide are also considered significant.

4. Notes

1. Many alternative reagents are commonly used to reduce nonspecific binding to the slide. Nonfat dry milk, normal serum to the host of the secondary antibody, and gelatin

are a few examples. Different microarray surface chemistries may perform differently with different blocking solutions. Protease-free BSA is highly recommended.

2. Always wear powder-free gloves when handling the slides and do not touch the printed surface of the array. Also, do not wear clothing that is capable of leaving fibers on the array. Hair coverings are recommended. Pay special attention to the cleanliness of the experiment.

3. Do not allow slides to dry at any point in the experiment, particularly between buffer changes. If slides do dry during the course of the experiment, it is likely that background will be very high.

4. Maintaining BSA in the buffers throughout the process will help keep background low and also increase the specificity of protein interactions.

5. A variety of other dyes can be scanned by these lasers. Refer to the scanner manufacturer's recommendations for dye compatibility. Many scanners also come equipped with added lasers for excitation of other fluorescent dyes commonly used for microarray imaging.

6. These settings have been optimized for glass and nitrocellulose-coated slides. If other microarray substrates are used; these settings will probably differ and should be optimized.

7. If no signals are observed on the array, particularly dye-labeled positive control spots, the scanner may not be set to the correct wavelength.

8. If the PMT is lowered to 400 and the image is still too bright, we recommend leaving the PMT at 400 and beginning to reduce the laser power.

9. Dyes can be light sensitive and thus susceptible to photobleaching. It is not recommended to scan any part of the array numerous times prior to data extraction.

10. The grid may not align accurately throughout the slide because of minor variations that occur during the printing process. The blocks should be individually repositioned before allowing the software to find features.

11. The method of analysis illustrated above is only for consideration. The user should determine the best method for data analysis and assignment of statistical significance.

Acknowledgments

We are grateful to Michael Salcius, Jaclyn Bonin, and Fang Zhou for their expertise in the process of manufacturing the yeast protein arrays that made all of these experiments possible.

References

1. Michaud, G. A. and Snyder, M. (2002) Proteomic approaches for the global analysis of proteins. *Biotechniques* **33**, 1308–1316.

2. MacBeath, G. and Schreiber, S. L. (2000) Printing proteins as microarrays for high-throughput function determination. *Science* **289**, 1760–1763.

3. Zhu, H., Bilgin, M., Bangham, R., et al. (2001) Global analysis of protein activities using proteome chips. *Science* **293**, 2101–2105.

4. Zhu, H., Klemic, J. F., Chang, S., et al. (2000) Analysis of yeast protein kinases using protein chips. *Nat. Genet.* **26**, 283–289.

5. Mendoza, L. G., McQuary, P., Mongan, A., Gangadharan, R., Brignac, S., and Eggers M. (1999) High-throughput microarray-based enzyme-linked immunosorbent assay (ELISA). *Biotechniques.* **27,** 778–780, 782–786, 788.

6. Kukar, T., Eckenrode, S., Gu, Y., et al. (2002) Protein microarrays to detect protein-protein interactions using red and green fluorescent proteins. *Anal. Biochem.* **306,** 50–54.

7. Lesaicherre, M. L., Jue, R. Y., Chen, G. Y., Zhu, Q., and Jao, S. Q. (2002) Intein-mediated biotinylation of proteins and its application in a protein microarray. *J. Am. Chem. Soc.* **124,** 8768–8769.

8. Arenkov, P., Kukhtin, A., Gemmell, A., Voloshchuk, S., Chupeeva, V., and Mirzabekov, A. (2000) Protein microchips: use for immunoassay and enzymatic reactions. *Anal. Biochem.* **278,** 123–131.

9. Predki, P. F. (2004) Functional protein microarrays: ripe for discovery. *Curr. Opin. Chem. Biol.* **8,** 8–13.

10. Zhu, H. and Snyder, M. (2003) Protein chip technology. *Curr. Opin. Chem. Biol.* **7,** 55–63.

11. Robinson, W. H., Steinman, L., and Utz, P. J. (2003) Protein arrays for autoantibody profiling and fine-specificity mapping. *Proteomics* **3,** 2077–2084.

12. Joos, T. O., Schrenk, M., Hopfl, P., et al. (2000) A microarray enzyme-linked immunosorbent assay for autoimmune diagnostics. *Electrophoresis* **21,** 2641–2650.

13. Sreekuman, A. and Chinnaiyan, A. M. (2002) Using protein microarrays to study cancer. *Biotechniques* **33,** S46–S53.

14. Haab, B. B. (2003) Methods and applications of antibody microarrays in cancer research. *Proteomics* **3,** 2116–2122.

15. Haab, B. B., Dunham, M. J., and Brown, P. O. (2001) Protein microarrays for highly parallel detection and quantitation of specific proteins and antibodies in complex solutions. *Genome Biol.* **2,** 1–13.

16. Schweitzer, B., Roberts, S., Grimwade, B., et al. (2002) Multiplexed protein profiling on microarrays by rolling-circle amplification. *Nat. Biotechnol.* **20,** 359–365.

17. Forero, A. and Lobuglio, A. F. (2003) History of antibody therapy for non-Hodgkin's lymphoma. *Semin. Oncol.* **30(6 suppl. 17),** 1–5.

18. Michaud, G. A., Salcius, M., Zhou, F., et al. (2003) Analyzing antibody specificity with whole proteome microarrays. *Nat. Biotechnol.* **21,** 1509–1512.

10

Multiplexed Protein Analysis Using Antibody Microarrays and Label-Based Detection

Brian B. Haab

Summary

This chapter describes methods for the production and use of antibody microarrays. The experimental methods are divided into three sections. The first gives information relating to the preparation and handling of antibodies and the production of microarrays. The second relates to the preparation of the samples, from either blood, tissue, or cell lysates. The third section describes methods for using the samples on antibody microarrays. Two related detection methods are described in which the proteins of a sample to be analyzed are labeled with a tag to allow detection after capture on antibody microarrays.

Key Words: Antibody microarrays; label-based detection; protein profiling.

1. Introduction

Antibody microarrays continue to be developed as a useful tool for multiplexed protein analysis. The benefits of the technology include highly parallel protein measurements, rapid experiments and analysis, quantitative and sensitive detection, and low volume assays. Recent publications have demonstrated the application of the technology to the study of proteins from serum *(1,2)*, cell culture *(3)*, tissue *(4)*, and culture media *(1,5)*. To facilitate the broad dissemination and more routine use of antibody microarray methods, this chapter describes practical and validated techniques that can be implemented by most labs. All aspects of the experimental process are described, including antibody handling, sample handling, and microarray production and use. Information on data analysis is not presented here but can be found in the above referenced citations.

The antibody microarray method described here uses label-based detection, in which covalently attached tags (such as biotin or the fluorophores Cy3 and Cy5) on the target proteins allow detection after proteins bind to the array.

From: *Methods in Molecular Medicine, Vol. 144, Microarrays in Clinical Diagnostics*
Edited by: T. Joos and P. Fortina © Humana Press Inc., Totowa, NJ

Label-based detection is an attractive complementary alternative to the sandwich assay, which employs a matched pair of antibodies for every protein target. Advantages of label-based detection are ease in assay development, since only one antibody per target is required, as opposed to a pair of antibodies for a sandwich assay, and the possibility for multicolor detection. Since different samples may be labeled with different tags, a reference sample may be coincubated with a test sample to provide internal normalization to account for concentration differences between spots. The two-color strategy is broadly used in DNA microarray experiments and has been used in antibody microarray experiments to detect multiple proteins in serum *(2,6)*, cell culture *(3,7,8)*, and tissue lysates *(9)*. Two-color methods are described here using labeling by either fluorophores (e.g., Cy3 and Cy5) or haptens (e.g., biotin and digoxigenin).

2. Materials

1. Robotic microarrayer (several commercial models available).
2. Microarray scanner (several commercial models available).
3. Clinical centrifuge with flat swinging buckets for holding slide racks (Beckman Coulter, among others).
4. HydroGel-coated glass microscope slides (PerkinElmer Life Sciences).
5. Nitrocellulose-coated glass microscope slides (FAST, Schleicher & Schuell).
6. NHS-linked Cy3 and Cy5 protein labeling reagents (Amersham, cat. nos. PA23001 and PA25001).
7. NHS-linked biotin and digoxigenin protein labeling reagents (Molecular Probes, cat. nos. B-1582 and A-2952).
8. Microscope slide staining chambers with slide racks (Shandon Lipshaw, cat. no. 121).
9. Polypropylene 384-well microtiter plates (Genetix or MJ Research).
10. Vacuum sealer (Tilia International, FoodSaver Vac 360).
11. Diamond scriber (VWR, cat. no. 52865-005).
12. Hydrophobic marker (PAP pen, Sigma, cat. no. Z37782-1).
13. Aluminum foil tape (R.S. Hughes, cat. no. 425-3).
14. Wafer handling tweezers (Technitool, cat. no. 758TW178, style 4WF).
15. Gel filtration columns for protein cleanup (Bio-Rad Micro Bio-Spin P-6, cat. no. 732-6222).
16. Kit for Protein A cleanup of antibodies (Bio-Rad Affi-gel Protein A MAPS kit, cat. no. 153-6159).
17. BCA protein assay kit (Pierce, cat. no. 23226).
18. Microcon YM-50 (Millipore, cat. no. 42423).
19. Phosphate-buffered saline (PBS), pH 7.4: 137 mM NaCl, 2.7 mM KCl, 4.3 mM Na_2HPO_4, 1.4 mM KH_2PO_4.
20. Carbonate buffer, pH 8.5: 50 mM Na_2CO_3.
21. PBST0.5: PBS + 0.5% Tween 20.
22. NP40 lysis buffer: 50 mM HEPES, pH 7.0, 5 mM EDTA, 50 mM NaCl, 10 mM NaPPi (sodium pyrophosphate), 50 mM NaF, 1% NP40, 10 mM sodium vanadate, and complete protease inhibitors (Roche, cat. no. 1 696 498).

3. Methods

The methods are divided into three sections: (1) antibody handling and microarray production, (2) sample preparation, and (3) microarray processing.

3.1. Antibody Handling and Microarray Production

The success of this method depends in part on the quality of the antibodies used on the microarrays. Each antibody has different performance characteristics in the microarray assay, and each needs to be evaluated independently. Antibody performance can be evaluated using standard immunological methods, which will not be discussed here.

3.1.1. Antibody Selection and Preparation

3.1.1.1. CHOOSING THE TARGETS AND ANTIBODIES

The first step in the project preparation is to determine the protein targets to be measured, which depend on the goals of the research. Not all proteins are suitable for measurement in this assay; the size of the target proteins and their estimated abundances in the samples need to be considered. If a protein is very small, it may not be compatible with label-based detection methods (discussed in **Subheading 3.3.**) that use a size-based separation of labeled products from unincorporated labeling reagents. If a protein is in very low abundance (*see* **Note 1**), it may fall outside the detection limit of the assay. We recommend choosing monoclonal antibodies that work in enzyme-linked immunosorbent assays, but polyclonal antibodies can also work well.

3.1.1.2. PURITY OF ANTIBODIES

Antibodies work best in the microarray assay when they are highly purified. A high concentration of other proteins in the antibody solution usually results in a weakened or nonspecific signal, since many binding sites on the microarray are occupied by the other proteins. Polyclonal antibodies collected from antisera should be antigen-affinity purified. IgG purification from antisera (e.g., using Protein A beads) is only good enough if the antibody is targeting a high-abundance protein. Monoclonal antibodies that are provided in ascites fluid should be further purified. The simplest method is to isolate the IgG fraction of the sample using a kit such as the Bio-Rad Affigel Protein A MAPS kit. Some antibodies come in a high concentration (up to 50%) of glycerol to improve stability. Although glycerol will not interfere with the assay, the added viscosity may negatively affect the printing process. Glycerol concentrations above approx 20% should be avoided. To change the buffer of an antibody, we recommend the Bio-Rad Micro Bio-Spin P30 column (*see* **Note 2**). If the antibody is subsequently to be labeled, do not put the antibody in a Tris-or amine-containing buffer, which will interfere with primary amine-based labeling reaction.

3.1.1.3. Buffer, Concentration, and Storage

Antibodies are stable refrigerated in a standard buffer such as PBS. The optimal spotting concentration is around 500 µg/mL. Higher concentrations could yield higher signal intensities and lower detection limits and may be desirable if consumption of antibody is not a concern. Most antibodies can be stored refrigerated for up to a year. New antibodies should be divided into aliquots, using one as a refrigerated working stock and freezing the others at −70°C, to avoid repeated freeze/thaw cycles that can damage proteins. When retrieving antibodies from a freezer stock, thaw the solution slowly on ice to reduce damage from the thawing process.

3.1.2. Preparation of Coated Microscope Slides

Various substrates for antibody microarrays have been demonstrated, such as poly-L-lysine-coated glass *(6)*, aldehyde-coated glass *(10)*, nitrocellulose *(4,11)*, and a poly-acrylamide-based hydrogel *(12,13)*. Microscope slides with these various coatings can be purchased commercially. Since each application is unique in some aspects, the choice of which to use should be determined empirically by each user. We recommend simply preparing arrays on several different substrates and running them in parallel. Several criteria could be used to determine which surface type is best. The signal-to-background ratio at each antibody is a good criterion, but one could also look at the reproducibility between replicate arrays and the consistency in the background level within each array.

3.1.3. Printing Microarrays

After the antibodies have been prepared at the proper purity and concentration, they are assembled into a "print plate"—a microtiter plate used in the robotic printing of the microarrays. Polypropylene microtiter plates are preferable to polystyrene because of lower protein adsorption. The plate should be rigid and precisely machined for optimal functioning with printing robots. The amount of antibody solution to load into each well of the print plate depends on the requirements of the printing robot—usually 10–15 µL is sufficient (*see* **Note 3**). If printing is sometimes inconsistent or variable between printing pins, it is desirable to fill multiple wells with the same antibody solution, so that different printing pins spot the same antibody. Store the 384-well print plates sealed in the refrigerator until ready to use. Aluminum foil tape provides a good seal (*see* **Subheading 2., Materials**). Long-term evaporation-free storage is ensured by enclosing the covered plate in a sealed plastic bag (*see* **Note 4**). Prepare a spreadsheet containing the well identities for use in downstream data processing applications.

The details of the printing process will depend on the type of printing robot used, but we give here some general notes.

1. Minimize the time that the print plates are unsealed and exposed in order to keep evaporation of the antibody solutions low.
2. Evaporation may be minimized by cooling the print plate (if the robot has that feature) and maintaining a moderately high humidity in the printing environment (~45%).
3. The proper printing of the robot should be confirmed with test prints on dummy slides before starting the microarray production. Use 500 μg/ml bovine serum albumin (BSA) in 1X PBS for the test prints.
4. Make sure the water in the tip wash bath is changed regularly to prevent contamination of the tips.
5. It is desirable to confirm sufficient washing of the pins between loads. This test can be done by loading labeled protein into one of the print plate wells in a dummy print, followed by scanning the slide. If fluorescence is seen in spots after the fluorescently labeled material, the pins need to be washed more stringently.
6. Most microarrayers will allow the printing of replicate spots on each array, which are useful to obtain more precise data through averaging and to ensure the acquisition of data if a portion of the array is unusable. Four to six spots per array per antibody are usually sufficient.

3.1.4. Postprint Processing of Microarrays

The handling of the arrays after printing depends on the surface used. Arrays printed on hydrogels should be incubated overnight in a humidified chamber to induce full binding of the antibodies to the hydrogel matrix. Microarrays printed on highly absorptive surfaces such as nitrocellulose will not require such a long incubation before blocking. We recommend vacuum sealing and refrigerating the arrays for storage before use (*see* **Note 4**) to minimize loss of antibody activity.

3.2. Sample Preparation

Here we describe the preparation of proteins for use in the microarray assay from either clinical specimens or cell culture. **Subheading 3.2.1.** concentrates on the use of serum or plasma (also applicable to other bodily fluids), and **Subheading 3.2.2.** describes the preparation of proteins from tissue specimens or cell culture.

3.2.1. Using Serum or Plasma Samples

The analysis of proteins from serum or plasma is convenient because all the proteins are soluble and only need to be diluted in the proper buffer (described in **Subheading 3.3.**).

Caution: Clinical samples should be handled as biohazards since they can be carriers of infectious agents. Tips and tubes that contact clinical samples should be disposed of in a biohazard bag. Samples should be aliquoted so that no more than three thaws are necessary for any experiment, as some researchers have observed measurable breakdown in proteins after three thaws. Samples should be stored at −80°C. Thaw the samples on ice to minimize protein breakdown.

3.2.2. Preparing Proteins From Cell Culture or Tissue

3.2.2.1. PREPARATION OF PROTEIN EXTRACTS FROM CELL CULTURE

1. Wash cells cultured in a 10-cm Petri dish at 80% confluency with ice-cold PBS three times.
2. Add 1 mL of NP40 lysis buffer and keep on ice for 15 min.
3. Scrape the lysate with a rubber policeman and transfer into a 1.5-mL Eppendorf tube.
4. Centrifuge at 10,000g for 10 min.
5. Transfer the supernatant into a fresh 1.5-mL tube.
6. Measure protein concentration using a Pierce BCA protein assay kit.
7. Bring the cellular extracts to the same concentration (~2 mg/mL) with NP40 lysis buffer.
8. Aliquot into working stocks and freeze at −80°C.

3.2.2.2. PREPARATION OF PROTEIN EXTRACTS FROM TISSUE

Tissue specimens should be handled as biohazards. Tissue samples fixed with formaldehyde and embedded in paraffin are not suitable for protein extraction for microarrays. Tissue samples either fresh frozen in liquid nitrogen or frozen embedded in Optimal Cutting Temperature (OCT) compound are suitable for this process. To make optimal use of the specimen, one may cut sections with a cryostat as needed for protein extraction, saving the rest of the specimen for later experiments. A 50-µm-thick section of a 1–2 cm^2 tissue sample yields approx 100–200 µg of protein (depending on the tissue type), plenty for several microarray experiments since about 20 µg is used per experiment.

1. Prepare 1.5-mL Eppendorf tubes with 70 µl of NP40 lysis buffer on ice.
2. Collect 50-µm tissue sections and put each section into a different tube.
3. Homogenize the tissue sections with a pellet pestle immediately. Keep on ice for 15 min.
4. Centrifuge at 10,000g for 10 min.
5. Transfer the supernatant into a fresh 1.5-mL tube.
6. Measure protein concentration using a Pierce BCA protein assay kit.
7. Bring the cellular extracts to same concentration (~2 mg/mL) with NP40 lysis buffer.
8. Aliquot into working stocks and freeze at −80°C.

3.3. Microarray Use

Figure 1 presents the types of detection methods described here: direct labeling (**Fig. 1A**) and indirect detection (**Fig. 1B**). In the direct labeling method, all

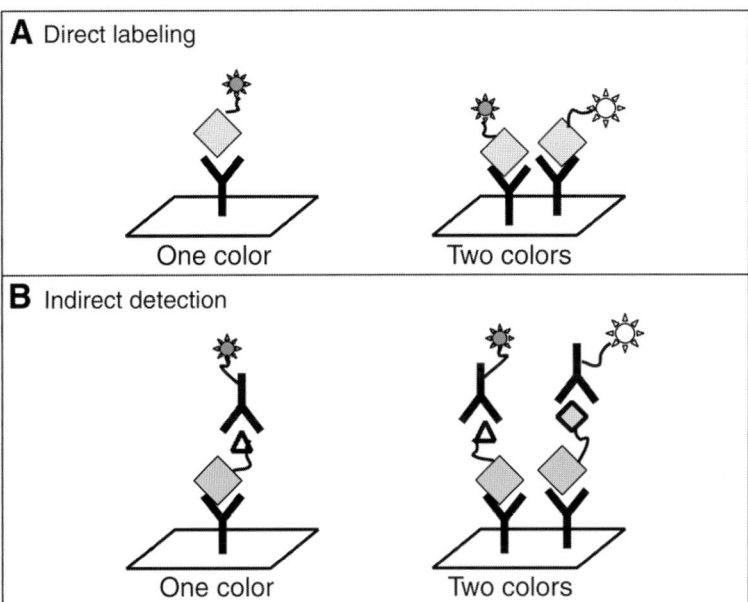

Fig. 1. Schematic representation of the detection methods described. **(A)** One- and two-color direct fluorescent labeling. Proteins are directly labeled with a fluorescent tag. In the two-color case, two pools of proteins are labeled with distinct fluorescent tags and incubated together on an antibody array. **(B)** One- and two-color indirect detection. Proteins are directly labeled with a hapten (such as biotin, represented by the triangle). In the two-color case, two pools of proteins are labeled with distinct haptens (represented by the triangle and the diamond). Fluorescently labeled antibodies that target the haptens are then incubated on the array.

proteins in a complex mixture are labeled with fluorophores that allow subsequent detection by fluorescence scanning. In indirect detection, the proteins are labeled with a tag that is detected by a fluorescently labeled antibody directed against the tag. **Figure 2** presents a schematic representation of the steps in two-color detection of labeled samples. Both detection methods start with the labeling of proteins in complex mixtures. The method described below uses commercially available reagents that react with primary amine groups.

3.3.1. Protein Labeling

1. Determine the volume to be used on each array. The volume should be enough to cover the array easily without the possibility of drying at the edges. (Drying will cause severe increase in background.) A 12 × 12-mm HydroGel pad with a hydrophobic marker boundary 1 mm beyond the edge of the pad requires about 50 μL of sample. The use of a cover slip over the pad reduces the required volume to about 20 μL.

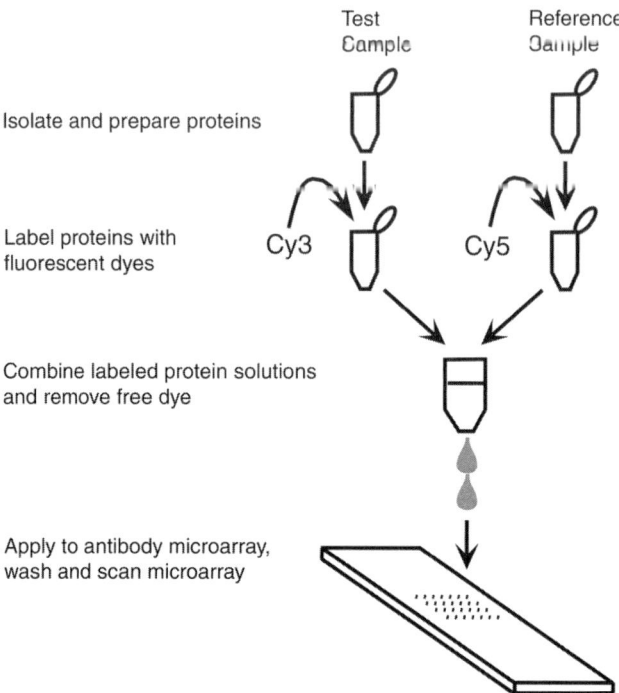

Fig. 2. Two-color detection of directly labeled samples. Proteins from a test sample and a reference sample are isolated and labeled with distinct fluorophores. The two samples are mixed, passed through a column to remove unincorporated dye, and incubated on an antibody microarray. After the array is washed and scanned, the relative sample-specific to reference-specific fluorescence at each antibody spot provides a measure of the protein binding from each sample.

2. Determine the volume to label of each sample.
   a. Serum samples: the amount to label of each sample is equal to $(V_a{*}A)/D$, where V_a is the volume per array determined above, A is the number of arrays on which the sample will be used, and D is the desired final dilution of the sample. A 50X final dilution usually works well for serum samples. Thus if the volume on the array is 50 µL with a dilution of 50X and two arrays per sample, the volume to use of each serum sample is 2.0 µL.
   b. Proteins extracted from cells: use enough to give about 500 µg/mL on the array, according to the formula $(500\ \mu g/mL \times V_a \times A)/C_s$, where V_a and A are as defined above, and C_s is the starting concentration of the sample. Thus if a protein extract solution is 10 mg/mL, and the volume on the array will be 50 µL, use 2.5 µL of sample per array.
   c. Samples to be used in a pooled reference: the amount to be labeled of each component of the reference is $(V_a \times A)/(D \times N_r)$, where V_a, A, and D are as defined

above, and N_r is the number of samples pooled in the reference. For example, if a pool of 10 samples at a 50X dilution will be used as the reference for 20 arrays of 50 µL each, the volume to be used of each sample in the reference will be $(50 \times 20)/(50 \times 10) = 2.0$ µL.

3. Dilute the samples with carbonate buffer spiked with a normalization standard (*see* **Note 5**). For serum samples, dilute by about 15X (e.g., 14 µL buffer per 1 µL sample). For cell lysates, dilute 2X, or to about 1 mg/mL.
4. Prepare an approx 7 mM stock of the labeling reagents (NHS-linked Cy3, Cy5, biotin, or digoxigenin) in dimethyl sulfoxide (DMSO) (*see* **Note 6**).
5. Add a 20th volume of the 7 mM labeling reagent stock to each sample (e.g., 0.75 µL for a 15 µL solution). Use different labels for the reference samples and the test samples. (e.g., use Cy3 for the test samples and Cy5 for the reference samples if using direct labeling, or use biotin for the test samples and digoxigenin for the reference samples if using indirect detection.)
6. Mix well, spin the tubes briefly to collect the solutions, and let the reactions proceed on ice in the dark for 1 h.
7. Add a 10th volume 1 M pH 7.5 Tris-HCI to quench the reaction. Let the reactions sit for 30 min to complete the quenching.
8. Prepare a Biorad Bio-spin 6 microcolumn for each sample (*see* **Note 2**).
9. Load each sample onto a Biorad Bio-spin 6 microcolumn, spin at 1000g for 2 min, and collect the flowthrough.
10. Mix each labeled sample with an appropriately labeled reference sample. If a pooled reference is being used, collect all the reference samples into one tube, mix, and distribute among the samples. Keep these solutions on ice and in the dark.
11. Make 10X blocking solution: 30% nonfat milk in PBS and 1% Tween-20 (e.g., add 3 g milk to 10 mL buffer).
12. Spin the blocking solution at 10,000g for 10 min. Use the supernatant in the next step.
13. To each labeled sample-reference mixture, add the 10X blocking solution. Use the volume of 10X blocking solution that will give 1X in the final volume. For example, for a final volume on the array of 50 µL, add 5 µL 10X blocking solution.
14. Add 1X PBS to bring each mixture to the final volume. The labeled samples may be stored overnight in the refrigerator.

3.3.2. Microarray Use

3.3.2.1. DIRECT LABELING

1. Carefully pipete the appropriate volume of each fluorophore-labeled sample (prepared according to **Subheading 3.3.1.**) onto each microarray. If using a cover slip, cover immediately (*see* **Note 7**).
2. Place the slides in a covered, humidified box (*see* **Note 8**). Place the box on an orbital shaker rotating about once per second.
3. Incubate the arrays for 2 h at room temperature with constant shaking.

4. Using wafer handling tweezers to hold each slide, dunk each slide briefly in PBST0.5 to remove the sample and cover slip (if used). Load each slide in a slide rack in a staining chamber filled with PBST0.1 (*see* **Note 9**).
5. Wash the slides for 10 min at room temperature in the PBST0.1.
6. Wash the slides twice more for 10 min each in fresh changes of PBST0.1.
7. Dry the slides by centrifugation (*see* **Note 10**). Keep the slides in the dark until ready to scan.

3.3.2.2. INDIRECT DETECTION

1. Prepare Cy3- and/or Cy5-labeled antihapten antibodies.

   a. Prior to labeling, the antibodies should be relatively pure (not in ascites fluid or antisera), at a concentration of at least 1 mg/mL, and not in a Tris- or amine-containing buffer. *See* **Subheading 3.1.** if the antibodies need to be purified or buffer-exchanged.
   b. Determine the amount of labeled antibody needed: equal to the volume per array (*see* **Subheading 3.3.1.**) times the final concentration times the number of arrays.
   c. Use the labeling and cleanup method described in **Subheading 3.3.1.**

2. Prepare the detection antibody solution.

   a. Mix the appropriate amounts of Cy3-labeled and Cy5-labeled antihapten antibodies, based on the solution volume per array times the concentration of each antibody times the number of arrays. The optimal concentration of the antibodies can be determined by titration (often around 10 µg/mL).
   b. Prepare a 10X blocker solution as described in **Subheading 3.3.1.**
   c. Add 10X blocker. Use 1/10th of the intended final volume of the antibody mix.
   d. Add 1X PBS to the final volume.

3. Incubate hapten-labeled samples (prepared according to **Subheading 3.3.1.**) on the arrays, and wash and dry the arrays according to **steps 1–7** in **Subheading 3.3.2.1.**
4. Incubate the detection antibody solution on the arrays for 1 h and wash and dry the arrays according to **steps 1–7** in **Subheading 3.3.2.2.**
5. Keep the slides in the dark until ready to scan.

4. Notes

1. Detection limits for the assay depend on the antibodies used, the protein background in the sample, and the detection conditions. In general, the direct labeling method (described in **Subheading 3.3.**) can give detection limits in the low ng/mL range for targets in a serum background, and the indirect detection method can give slightly lower detection limits. A modification of the indirect detection method, which involves amplification of the fluorescence signal from the tagged proteins, gives further reduced detection limits *(11)*.
2. The Biospin columns come prepacked with two types of buffers: sodium saline citrate (SSC) and Tris-HCl buffer. The packing buffer comes out of the column with the sample that was applied to it, that is, after a sample is run through the column, it

will be in the buffer with which the column was packed. The packing buffer can be changed by running a different buffer through the column three times. The P30 column removes solution components smaller than 30 KD, and the P6 column removes components smaller than 6 kDa. Thus the P30 column is better for purification of antibodies, and the P6 column is better for the purification of complex mixtures in which low-molecular-weight species should be preserved.

3. The volume may depend on the shape of the well and how far the print tips descend into the well. Too much volume can lead to droplets of antibody solution sticking to the outside of the print tip. The volume may also need to be optimized for particular applications, such as multiple draws from each well, which would require a greater volume.

4. Food storage sealers work well for this purpose. Some models have the option of applying vacuum, or simply sealing without vacuum. To prevent evaporation of fluid from the print plate completely, insert a moist piece of paper towel into the bag with the print plate before sealing—that will keep the humidity in the bag high.

5. As part of the analysis of each array, a normalization factor is calculated for each array that sets data from normalization antibodies to known values. A normalization antibody could detect a spiked-in standard such as flag-labeled BSA. A normalization standard could also be a protein normally found in the samples, such as IgG in serum.

6. The Amersham dyes come in tubes of about 200 nmol each, so an approx 7 mM solution can be achieved by dissolving in about 30 μL DMSO. The concentration of the stock should be checked by UV absorbance. The labeling stocks should be aliquoted and frozen at −80°C.

7. Use wafer handling tweezers to hold the cover slip. Lower it onto the array at an angle so one edge of the cover slip touches the solution first. Lower the rest of the cover slip onto the array so that no bubbles are trapped. We recommend using the Lifterslip-style cover slip, which has thin spacers attached to two edges of the cover slip. The spacers slightly elevate the cover slip above the array and allow more movement of the liquid under the cover slip. The use of a cover slip reduces the volume required per array and is useful if sample is limited. However, a cover slip greatly limits diffusion in the sample and can reduce signal strengths, so we recommend normally not using a cover slip.

8. A microscope slide box that holds 100 slides works well for this purpose. A paper towel soaked with 1X PBS can be placed in the bottom of the box, and the slides can be placed flat over the slide holding slots.

9. A microscope slide staining chamber is useful for the washing steps (*see* **Subheading 2., Materials**). The staining chambers come with slide racks that hold 10–30 slides. The racks can be transferred between staining chambers containing different washes and to a clinical centrifuge for drying the slides.

10. A clinical centrifuge with flat swinging bucket holders works well for this task. Place a paper towel layer on the bottom of the swinging bucket to absorb water that is removed from the slides. Place the slide rack on the paper towel and centrifuge at approx 300g for about 3 min.

Acknowledgments

This work was funded through the Van Andel Research Institute, the Early Detection Research Network (U01 CA84986), and the Michigan Economic Development Corporation.

References

1. Huang, R.-P., Huang, R., Fan, Y., and Lin, Y. (2001) Simultaneous detection of multiple cytokines from conditioned media and patient's sera by an antibody-based protein array system. *Anal. Biochem.* **294,** 55–62.
2. Miller, J. C., Zhou, H., Kwekel, J., et al. (2003) Antibody microarray profiling of human prostate cancer sera: antibody screening and identification of potential biomarkers. *Proteomics* **3,** 56–63.
3. Sreekumar, A., Nyati, M. K., Varambally, S., et al. (2001) Profiling of cancer cells using protein microarrays: discovery of novel radiation-regulated proteins. *Cancer Res.* **61,** 7585–7593.
4. Knezevic, V., Leethanakul, C., Bichsel, V. E., et al. (2001) Proteomic profiling of the cancer microenvironment by antibody arrays. *Proteomics* **1,** 1271–1278.
5. Schweitzer, B., Roberts, S., Grimwade, B., et al. (2002) Multiplexed protein profiling on microarrays by rolling-circle amplification. *Nat. Biotechnol.* **20,** 359–365.
6. Haab, B. B., Dunham, M. J., and Brown, P. O. (2001) Protein microarrays for highly parallel detection and quantitation of specific proteins and antibodies in complex solutions. *Genome Biol.* **2,** 1–13.
7. Nielsen, U. B., Cardone, M. H., Sinskey, A. J., MacBeath, G., and Sorger, P. K. (2003) Profiling receptor tyrosine kinase activation by using ab microarrays. *Proc. Natl. Acad. Sci. USA* **100,** 9330–9335.
8. Lin, Y., Huang, R., Chen, L. P., et al. (2003) Profiling of cytokine expression by biotin-labeled-based protein arrays. *Proteomics* **3,** 1750–1757.
9. Tannapfel, A., Anhalt, K., Hausermann, P., et al. (2003) Identification of novel proteins associated with hepatocellular carcinomas using protein microarrays. *J. Pathol.* **201,** 238–249.
10. MacBeath, G., and Schreiber, S. L. (2000) Printing proteins as microarrays for high-throughput function determination. *Science* **289,** 1760–1763.
11. Zhou, H., Bouwman, K., Schotanus, M., et al. (2004) Two-color, rolling-circle amplification on antibody microarrays for sensitive, multiplexed serum-protein measurements. *Genome Biol.* **5,** R28.
12. Guschin, D., Yershov, G., Zaslavsky, A., et al. (1997) Manual manufacturing of oligonucleotide, DNA, and protein microchips. *Anal. Biochem.* **250,** 203–211.
13. Arenkov, P., Kukhtin, A., Gemmell, A., Voloshchuk, S., Chupeeva, V., and Mirzabekov, A. (2000) Protein microchips: use for immunoassay and enzymatic reactions. *Anal. Biochem.* **278,** 123–131.

11

Allergen Microarrays

Tito Bacarese-Hamilton, Julian Gray, Andrea Ardizzoni, and Andrea Crisanti

Summary

Allergy affects more than 25% of Western populations (*1*) and is estimated to be the sixth leading cause of chronic disease in the United States and Western Europe. The complexity of the condition is such that hundreds of common allergens have been described, and in order to maximize diagnostic efficiency there is an urgent clinical requirement for assays to provide multiple-allergen determination in a timely and cost-effective manner. Miniaturized immunoassays that utilize protein microarray technology now offer the possibility of circumventing most of the current limitations in the serodiagnosis of allergic disease. The heterogeneous nature of allergens presents many challenges in all aspects of developing such arrays, from immobilization of the capture molecule to detection of the bound ligand. In addition, there is no simple method of protein amplification (such as PCR for nucleic acids), and stabilization is yet a further major consideration. Notwithstanding these challenges, protein microarrays have been developed for the serodiagnosis of allergies and other complex clinical conditions. These assays exhibit good analytical and clinical performance and deliver significant advantages in convenience and cost compared with traditional ELISA test formats. This chapter details the techniques employed in the construction and processing of an allergen array specific for the serodiagnosis of allergic disease. An overview of protein microarray technology is provided and the principles that underpin the suitability for use of this technology in the identification and measurement of particular proteins in patient sera (serum profiling) are discussed.

Key Words: Protein microarray; immunoassay; serodiagnosis; allergy; signal amplification.

1. Introduction

1.1. Protein Microarrays and Their Use in Serum Profiling

The term *microarray* refers to the deposition of minute volumes of aqueous biomolecules onto a solid surface in an ordered and highly dense manner. Microarray technology represents the culmination of novel technological advances that together permit the simultaneous, parallel analysis of thousands of

From: *Methods in Molecular Medicine, Vol. 144, Microarrays in Clinical Diagnostics*
Edited by: T. Joos and P. Fortina © Humana Press Inc., Totowa, NJ

parameters within a discrete microenvironment. With its innate capacity for miniaturization, the microarray format overcomes the spatial, logistical, and economic constraints that have previously made such wide-scale, truly parallel investigations impossible. Interest in the microarray platform has grown rapidly since its inception. Following the initial development of DNA-based arrays *(2)*, the number of publications documenting protein-based *(3,4)* and more recently carbohydrate-based *(5,6)* microarray technology has increased exponentially.

Protein microarrays can be conveniently divided into two classes on the basis of their application. Broadly speaking, these categories constitute those arrays that are functional and those that are analytical in their nature *(7)*. Functional microarrays are used to assess proteins for a variety of biochemical activities and provide valuable insight into the function of specific proteins. The high-throughput capacity of such arrays permits the simultaneous screening of multiple proteins, such that global functional analysis at almost the entire proteome level has recently been demonstrated *(8)*. These high-density functional arrays have huge applications in protein characterization as well as in drug and drug target identification. Future improvements in protein expression technology will enormously benefit the ability to scale up these arrays *(9)*. Analytical arrays, on the other hand, are implicitly designed to detect the presence and measure the concentration of particular proteins of interest (target proteins) from a complex protein mixture. Here immobilized capture molecules (typically antibodies or antigens) with characterized affinity for target protein(s) are used to screen sera or lysate for the presence of the target protein(s). Analytical arrays are typically of low to medium density owing to limited numbers of characterized capture molecules and have application in many fields including diagnostics, prognostics, and protein expression profiling *(10)*.

Key products for functional and analytical applications for protein arrays are whole proteome array slides, slides for diagnostic applications, and slides for ligand and/or small molecule screening. All applications have the same general requirements;

1. Immobilization of a capture molecule on the slide.
2. Getting the target molecule to the capture site.
3. Detection of the target molecule when bound.

The technology for printing high-density protein arrays has been lifted almost entirely from DNA chip construction. There are numerous attachment strategies and many commercial suppliers for these products *(11)*. The capture surface plays a critical role in the quality of the microarray since it influences the efficiency of antibody attachment, the degree of nonspecific binding, and the accessibility of the capture molecules to the target proteins. Glass slides have been adopted as the substrate of choice owing to their ease of use, compatibility with laboratory equipment, and robustness. However, many new surfaces such as hydrogels and a variety of membranes are in development for specific applications. The accurate and precise high-density spotting of proteins is performed by robotic arrayers mainly based on either contact or ink-jet printing techniques *(12,13)*. The binding of the target molecules to the capture surface is performed using traditional solid-phase immunoassay techniques such as, enzyme-linked immunosorbent assay (ELISA). The key consideration is to bind the target molecules specifically and avidly and to minimize nonspecific binding. Nonspecific binding slows the equilibration of the target molecules with the capture surface and is one of the major constraints on the sensitivity of the system. The glass slide allows for good washing efficiency since most of the protein array assays developed to date are of the "heterogeneous" type in which various wash steps are performed to remove the unbound label *(14)*. At least one "homogeneous" assay format incorporating the optical phenomenon of evanescent waves is also in development *(15)*. Other "homogeneous" assay formats, e.g., using microcantilever technology, have also been described *(16)*. The main detection method used in both DNA and protein array technology is fluorescence. This is because fluorophores are in the main small molecules that produce high specific affinity labels (i.e., high dye: protein incorporation ratios), and fluorescence offers a large dynamic range and lends itself to scanning techniques. Fluorescence is detected by either confocal microscopy or by cameras incorporating charged coupled devices (CCDs) *(17)*. Other detection methods described include chemiluminescence, colorimetry, mass spectrometry and surface plasmon resonance *(18–21)*.

In recent years, much interest has focused on the use of protein microarrays for immunoassay development. Ekins developed the theoretical background for microarray-based ligand-binding assays in the late 1980s *(22–24)*. He formulated the concept of the ambient analyte immunoassay, which he predicted would not only allow simultaneous detection of an analyte panel but would also have a sensitivity equal to, or even greater than, standard immunoassay techniques. Validation of these theories has had to await the advent of robotic arrayers and fluorescent scanners, but they are now widely accepted and are the basis of all microarray assay applications.

The profiling of serum is fundamental to both diagnostic clinical procedure (through identification of serum antibodies that suggest the presence of disease agents) and prognostic clinical procedure (using biomarker and surrogate levels in serum to predict the risk of disease and monitor disease progression/regression during treatment). However, the assays currently used commercially (e.g., ELISA *[25]*) fail to meet the clinical demand for high-throughput and low sample requirements (**Tables 1** and **2**). Protein microarrays can overcome these limitations of current clinical assays *(26)* and facilitate the simultaneous, parallel determination of hundreds of target proteins within patient sera. In recent years huge progress has been made in the development of microarray immunoassays intended for the profiling of human serum. Microarrays have been developed

Table 1
Comparison of Microarray and ELISA for Serum Reactivity[a]

	ELISA		Microarray			
Allergen	Positive	Negative	Positive	Negative	Sensitivity %	Specificity %
Dermatophagoides pteronyssinus	11/22	11/22	13/22	9/22	90.9	72.7
Dermatophagoides farinae	8/16	8/16	9/16	7/16	87.5	75
Olea europaea	7/22	15/22	8/22	14/22	71.4	80
Altemaria altemata	3/20	17/20	4/20	16/20	100	94.1
Artemisia vulgaris	2/22	20/22	2/22	20/22	50	95
Dactylis glomerata	6/22	16/22	7/22	15/22	83.3	87.5

Clinical sensitivity is calculated by expressing the number of samples that are positive by microarray immunoassay as a percent of the number of samples positive for the disease (i.e., positive by ELISA). Clinical specificity is calculated by expressing the number of samples that are negative by the microarray immunoassay as a percent of the number of samples negative for the disease (i.e., negative by ELISA). As an example, for *Dactylis glomerata*:

	Test positive	Test negative
Disease present: 6	5	1
Disease absent: 16	2	14

Therefore sensitivity = 5/6 = 83.3% and specificity = 14/16 = 87.5%.

[a] For serodiagnosis of allergy.
From **ref. 27.**

that permit the detection of antibodies directed against antigens implicated in allergy *(1,18,27)*, autoimmune disease *(28,29)*, and parasitic/viral infections *(30)*. Similarly, protein microarrays have been developed that permit the detection and accurate quantification of prognostic biomarkers from human serum *(31,32)*.

In our laboratory, the main application of protein arrays is for serodiagnosis of disease. Such assays, because of their ability to detect in parallel many antibodies with different specificities, have great potential in epidemiological research and vaccine development as well as in the diagnosis of complex clinical conditions such as allergy and autoimmune and infectious diseases. In the serodiagnosis of autoimmune, infectious diseases, IgG and IgM immunoglobulins are the antibody responses determined in patient sera. These antibodies circulate at µm/mL concentrations, and their determination is made directly, using fluorescently labeled anti-IgG and anti-IgM reagents. However, the diagnosis of allergy necessitates the measurement of immunoglobulin IgE, which circulates at ng/mL concentrations in human serum. Consequently, techniques have been developed to increase the sensitivity of protein arrays for such purposes.

Table 2
Comparison of Microarray and ELISA Test Protocols[a]

	ELISA	Microarray
Sample		
Dilution	1:300	1:200
Volume (μL)	100	80
Incubation (min)	60	15
Temperature	37°C	Room temperature
Wash (μL)	1050	5000
Conjugate		
Volume (μL)	100	80
Incubation (min)	30	5
Temperature	37°C	Room temperature
Wash (μL)	1050	5000
Substrate/Stop		
Volume (μL)	100 + 100	None
Incubation (min)	10	None
Temperature	37°C	N/A
Read time (min/sample)	2	2
Signal stability	15 min	>3 mo
Total time (min)		
1 parameter	110	20
10 parameter	1100	20
Estimated cost (US$)		
1 parameter	1.5	5
10 parameter	15	5

[a] For serodiagnosis of infectious disease. NA, not applicable.
From ref. *30*.

One such technique is tyramide signal amplification *(27)*; another uses rolling circle amplification *(18)*.

1.2. Arrays for Serodiagnosis of Allergy

Allergy is a state of immune dysregulation that results in an overproduction of IgE *(33)*. Identification of the causative agents that provoke allergies in individuals is a challenging task owing to the large numbers of such agents. Traditionally, techniques such as skin prick testing and immunoassay determination of single allergens at a time have been performed to identify these causative agents *(34)*. Skin prick testing is uncomfortable for the patient and requires observation in case the patient develops anaphylactic shock; traditional

ELISAs do not address the full spectrum of allergens because of the quantities of blood required and the expense of the kits. Allergen arrays offer the possibility of determining and monitoring the IgE reactivity profile of allergy sufferers to very large numbers of disease-causing allergens simultaneously and use minute volumes of serum. This is particularly important for the diagnosis of allergy in children, in whom taking a blood sample can be difficult. The following sections detail our experiences in the development of such arrays.

2. Materials

2.1. Equipment

1. Computer-controlled high-speed robot (Total Array System, Biorobotics, www. biorobotics.com).
2. S5000. Scanner including ScanArray™ and QuantArray™ software (Perkin Elmer).
3. Equipment for ELISA:
   a. Microtiter plate washer: Autura Plate Washer (Mikura, www.mikura.co.uk).
   b. Microtiter plate reader: Kinetic Microplate Reader, v max (Molecular Devices).
4. Spectrophotometer: DU 640 (Beckman Coulter).
5. Silanized glass microscope slides (CEL Associates, www.cel-1.com).
6. 96- or 384-Well microtiter plates (Thermo Life Sciences, www.thermols.com).
7. Gene-frames (65 µL, 1.5 × 1.6 cm) (Abgene, www.abgene.com).

2.2. Reagents

1. Allergen extracts: *Dermatophagoides pteronyssinus* (D1), *Dermatophagoides farinae* (D2), *Alternaria alternata* (M6), *Olea europaea* (T9), *Artemisia vulgaris* (W6), *Dactylis glomerata* (G3), and house dust (H2; Allergon AB, www.allergon.com).
2. Human IgE purified from myeloma plasma (Calbiochem).
3. Horseradish peroxidase-streptavidin conjugate (0.1 mg/mL), Alexa 546-tyramide conjugate, amplification buffer, and 300 mL/L hydrogen peroxide as provided in the TSA™ kit # 23 (Molecular Probes).
4. Biotinylated anti-human IgE at a final concentration of 1 mg/mL (KPL).
5. Slide blocking solution: 20 g/L bovine serum albumin (BSA) in 1X phosphate-buffered saline (PBS).
6. Signal control: Alexa 546 dye (Molecular Probes.) in 2X PBS containing 0.1 mL/L Tween-20.
7. Negative control: 10 g/L BSA in 2X PBS containing 0.1 mL/L Tween-20.
8. Dilution buffers.
   a. Human IgE, allergen and slide blocking solution diluent: Phosphate-buffered saline (PBS) 1X (0.2 g/L KCl, 1.44 g/L Na_2HPO_4, 0.24 g/L KH_2PO_4, 8 g/L NaCl, pH 7.4).
   b. Serum sample, horseradish peroxidase-streptavidin conjugate and conjugated antibody diluent: 2X PBS containing 10 g/L BSA and 0.1 mL/L Tween 20.
9. Wash buffer slide wash buffer: 1X PBS with 0.1 mL/L Tween-20.

10. Print buffer human IgE print buffer: 1X PBS containing 0.1 mL/L Tween-20 and 1 g/L sodium dodecyl sulfate (SDS).

3. Methods

3.1. Construction of the Array

3.1.1. Preparation of the Slide Surface

Commercial aminosilane-coated slides provided by Cel Associates are routinely used in our laboratory. Some commonly used silanization methods have been documented *(35)*.

3.1.2. Preparation of Biomolecules for Printing

1. The arrays employed in our investigations are usually designed as 7×7 matrices. This format incorporates human IgE antibodies, allergens, and negative controls (**Fig. 1A**; *see* Color Plate 18 following p. 178). Human IgE antibodies included in the array are printed at known concentrations and serve as internal calibration curves.
2. Human IgE, provided by Calbiochem, is diluted in 1X PBS to a concentration of 100 µg/mL and stored at −20°C.
3. According to the manufacturer's instructions, the resulting solution should not undergo freeze/thaw cycles; to avoid this problem, the solution is aliquoted prior to storage.
4. The IgE calibration curve (2–200 ng/mL) is prepared by diluting the human IgE preparation in 1X PBS containing Tween-20 (0.1 mL/L) and SDS (1 g/L). It is our experience that once diluted in its printing buffer, the human IgE should not be stored and reused for further print runs.
5. Importantly, 0.01% Tween-20 is routinely added to all print solutions (including allergen preparations; *see* **Note 1**).
6. Allergens are obtained as 1-g samples, reconstituted in 1X PBS, pH 7.2, to a concentration of 10 mg/mL and stirred overnight at room temperature.
7. After centrifugation at 503*g* for 30 min at 10°C, the allergen extracts are filtered in four stages: first through Whatman paper (125-mm diameter, cat. no. 1001125), then through 5-µm filter disks (Pall, cat. no. 4199; www.pall.com), then through 0.45-µm filter units (Schleicher & Schuell, cat. no. 10462 450; www.S-and-Side), and finally through 0.2-µm filter units (Schleicher & Schuell, cat. no. 10462 200).
8. Extracts are stored at −20°C as 1.5-mL aliquots.
9. A solution of 10 g/L BSA in 2X PBS containing 0.1 mL/L Tween-20 is employed as a negative control for accurate, parallel assessment of signal specificity within the array. A preparation of Alexa 546 dye (described in **Subheading 2.2.**) can also be employed as a signal control. It should be noted that such a control is not included within the array depicted in this chapter. Alexa 546 is chosen for use as the signal control because the signal amplification technique used to process the slides makes use of this same fluorophore coupled to the tyramide (*see* **Note 2**).

A Array

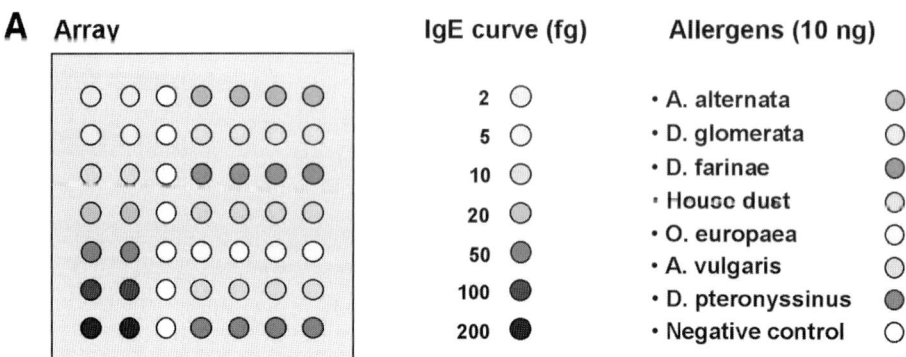

B

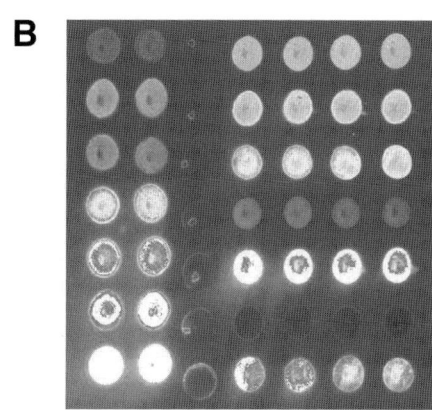

Fig. 1. Schematic array format and fluorescent scan for the allergen array (*see* Color Plate 18 following p. 178). (**A**) Schematic representation of the allergen array format. Arrays are printed on silanated glass slides using a robotic contact-arrayer. Printed arrays consist of various allergens printed in replicate alongside a standard curve of human IgE for quantitation and a negative control. (**B**) Fluorescent scan of allergen array incubated with serum and developed using the tyramide amplification protocol. For processing, the slide surface is blocked before the array is incubated with serum. Slides are then washed, and bound human IgE is revealed utilizing the tyramide single amplification system. Slides are incubated with biotinylated antihuman IgE, then with HRP-labeled streptavidin, and finally with activated Alexa 546-labeled tyramide. After a final washing step, slides are dried before being scanned with a confocal laser. The signal from the scanned digital image is then quantified.

3.1.3. Printing of Biomolecules

1. In order for printing of the array to take place, the solutions for printing (antibodies/allergens) must be transferred to the source plate (a 96- or 384-well microtiter plate). In our laboratory we commonly employ 384-well plates.

2. The solutions are prepared, and at least 50 μL of each are transferred, ensuring thorough mixing, to the source plate just prior to the start of the print run.
3. Depending on the number of replicate arrays required, a corresponding number of pins is loaded into the pinhead, and the appropriate number of wells is filled with each printing solution.
4. Biomolecules are contact-printed on silanized glass slides by means of computer-controlled high-speed robotics.
5. The samples are transferred from the microtiter plate to immobilized glass slides by use of stainless-steel solid pins (200 μm diameter). Each pin is estimated to transfer approx 1 nL of sample to the slide and produces spots with a pitch of 0.6 mm.
6. After deposition of each print solution, pins are washed first with distilled water and then with 70% ethanol and finally dried as part of the printing program.

3.1.4. Storage of Printed Slides

Once the printing process has been completed, the slides remain inside the robot printer cabinet for at least 12 h to "mature." This stable and controlled environment (approx 55% humidity and 25°C) is conducive to the binding of the antibodies and the allergens to the slide surface. Printed slides are stored in the dark in boxes containing silica gel bags. It is critical that the slides be kept dry; slides that have been exposed to moisture during storage give imprecise and varying signals.

3.2. Processing of the Printed Array for Determination of Serum IgE

A protocol incorporating signal amplification through tyramide reagentry has been identified (*see* **Note 3**) and is described here. Use of this type of chemistry results in significant amplification of the bound signal (50–100-fold). Amplification is required to increase test sensitivity since human IgE circulates at very low concentrations.

1. First, undiluted serum samples are analyzed with a commercial ELISA (e.g., Radim). ELISAs are performed and samples categorized according to the manufacturer's instructions.
2. Slides are then prepared for processing by the attachment of an adhesive frame (Gene-Frame; *see* **Subheading 2.2.**) around the printed array. This frame serves two purposes: first, this ensures that samples and reagents are contained within the array area; second, the frame limits the development area to the direct vicinity of the printed array, thereby minimizing reagent use.
3. To block nonspecific binding to the activated slide surface, the printed slides are incubated for 60 min at room temperature (RT) with the slide blocking solution (*see* **Subheading 2.2.**). Staining jars with at least 50 mL of blocking solution are commonly used for this purpose.
4. Slides are then incubated with serum (150 μL) for 60 min at RT and then thoroughly rinsed with wash buffer (5 × 1 mL).

5. A 60-min incubation with a biotinylated antihuman IgE (150 µL) at a final concentration of 1 mg/L in the appropriate diluent (*see* **Subheading 2.3.**) is then performed at RT.

6. After rinsing with wash buffer (5 × 1 mL), slides are incubated for a further 60 min at RT with a horseradish peroxidase-streptavidin conjugate diluted 1:100 in the appropriate diluent (*see* **Subheading 2.3.**).

7. Slides are rinsed again (5 × 1 mL), and the resulting streptavidin-biotinylated antibody complexes are detected by incubating the slides for 15 min at RT with Alexa 546-tyramide conjugate, diluted 1:100 with Molecular Probes diluent (namely, Amplification Buffer) containing 0.015 mL/L hydrogen peroxide.

8. Upon completion of the final processing step, slides are rinsed for the final time (5 × 1 mL), and the gene-frame is removed.

9. After drying at 37°C for 10–15 min, slides are then ready to be scanned.

3.3. Collection and Analysis of Array Data

3.3.1. Scanning of Processed Slides

Scanning is performed using ScanArray software, according to the manufacturer's instructions. Images (**Fig. 1B**) are optimized using the fine-tuning capabilities provided (notably, adjustment of laser/PMT settings) and stored as Bitmap files.

3.3.2. Quantitative Analysis of Image Data

Quantification is achieved using QuantArray software according to the manufacturer's instructions. Background signal is automatically subtracted, and data are stored as Excel files. Calibration curves incorporated in each protein array are fitted using the appropriate curve-fitting routine and provide a basis on which to correlate the signal generated against a concentration of the analyte under determination. Signals from printed antigens are then interpolated from these calibration curves.

4. Notes

1. Printing solutions for allergens: the heterogeneous nature of allergens demands careful optimization of print conditions for individual allergens. Useful additions include Tween-20 as the surfactant of choice; print quality can be much improved by addition of 0.01% Tween-20 to print solutions. SDS can be beneficial to the printing of certain allergens owing to its denaturing properties, which can enhance binding to the substrate and/or expose epitopes for binding to the serum antibody. Sucrose can be added to print solutions as a means of extending shelf life; sucrose retains a hydration shell around the immobilized material, thereby maintaining its 3D structure.

2. Signal generation: fluorescent end points are used as the signal generation option of choice. Other end points have been described (such as colorimetric

and chemiluminescent), but direct labeling of proteins with fluorophores to generate high specific activity and stable labels is easy to perform and instrumentation is widely available for the determination of the emitted signal. It is preferable to choose fluorophores with spectral properties >500 nm since many proteins in blood fluoresce at <500 nm, and background signals are much reduced at these wavelengths. The Alexa range of dyes (Molecular Probes) and the Cy dyes (Amersham Biosciences) can be recommended for this type of protein array immunoassay.

3. Signal amplification using tyramide reagentry: In the presence of hydrogen peroxide, horseradish peroxidase activates the dye-labeled tyramide derivatives, creating extremely reactive short-lived tyramide radicals that bind to nucleophilic residues on the slide surface. This results in significant amplification of the bound signal. Use of this type of chemistry can increase signal generation by 50- to 100-fold.

References

1. Hiller, R., Laffer, S., Harwanegg, C., et al. (2002) Microarrayed allergen molecules: diagnostic gatekeepers for allergy treatment. *FASEB J.* **16**, 414–416.
2. Brown, P. O. and Botstein, D. (1999) Exploring the new world of the genome with DNA microarrays. *Nat. Genet.* **21**, 33–37.
3. Cutler, P. (2003) Protein arrays: the current state-of-the-art. *Proteomics* **3**, 3–18.
4. MacBeath, G. (2002) Protein microarrays and proteomics. *Nat. Genet.* **32**, 526–532.
5. Wang, D., Liu, S., Trummer, B. J., et al. (2002) Carbohydrate microarrays for the recognition of cross-reactive molecular markers of microbes and host cells. *Nat. Biotechnol.* **20**, 275–281.
6. Willats, W. G., Rasmussen, S. E., Kristensen, T., et al. (2002) Sugar-coated microarrays: a novel slide surface for the high-throughput analysis of glycans. *Proteomics* **2**, 1666–1671.
7. Zhu, H. and Snyder, M. (2003) Protein chip technology. *Curr. Opin. Chem. Biol.* **7**, 55–63.
8. Zhu, H., Biglin, M., Bangham, R., et al. (2001) Global analysis of protein activities using proteome chips. *Science* **293**, 2101–2105.
9. Bacarese-Hamilton, T., Bistoni, F., and Crisanti, A. (2002) Protein microarrays: from serodiagnosis to whole proteome scale analysis of the immune response against pathogenic microorganisms. *Biotechniques* **33**, S24–S29.
10. Stoll, D., Templin, M. F., Schrenk, M., et al. (2002) Protein microarray technology. *Front. Biosci.* **7**, 13–32.
11. Kusnezow, W. and Hoheisel, J. (2002) Antibody microarrays: promises and problems. *Biotechniques* **33**, S14–S23.
12. Schena, M., Shalon, D., Davis, R. W., et al. (1995) Quantitative monitoring of gene expression patterns with a complementary DNA microarray. *Science* **270**, 467–470.
13. Okamoto, T., Suzuki, T. and Yamamoto, N. (2000) Microarray fabrication with covalent attachment of DNA using bubble jet technology. *Nat. Biotechnol.* **18**, 438–441.

14. Davies, C. (2001) Introduction to immunoassay principles, in *The Immunoassay Handbook, 2nd ed.* (Wild, D. cd.), Nature Publishing Group, London, pp. 3–40.

15. Pawlak, M., Schick, E., Bopp, M. A., et al. (2002) Zeptosens' protein microarrays: a novel high performance microarray platform for low abundance protein analysis. *Proteomics* **2**, 383–393.

16. Wu, G., Datar, R. H., Hansen, K. M., et al. (2001) Bioassay of prostate-specific antigen (PSA) using microcantilevers. *Nat. Biotechnol.* **19**, 856–860.

17. Pastinen, T. (2003) Single nucleotide polymorphism genotyping using microarrays, in *Microarrays & Microplates: Applications in Biomedical Sciences* (Ye, S. and Day, I. N. M., eds.), BIOS, Oxford, UK, pp. 89–108.

18. Wiltshire, S., O'Malley, S., Lambert, J., et al. (2000) Detection of multiple allergen-specific IgE's on microarrays by immunoassay with rolling circle amplification. *Clin. Chem.* **46**, 1990–1993.

19. Mendoza, L. G., McQuary, P., Mongan, A., et al. (1999) High-throughput microarray-based enzyme-linked immunosorbent assay (ELISA). *Biotechniques* **27**, 778–788.

20. Weinberger, S. R., Morris, T. S., and Pawlak, M. (2000) Recent trends in protein biochip technology. *Pharmacogenetics* **1**, 395–416.

21. Brockman, J. M., Nelson, B. P., and Corn, R. M. (2000) Surface plasmon resonance imaging measurements of ultrathin organic films. *Annu. Rev. Phys. Chem.* **51**, 41–63.

22. Ekins, R. P. (1989) Multi-analyte immunoassay. *J. Pharmacol. Biomed. Anal.* **7**, 155–168.

23. Ekins, R. P., Chu, F. W., and Biggart, E. (1990) Multispot, multianalyte, immunoassay. *Ann. Biol. Clin. (Paris)* **48**, 655–666.

24. Ekins, R. P. and Chu, F. W. (1991) Multianalyte microspot immunoassay—microanalytical "compact disk" of the future. *Clin. Chem.* **37**, 1995–1967.

25. Engvall, E. and Perlmann, P. (1971) Enzyme linked immunosorbent assay (ELISA). Quantitative assay of immunoglobulin G. *Immunochemistry* **8**, 871–874.

26. Walter, G., Bussow, K., Cahill, D., et al. (2000) Protein arrays for gene expression and molecular interaction screening. *Curr. Opin. Microbiol.* **3**, 298–302.

27. Bacarese-Hamilton, T., Mezzasoma, L., Ingham, C., et al. (2002) Detection of allergen-specific IgE on microarrays by use of signal amplification techniques. *Clin. Chem.* **48**, 1367–1370.

28. Joos, T. O., Schrenk, M., Hopfl, P., et al. (2000) A microarray enzyme-linked immunosorbent assay for autoimmune diagnostics. *Electrophoresis* **21**, 2641–2650.

29. Robinson, W. H., DiGennaro, C., Hueber, W., et al. (2002) Autoantigen microarrays for multiplex characterization of autoantibody responses. *Nat. Med.* **8**, 295–301.

30. Mezzasoma, L., Bacarese-Hamilton, T., Ingham, C., et al. (2002) Antigen microarrays for serodiagnosis of infectious diseases. *Clin. Chem.* **48**, 121–130.

31. Wiese, R., Belosludtsev, Y., Powdrill, T., et al. (2001) Simultaneous multianalyte ELISA performed on a microarray platform. *Clin. Chem.* **47**, 1451–1457.

32. Miller, J. C., Zhou, H., Kwekel, J., et al. (2003) Antibody microarray profiling of human prostate cancer sera: Antibody screening and identification of potential biomarkers. *Proteomics* **3**, 56–63.

33. Lacour, M. (1994) Acute infections in atopic dermatitis: a clue for a pathogenic role of a Th1/Th2 imbalance? *Dermatology* **188,** 255–257.
34. Brown, W. G., Halonen, M. J., Kaltenborn, W. T., et al. (1979) The relationship of respiratory allergy, skin test reactivity, and serum IgE in a community population sample. *J. Allergy Clin. Immunol.* **63,** 328–335.
35. Shriver-Lake, L. (1998) Silane-modified surfaces for biomaterial immobilization, in *Immobilized Biomolecules in Analysis* (Cass, T. and Ligler, F., eds.), Oxford University Press, Oxford, pp. 1–14.

12

Enhanced Protein Profiling Arrays for Quantitative Measurement of Protein Expression in Multiple Samples

Ruochun Huang, Qian Shi, Weimin Yang, and Ruo-Pan Huang

Summary

Simultaneous measurement of molecules from many samples is particularly useful for screening of certain molecules from different samples, for targeting verification of results obtained from protein and cDNA arrays, and for molecular epidemiology-based investigations. Described herein is a sensitive and quantitative approach, referred to as enhanced protein profiling arrays, that measures molecular expression levels from numerous samples. Enhanced protein profiling arrays can easily be constructed from tissue lysates, cell lysates, conditioned media, and body fluids, along with standard proteins and appropriate controls onto antibody-coated surfaces of solid supports. Coupled with many common detection techniques, the expression levels of certain molecules can be determined quantitatively. An account of the successful development and application of this enhanced protein profiling array technology is presented.

Key Words: Protein arrays; expression; ELISA; quantitation; high-throughput screening.

1. Introduction

Simultaneous detection of multiple protein expression levels and functions using protein array technology is a powerful tool in biomedical research *(1,2)*. Equally important is the ability to measure these expression levels and functions from multiple samples quantitatively. Protein array technology allows one to screen a number of important factors associated with a variety of cell events, disease phenotypes, and responses to certain treatment. To validate the role of these key molecules further, researchers need to examine expression levels in many samples. A challenge that many scientists face is dealing with data obtained from different samples and the costs associated with the experiments. At this stage of study, only limited molecules will be examined; thus, it is much more efficient to quantitatively measure the expression levels or functions of particular molecules from as many samples as possible at one time. One

From: *Methods in Molecular Medicine, Vol. 144, Microarrays in Clinical Diagnostics*
Edited by: T. Joos and P. Fortina © Humana Press Inc., Totowa, NJ

Table 1
Detection Sensitivity of Enhanced Protein Profiling Arrays

	Coated (ng/mL)	Noncoated (ng/mL)	Increased sensitivity (fold)
II-8	0.2	10,000	50,000
EGF	0.4	1000	2500
MCP-1	25	10,000	400
MIP-1α	25	10,000	400
VEGF	4	1000	400
GRO-α	8	1000	125
Ang	10	1000	100
GRO-β	5	>200	ND
SCF	25	1000	40
GRO-γ	5000	>5000	ND

Abbreviations: IL-8, interleukin-8; EGF, epidermal growth factor; MCP-1, macrophage chemtractaus protein-1; MIP-1α, macrophage inflammatory protein-1α; VEGF, vascular eudothelial growth factor; GRP, growth-regulated protein; Ang, angiogenin; SCF, stem cell factor.
From **ref. 9**.

approach currently used is the application of tissue array technology *(3,4)*, which allows one to screen expression levels of certain factors from many tissues. The major advantage of tissue arrays is that the protein's location can be determined. However, this technology requires highly trained professionals to evaluate the results, and the lack of quantitative measurements limits its access by the wider research community. Another approach to protein detection is the use of reversed-phase protein arrays. This technology can easily be performed by investigators in most general laboratories *(5)* and has been demonstrated to be a useful tool to determine protein levels and functions from tissue lysates *(6)* and cell lysates *(7)*.

One of the difficulties associated with tissue lysate and serum protein array technology is low detection sensitivity when the sample volume is significantly scaled down, particularly when detecting proteins from body fluid samples. Another potential problem is the variation of antigens, which have considerably distinct binding abilities. To overcome both problems, an enhanced protein profiling array methodology has been developed *(8,9)*. The unique features of this technology include precoating the surface of the solid support with specific antibodies before spotting samples onto the surface. Because this technique greatly increases detection sensitivity (**Table 1**) and allows the use of standard proteins, the expression levels can be measured quantitatively. In this chapter, enhanced protein profiling array technology is discussed, and an example of its use is presented.

2. Materials

The following materials are required to perform experiments using enhanced protein profiling array technology.

1. Pairs of antibodies against cytokine.
2. Horseradish peroxidase (HRP)-conjugated streptavidine (BD PharMingen, San Diego, CA).
3. Substrates, including Hybond enhanced chemiluminescence (ECL) membrane (Amersham, UK) and nitrocellulose slides (Schleicher & Schuell, Keene, NH).
4. Biotinylated bovine IgG.
5. Coating buffer: 0.1 M Na$_2$HPO$_4$, pH 9.0.
6. Tris-buffered saline (TBS): 0.01 M Tris-HCl, pH 7.6, 0.15 M NaCl.
7. TBS/0.1% Tween-20: 0.01 M Tris-HCl, pH 7.6, 0.15 M NaCl, 0.1% Tween-20.
8. Blocking buffer: 5% bovine serum albumin (BSA)/1X TBS (BSA: Roche Molecular Biochemistry).
9. Radioactive immunoprecipitation assay (RIPA) buffer: 10 mM Tris-HCl, pH 7.5, 0.15 M NaCl, 1% sodium deoxycholate, 1% Triton X-100.
10. ECL reagents A and B (Amersham).
11. Cy3-conjugated streptavidin (Rockland, Gilbertsville, PA).
12. Laser scanner (Perkin Elmer Life Science, Meriden, CT).
13. Densitometry (Bio-Rad, Hercules, CA).
14. Sigma plot, SPSS 8.0 computer program (SPSS, Chicago, IL) and Clusfavor 6.0 (http://mbcr.bcm.tmc.edu/genepi/).
15. Kodak X-OMAT film and film processor or chemiluminescence imaging system.
16. Centrifugal filter tube (Millipore, Bedford, MA).
17. Arrayer.
18. Orbital shaker.
19. Pipetman (0.1–2 µL).
20. Small plastic boxes or containers or chamber for washing steps.
21. Frame and cover slip (Continental Lab Products, San Diego, CA).
22. Other general laboratory equipment, such as a centrifuge.

3. Methods

3.1. Principle

The principle of enhanced protein profiling arrays is based on sandwich enzyme-linked immunosorbent assay (ELISA), as shown in **Fig. 1**. The surface of solid supports, such as membrane or glass slides, is coated with an antibody functioning as a captor. Samples, along with a series of dilutions of standard protein, are then spotted on the coated surface in an array format. The array membranes or array slides are incubated with a biotinylated antibody serving as a detector. Both capture and detection antibodies recognize two different epitopes of the same antigen. Signals can be visualized with HRP-conjugated streptavidin or Cy3- or Cy5-conjugated streptavidin. By comparing signal

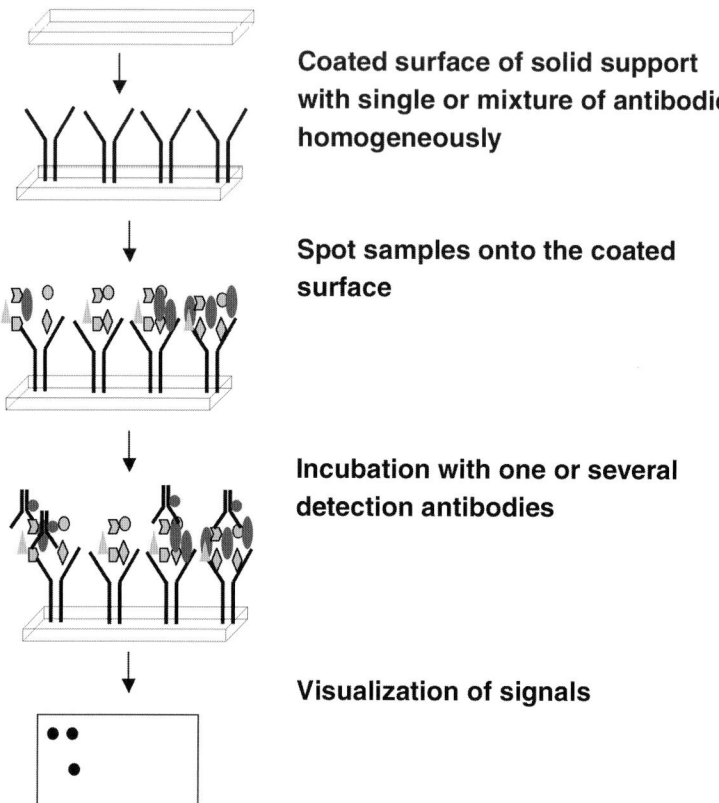

**Coated surface of solid support
with single or mixture of antibodies
homogeneously**

**Spot samples onto the coated
surface**

**Incubation with one or several
detection antibodies**

Visualization of signals

Fig. 1. Schematic representation of enhanced protein profiling arrays. The surface of the solid support is coated with antibody against the corresponding protein. Samples, together with standard protein and appropriate controls, are spotted onto the surface of the solid support. The expression levels of corresponding proteins from numerous samples are detected by specific antibody and quantitated according to a standard curve. (From **ref. 9**.)

intensities generated from standard curves, the exact amounts of a protein can be measured quantitatively.

3.2. Basic Steps

The prerequisites for enhanced protein profiling arrays are a pair of good antibodies and experimental samples. The experimental procedures, described below, are relatively easy.

1. Coat the surface of the solid support with capture antibody overnight at 4°C.
2. Dry the coated membrane or slide.

3. Spot samples and diluted standard protein.
4. Block the membranes or slides with BSA for 30 min at room temperature.
5. Incubate membranes or slides with detection antibody for 1–2 h.
6. Wash array membranes or array slides.
7. Incubate array membranes or slides with HRP-streptavidin, Cy3-streptavidin, or Cy5-strepatavidin.
8. Wash array membranes or slides.
9. Visualize the signals.
10. Generate standard curves and determine expression levels of protein in different samples.

3.3. Array Design

3.3.1. Antibody

Using a good antibody is the most important factor for performing experiments successfully with enhanced protein profiling arrays. Great attention must be paid to selecting a pair of antibodies with high specificity and sensitivity (*see* **Note 1**). The specificity and sensitivity of antibodies can be tested using Western blot analysis, protein array assays, or ELISA. Generally, commercially available antibodies are a good choice. If the appropriate antibodies are not available commercially, then one can use a monoclonal antibody as capture and a polyclonal antibody as detection. We have found that such combinations are successful in most cases.

3.3.2. Standard Protein

To measure protein expression levels quantitatively, one needs to have purified protein as a standard. Recombinant proteins are the simplest approach to achieving this standard. Because some proteins are easily degraded, appropriate storage methods must be followed. It is also essential that protein be aliquoted when dissolved in buffer and kept at −80°C.

3.3.3. Positive Control

The enhanced protein profiling array system must include a series of positive and negative controls, in addition to samples and standard proteins. Positive controls may consist of one or all of the following:

1. HRP-conjugated bovine IgG or other appropriate HRP-conjugated protein if signals are detected using an ECL approach.
2. Cy3- and Cy5-conjugated bovine IgG or other appropriate fluorescence-labeled protein if signals are detected using a laser scanner.
3. Biotin-conjugated bovine IgG (BIgG), which serves to normalize streptavidin incubation efficiency.
4. Standard proteins.

Negative controls include:

1. BSA.
2. TBS.
3. Other proteins that are not recognized by capture and detection antibodies.

3.3.4. Selection of Solid Support

Glass slides, membranes, and plates are the three major supports currently being used for protein arrays. The authors prefer Hybond ECL membranes and Schleicher & Schuell nitrocellulose slides. Both of these array supports provide low background and high sensitivity. Although we have used HydroGel slides in our development of cytokine antibody arrays, these slides are not suitable for enhanced protein profiling arrays owing to high background.

3.4. Sample Preparation

3.4.1. Serum or Plasma

Serum and plasma samples should be stored at −80°C and aliquoted to avoid a repeat thaw of samples. Serum and plasma can be directly spotted onto coated membranes or glass slides without any further treatment. If stored appropriately, the samples can be used for years.

3.4.2. Conditioned Medium

It is better to prepare serum-free or low-serum medium for conditioned medium because bovine serum also contains numerous proteins, which may be detected by antihuman protein antibodies. If serum is absolutely necessary for cell growth, the control medium should be prepared at the same condition and spotted onto the same array membranes.

1. Day 0: seed 1×10^5 cells in a 24-well tissue culture plate with complete medium.
2. Days 2–3: remove complete medium and replace with low-serum medium (e.g., medium containing 0.2% calf serum). Keep medium to a minimum to obtain high concentrations of secreted protein. Treatment can be given at this time.
3. Days 4–5: collect media into 15-mL tubes. Spin at 2000 rpm in a benchtop centrifuge at 4°C for 10 min. Save and aliquot the supernatant into 1.5-mL Eppendorf tubes. Store supernatant at −80°C until time for the experiment (up to 1 y). Alternatively, conditioned medium can be concentrated using a centrifuge filter tube or lyophilization. The lyophilized samples can be stored at −20°C or −80°C for several years.
4. Harvest cells and determine protein concentrations using a routine procedure or count cell numbers. Protein concentrations or cell numbers are used to normalize the amounts of samples. Adjust the conditioned media with 5% BSA to maintain equal amounts of proteins or cell numbers in different conditioned media. Directly spot the cell lysates onto membranes or slides.

3.4.3. Cell Lysate

The following procedure can be used to produce cell lysates:

1. Harvest cells into Eppendorf tube.
2. Add RIPA buffer to sample (e.g., add 20–100 μL to L × 10^5 cells), and vortex for approx 1 min. Extract protein at minimal volume of RIPA buffer to maintain high concentrations of proteins in the sample.
3. Spin down tube in Eppendorf centrifuge at maximal speed for 5 min at 4°C.
4. Save, aliquot, and store supernatant at −80°C until experiment.
5. Determine protein concentrations. Adjust the cell lysates with 5% BSA to maintain equal amounts of proteins or cell number in different cell lysates. Directly spot the cell lysate onto membranes or slides.

3.4.4. Tissue Lysate

Tissue lysates can be produced as follows:

1. Obtain patient or animal tissues, cut into small pieces, and add RIPA (e.g., for 1 mg tissue, add 200 μL of RIPA).
2. Homogenize samples until tissue is dissolved well.
3. Spin down sample in Eppendorf tube at maximal speed for 5 min at 4°C.
4. Save, aliquot, and store supernatant at −80°C until time for the experiment.
5. Determine protein concentrations and directly spot the cell lysates onto membranes or slides.

3.4.5. Paraffin-Embedded Tissue Lysate

Protein extracted from paraffin-embedded tissues using laser capture-dissection can also be used for enhanced protein profiling arrays. For extraction of protein from paraffin-embedded tissues, please refer to **refs. (*6* and *10*)**.

3.5. Array Construction

After validation of antibodies, preparation of samples, and selection of suitable solid supports (either membranes or slides), the arrays can be constructed.

3.5.1. Layout

The layout of the arrays will depend on the number of samples. Several factors should be taken into consideration when determining the layout. Positive controls should be spotted in such a way that the orientation of membranes or slides can be easily identified. Spots should be spaced far enough apart to prevent overlapping of neighboring spots. Care should be taken to not spot the samples next to positive controls or high amounts of standard protein. An easier overall impression of expression data is more attainable if the same group of samples is spotted in the same area.

3.5.2. Coating

The procedure for coating the surface support follows:

1. Prepare antibody in 10–40 µg/mL coating buffer (*see* **Note 2**).
2. Incubate membranes or one side of the glass slides with antibody at 4°C for overnight. The antibody solution must cover the entire area that will be used for printing of samples.
3. Dry the membranes or slides at room temperature for 10–30 min.
4. The coated membranes or slides can be stored at −20°C for several weeks.

3.5.3. Printing

The next step is to array the samples onto membranes or slides. The authors have tried three different methods: pin tool-based contact arrayer, noncontact arrayer, and pipeitor.

The pin tool-based contact arrayer deposits tiny amounts of protein onto a surface through direct contact. The amount of proteins deposited can be controlled by the size of the pins and the length of contact and is greatly affected by different substrates. The major advantages of pin tool-based contact arrayer are relatively low cost and high speed. However, it is difficult to control the precise amounts of proteins to be deposited onto the surfaces, and this method usually has higher CVs than noncontact arrayers. Noncontact arrayers, on the other hand, can deposit precise amounts of solution onto the surface of a solid support. Two types of noncontact arrayers are commercially available. The first, peizoeletronic robotic dispensers from Perkin Elmer, can deliver as little as 325 pL of reagents. A potential problem with this type of arrayer is the low printing speed . Genomics Solution produces a dispenser that can print at a high speed, but the smallest dispensed volume is more than 10 nL. A disadvantage of non-contact arrayers is that they are more expensive than contact arrayers.

3.6. Array Assay

The procedure below describes the preparation of array assays using membranes:

1. Put membranes in suitable containers such as 8-well tissue culture plates or other small containers or plastic bags to save samples and reagents, depending on the size of array.
2. Block nonspecific binding by incubating membranes with 5% BSA/TBS at room temperature for 30 min (*see* **Note 3**).
3. Add 100 µL to 2 mL of biotin-conjugated antibody. (The volume used depends on the size of the array membranes.) Dilute biotin-conjugated antibody into 50–500 ng/mL with 5% BSA/TBS, and incubate at room temperature.
4. Wash membranes three times with TBS/0.1% Tween-20 at room temperature, 5 min per wash.

5. Wash membranes two times with TBS at room temperature, 5 min per wash.
6. Add 100 μL to 2 mL of 2.5 ng/mL of HRP-conjugated streptavidin (*see* **Note 4**). (The volume depends on the size of the anay membranes.) Dilute HRP-conjugated streptavidin with 5% BSA.
7. Incubate at room temperature for 30–60 min.
8. Wash as directed in **steps 5** and **6**.
9. Visualize signals using ECL according to the manufacturer's instructions coupled with film development, or a chemiluminescent imaging system. Expose membranes several times for 30 s to 10 min (*see* **Notes 5** and **6**).

To prepare array assays for slides, use frames and cover slips during the incubation steps and a chamber for washes.

3.7. Multiple Protein Detection

To detect multiple proteins, multiple antibodies, or an antibody against a common domain of several proteins, samples can be coated onto membranes or slides. The samples are spotted onto the coated membranes or slides as described. The arrays are incubated with different biotinylated antibodies, and different protein expression levels can be detected.

3.8. Illustrative Case

The primary application of enhanced protein profiling arrays is to measure certain protein expression levels quantitatively from multiple samples. The authors have successfully applied this technology in several projects. Following is an example of the application of the enhanced protein profiling array system to identify potential biomarkers for gynecological cancers (*9*).

1. Collect patient's plasma samples and store at −80°C in aliquots.
2. Enhanced protein profiling arrays.
   a. Incubate membranes with 10 μg/mL of anti-vascular endothelial growth factor, anti-interleukin-8 (IL-8), anti-platelet-derived growth factor B, anti-angiogenin, and anti-epidermal growth factor, respectively, overnight at 4°C.
   b. Spot 0.25 μL of plasma along with a series of diluted recombinant protein (0.05–2000 ng/mL) and positive and negative controls onto membranes.
   c. Probe the array membranes with biotinylated antibody (50–500 ng/mL).
   d. Incubate the array membranes with HRP-conjugated streptavidin.
   e. Visualize the signals with a chemiluminescence kit (Amersham Biosciences).
3. Data analysis.
   a. Determine the signal intensities using Bio-Rad densitometry.
   b. Generate standard curves of recombinant proteins using a Sigma plot (*see* **Note 7**). From the standard curve, the exact amounts of protein from different samples can be determined, as shown in **Fig. 2**.
   c. Confirm the results further confirmed by conventional ELISA.

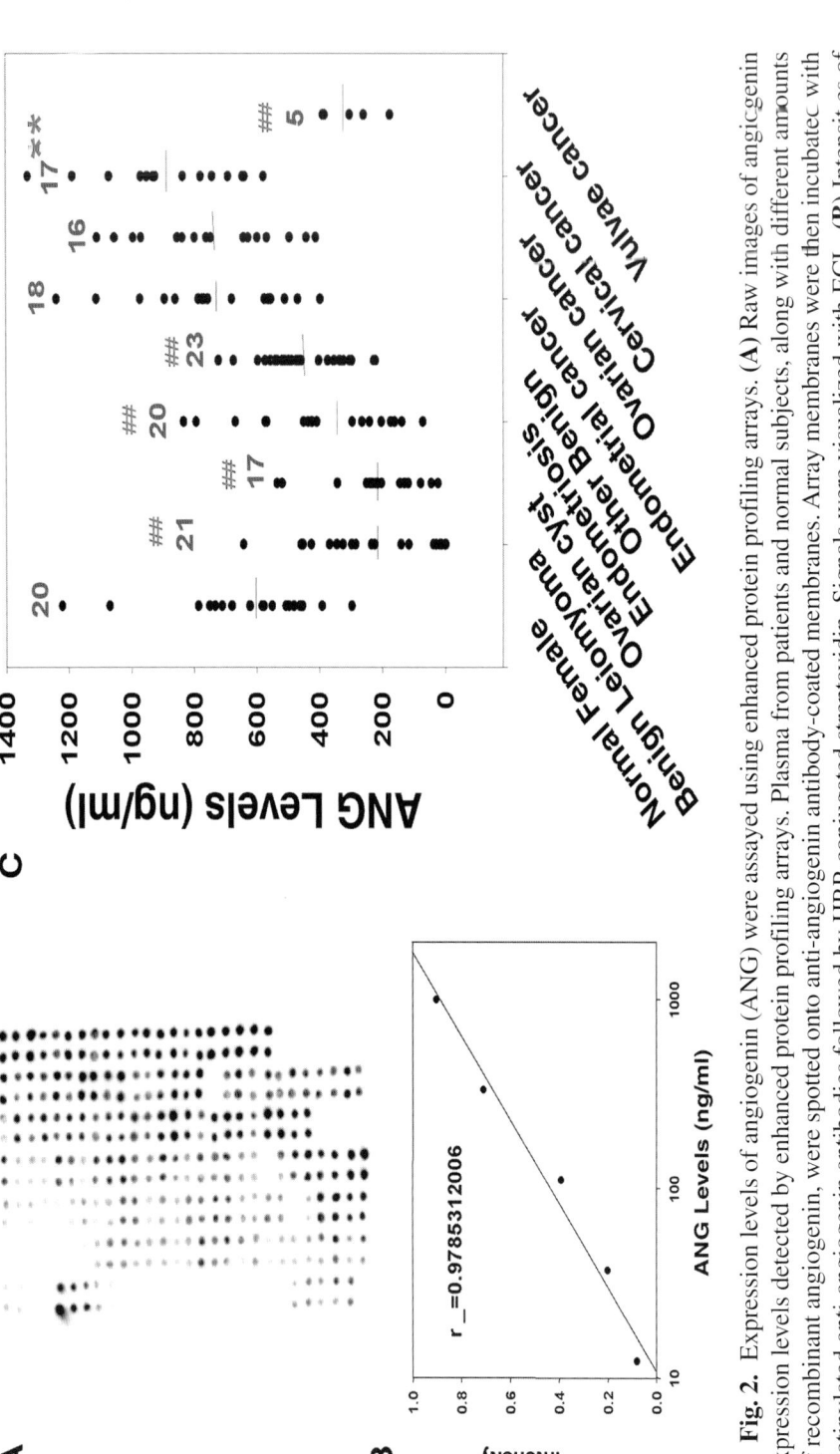

Fig. 2. Expression levels of angiogenin (ANG) were assayed using enhanced protein profiling arrays. (A) Raw images of angiogenin expression levels detected by enhanced protein profiling arrays. Plasma from patients and normal subjects, along with different amounts of recombinant angiogenin, were spotted onto anti-angiogenin antibodies, were then incubated with biotinylated anti-angiogenin antibodies followed by HRP-conjugated streptavidin. Signals were visualized with ECL. (B) Intensities of signals derived from standard angiogenin were determined by densitometry and plotted against known concentrations of angiogenin. (C) Expression levels of angiogenin in control and patient samples. Comparisons were made between normal subjects and patients, by Student's t-test. Significance is shown at the levels of $*p \leq 0.05$, $**p \leq 0.01$ for high expression in patient compared with normal and at the levels of $^{\#}p \leq 0.05$, $^{\#\#}p \leq 0.01$ for low expression in patients compared with normal. (From **ref. 9.**)

Color based on sorted raw data
Distance function: Euclidean distance

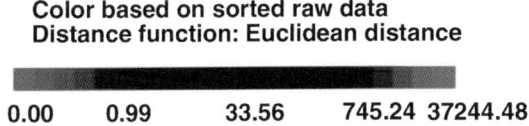

0.00   0.99   33.56   745.24 37244.48

3 normal/11 ovarian cancer   14 normal

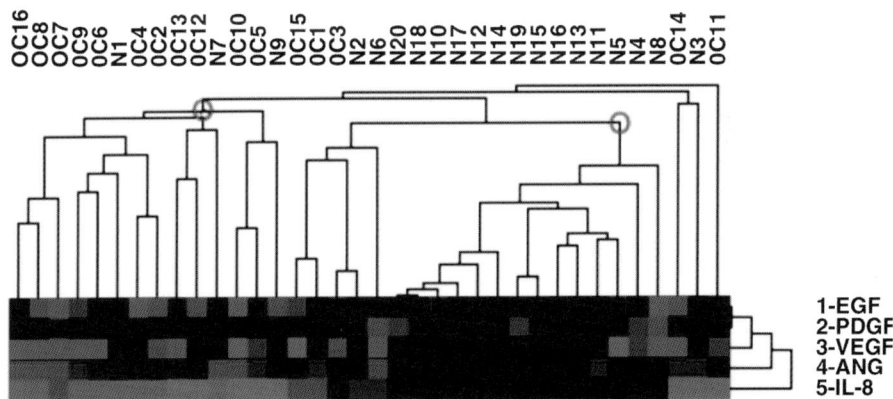

1-EGF
2-PDGF
3-VEGF
4-ANG
5-IL-8

Fig. 3. Clustering analysis of angiogenic factors measured by enhanced protein profiling arrays between normal subjects and patients with gynecological diseases. (*See* Color Plate 19 following p. 178.) Hierarchical clustering based on the expression levels of five angiogenic factors was performed on the 20 normal subjects and 16 ovarian cancer patients. EGF, epidermal growth factor; PDGF, platelet-derived growth factor; VEGF, vascular eudothelial growth factor; ANG, angiogenin; IL-8, interleukin 8.

    d. The expression levels of five angiogenic factors were clustered against different samples. As shown in **Fig. 3** (*see* Color Plate 19 following p. 178), most of the ovarian cancer cells were clustered together, as were normal cells.

4. Notes

1. Limitations of this assay greatly depend on the antibodies used. Good antibodies with high affinity and specificity are the most important factor for successful assays. Great care must be taken in the selection and development of good antibodies.
2. In general, 20 µg/mL of capture antibody and 50 ng/mL of biotinylated antibody are used. However, to achieve the highest sensitivity and lowest background possible, titration for individual antibody concentrations may be required.
3. Since only a minute amount of sample is spotted, it is better to prepare as high a concentration of sample as possible. If the detection sensitivity for IL-8 is 200 pg/mL,

and 1 nL of sample is deposited onto the membranes or slides, the absolute amount of IL-8 detected is 1 fg.

4. Signal detection sensitivity may be further increased by tyramide signal amplification (11) or rolling cycle amplification (12).

5. In some cases, the detection ranges of proteins exceed the practical dynamic range, particularly when film is used. Legitimate data at higher and lower ends of the concentration ranges may be obtained by exposure at different times.

6. If HRP-conjugated streptavidin used, signals can be detected by chemiluminescence, as described above, or colorimetry. Furthermore, If fluorescence-labeled streptavidin used, the signal can be visualized by a laser scanner.

Acknowledgments

This work was supported by NIH/NCI grant CA89273 to (R.P.H.) and NIH/NCI grant (CA107783) (R.P.H.). This research was also supported in part by the University Research Committee of Emory University. Dr. Ruo-Pan Huang may be entitled to royalties derived from RayBiotech, Inc., which develops and produces protein array technology. The terms of this arrangement have been reviewed and approved by Emory University in accordance with its conflict of interest policies.

References

1. Huang, R. P. (2003) Protein arrays, an excellent tool in biomedical research. *Front. Biosci.* **8,** D559–D576.

2. Huang, R. P. (2003) Cytokine antibody arrays: a promising tool to identify molecular targets for drug discovery. *Comb. Chem. High Throughput Screen.* **6,** 769–775.

3. Kononen, J., Bubendorf, L., Kallioniemi, A., et al. (1998) Tissue microarrays for high-throughput molecular profiling of tumor specimens. *Nat. Med.* **4,** 844–847.

4. Simon, R., Nocito, A., Hubscher, T., et al. (2001) Patterns of her-2/neu amplification and overexpression in primary and metastatic breast cancer. *J. Natl. Cancer Inst.* **93,** 1141–1146.

5. Paweletz, C. P., Charboneau, L., Bichsel, V. E., et al. (2001) Reverse phase protein microarrays which capture disease progression show activation of prosurvival pathways at the cancer invasion front. *Oncogene,* **20,** 1981–1989.

6. Grubb, R. L., Calvert, V. S., Wulkuhle, J. D., et al. (2003) Signal pathway profiling of prostate cancer using reverse phase protein arrays. *Proteomics* **3,** 2142–2146.

7. Nishizuka, S., Charboneau, L., Young, L., et al. (2003) Proteomic profiling of the NCI-60 cancer cell lines using new high-density reverse-phase lysate microarrays. *Proc. Natl. Acad. Sci. USA* **100,** 14229–14234.

8. Huang, R. P., Huang, R., Fan, Y., and Lin, Y. (2001) Simultaneous detection of multiple cytokines from conditioned media and patient's sera by an antibody-based protein array system. *Anal. Biochem.* **294,** 55–62.

9. Huang, R., Lin, Y., Shi, Q., et al. (2004) Enhanced protein profiling arrays with ELISA-based amplification for high-throughput molecular changes of tumor patients' plasma. *Clin. Cancer Res.* **10,** 598–609.

10. Celis, J. E., Gromov, P., Cabezon, T., et al. (2004) Proteomic characterization of the interstitial fluid perfusing the breast tumor microenvironment: a novel resource for biomarker and therapeutic target discovery. *Mol. Cell. Proteomics* **3,** 327–344.

11. Woodbury, R. L., Varnum, S. M., and Zangar, R. C. (2002) Elevated HGF levels in sera from breast cancer patients detected using a protein microarray ELISA. *J. Proteome Res.* **1,** 233–237.

12. Schweitzer, B., Wiltshire, S., Lambert, J., et al. (2000) Inaugural article: immunoassays with rolling circle DNA amplification: a versatile platform for ultrasensitive antigen detection. *Proc. Natl. Acad. Sci .USA* **97,** 10113–10119.

13

Multiplexed Cytokine Sandwich Immunoassays

Clinical Applications

Uma Prabhakar, Edward Eirikis, Bruce E. Miller, and Hugh M. Davis

Summary

Flow cytometric, microsphere-based immunoassays have been developed for the simultaneous detection of soluble analytes in a variety of sample types. The ability to discriminate between individual microspheres on the basis of size, fluorescent intensity, and/or wavelength has allowed the simultaneous analysis of multiple analytes from the same sample. Cytokines are particularly good candidates for multiplexed analysis. Through intricate networks and complex feedback mechanisms, cytokines modulate each other as well as a multitude of cellular events and play an important role in health and disease. The simultaneous measurement of multiple cytokines in a single biological sample has tremendous potential value to further our understanding of the role of these immunomodulators in health and disease. This chapter describes a multiplexed, microsphere-based flow cytometric method to quantitate and compare multiple cytokines simultaneously in human serum. Serum poses a challenge for multiplexed cytokine analysis owing to the presence of numerous proteins and other potentially interfering factors. Assessment of the "real world" performance of the multiplexed, microsphere-based flow cytometry assay with clinically relevant sample types is critical for determining the proper application of these methods in deciphering the role of cytokines in disease pathogenesis and for evaluating drug action.

Key Words: Autoimmune disease; cytokines; immunomodulators; multiplex analysis; immunoassays.

1. Introduction

Cytokines are small to medium-sized proteins and glycoproteins that play important roles in cell-to cell communication and immunoregulation; they have been implicated in inflammatory, neoplastic, and infectious disease processes. *(1)* They form intricate regulatory and homeostatic networks that serve to regulate production of each other and to modulate a multitude of cellular functions. Pleiotropy, redundancy, synergistic activity, and antagonistic effects are common features

From: *Methods in Molecular Medicine, Vol. 144, Microarrays in Clinical Diagnostics*
Edited by: T. Joos and P. Fortina © Humana Press Inc., Totowa, NJ

of these cytokine cascades. A deeper understanding of the normal operation of these networks in health and disease is needed to unravel how cytokines mediate their diverse effects on biological systems. Such knowledge will also be important for developing and understanding the action of therapeutic agents.

For these reasons, simultaneous analysis of the multiple cytokines expressed within biological fluids is of greater value for research in various disease states than analysis of a single cytokine. In recent years, several commercial vendors and/or independent laboratories have developed multiplexed microsphere/polystyrene bead-based arrays for measuring cytokines. The microspheres are internally dyed with a spectrally distinct fluorophore, and each dyed microsphere is conjugated with a monoclonal antibody specific for a target cytokine. Multiplexed flow cytometric-based immunoassays are created by mixing bead sets with different conjugated antibodies to permit simultaneous measurement of multiple cytokines (or other proteins) from a given sample. The antibody-conjugated microspheres are allowed to react with a biological sample and detection antibody in a microplate well to form a capture sandwich immunoassay. Flow cytometric-based hardware with dual lasers and associated optics detects the interactions that occur on the surface of the microspheres, and a high-speed digital signal processor manages the fluorescent output. The features and advantages of the multiplexed flow cytometric platform compared with conventional quantitative approaches such as enzyme-linked immunosorbent assay (ELISA) are discussed in greater detail by Dunbar and Jacobson in Chapter 5, and are reviewed by Kellar and Iannone *(2)*.

The performance of the multiplexed flow cytometry platform to quantify levels of multiple cytokines in culture supernatants of stimulated peripheral blood mononuclear cells (PBMCs) from normal volunteers has been fairly well characterized *(3–6)*. As many as 15 cytokines have been simultaneously measured in PBMC supernatants using the multiplexed technology *(5,7)*.

There are few published reports, however, that provide a thorough evaluation of the performance (i.e., accuracy, precision, and reproducibility) of these multiplexed immunoassays with the more complex sample matrices that would be used during preclinical and clinical studies *(3,4,8)*. Serum, plasma, tear fluids, and other bodily fluids are complex matrices, consisting of a large dynamic range of proteins, heterophilic antibodies, and other potentially interfering substrates. The applicability of the multiplexed, flow cytometric immunoassay platform to further our understanding of disease processes and drug actions, however, will also require that assay performance be well characterized in samples collected from patients as well as from healthy donors *(9)*.

This chapter describes our efforts to develop a microsphere-based multiplexed flow cytometric immunoassay to quantitate and compare simultaneously up to six cytokines in serum collected from healthy subjects and patients with different autoimmune diseases.

2. Materials

1. Human cytokine LINCO-plex Kit (Linco Research, St. Charles, MO).
2. Bioplex Protein Array System (Bio-Rad, Hercules, CA).
3. 9.5-mL Vacutainer Serum Separator tubes (Becton Dickinson, Franklin Lakes, NJ).
4. Filter plates (Cat. No. MABVN1250, Millipore, Bedford, MA).
5. Microplate shaker (Labline Instruments, Melrose Park, IL).
6. Filtration vacuum manifold (Millipore).
7. Eppendorf microcentrifuge tubes (VWR, Bridgeport, NJ).
8. Pipetors (Rainin Instruments, Woburn, MA).
9. 12 × 75-mm Polypropylene tubes (VWR, Bridgeport, NJ).
10. Vortex (VWR).
11. Sonicator (Branson, Danbury, CT).
12. Deionized water.
13. Wash buffer: the manufacturer's kit contains 1 bottle of 10 mM phosphate-buffered saline (PBS) with 0.08% sodium azide and 0.05% Tween-20 at a pH of 7.4. Prepare a 1:10 dilution of this 10X wash buffer with deionized water (e.g., 30 mL 10X wash buffer with 270 mL deionized water).

3. Methods

The methods described below outline (1) sera collection, (2) preparation of reagents, and cytokine and control standards, (3) cytokine detection, (4) determination of assay performance, and (5) data analysis (*see* **Note 1**).

3.1. Sera Collection

Whole blood is collected from healthy donor subjects or patients in 9.5-mL serum separator vacutainer tubes and is allowed to sit at room temperature for 30 min to clot. It is then centrifuged at 1300g for 10 min. Then 500-µL aliquots of the separated serum is stored at −70°C until analysis (*see* **Note 2**).

3.2. Preparation of Reagents and Cytokine and Control Standards

This section describes the steps involved in preparing each of the reagents, cytokine standards, and control standards for use in this assay (*see* **Note 3**). All preparations are done according to the manufacturer's instructions using the LINCO-plex kit.

3.2.1. Antibody-Immobilized Microspheres

Antibody-immobilized microspheres are supplied at a concentration of 25,000 beads/mL in a volume of 400 µL. The microspheres are vortexed for 30 s followed by sonication for 30 s. Then 350 µL of microspheres from each bead set are added to a vial containing 2.1 mL of bead diluent (supplied in the kit by Linco) to obtain the working bead concentration.

3.2.2. Cytokine Standards

Reconstitute the cytokine standard cocktail provided with the manufacturer's kit with 250 µL deionized water to yield a stock concentration of 10,000 pg/mL of each cytokine. This serves as the high concentration standard. Vortex the vial and transfer to an appropriately labeled polypropylene tube. Prepare the following standards from the 10,000 pg/mL stock using the assay buffer provided in the manufacturer's kit: 2000, 400, 80, 16, 3.2, and 0 pg/mL.

Each of the two vials of lypholized cytokine controls provided in the manufacturer's kit is reconstituted by adding 250 µL deionized water and gently vortexing.

3.3. Cytokine Detection

1. The assays are performed in filter plates that are first blocked with the assay buffer (50 m*M* PBS, 25 m*M* EDTA, 0.08% sodium azide, 0.05% Tween-20, and 1% bovine serum albumin [BSA]) provided in the manufacturer's kit. A total of 200 µL of the assay buffer is pipeted into each well of the microtiter plate.
2. Cover the filter plate with a plastic lid (or aluminum foil) and mix on the plate shaker at a setting of 4 for 10 min at room temperature.
3. Vacuum-filter the assay buffer through the wells using the vacuum manifold apparatus at a setting of 2.5–3.5 psi. Any excess assay buffer should be removed from the bottom of the plate by inverting and gently blotting on an absorbent pad or paper towels.
4. Add 25 µL of the assay buffer to the 0 standard and sample wells. Add 25 µL of the serum matrix diluent (provided in the kit) to the standard and control wells.
5. Add cytokine standards, controls, and serum samples (25 µL) to the appropriate wells.
6. Vortex and sonicate the vial containing the microspheres as described in **Subheading 3.2.1.** Add 25 µL of the mixed bead solution to each well.
7. Cover the plate with a plastic lid (or aluminum foil) and place on the plate shaker at a setting of 4 for 1 h at room temperature in the dark.
8. Gently remove fluid using the vacuum filtration manifold as described in **step 3** above.
9. Add 200 µL of the diluted wash buffer to each well and vacuum any excess wash buffer as in **step 3**. Repeat this washing process for a total of two times.
10. Add 50 µL of the detection antibody cocktail to each well (biotinylated detection antibodies to each target cytokine, provided in manufacturer's kit).
11. Cover the plate with a plastic lid (or aluminum foil) and place on the plate shaker at a setting of 4 for 30 min at room temperature in the dark.

12. Without washing, add 50 µL of streptavidin-phycoerythrin assay buffer (provided in manufacturer's kit) to each well.
13. Cover the plate with a plastic lid (or aluminum foil) and place on the plate shaker at a setting of 4 for 30 min at room temperature in the dark.
14. Wash the plate twice as described in **step 9** above.
15. Following the second wash, add 100 µL of the wash buffer to each well. Cover the plate with a plastic lid (or aluminum foil) and place on the plate shaker at a setting of 4 for 5 min at room temperature in the dark to resuspend the microspheres.
16. Place the plate into the Bioplex plate reader. Analyze 50 µL of sample per well using a minimum of 50 microspheres per region (*see* **Note 4**).

3.4. Determination of Assay Performance

In this section, standard methodology for determining the sensitivity, precision, and accuracy of the multiplexed flow cytometry analysis in simultaneously quantifying target cytokines in clinical samples is described. The criterion for determining assay performance conforms to the guidance provided by the Food and Drug Administration for bioanalytical method validation *(10)* and also to the guidance for immunoassay validation *(11)*.

3.4.1. Assay Sensitivity

The lowest cytokine concentration that can be detected in the clinical samples (limit of detection [LOD]) and the lowest cytokine concentration that can be reliably quantified (limit of quantitation [LOQ]) are based on analyzing replicate (n = 30–40) measurements of the 0 pg/mL standard. To obtain the LOD, we typically add 3 standard deviations to the average mean fluorescence intensity (MFI) of the 0 standard. For the LOQ, we add 6 standard deviations to the mean MFI. Values expressed as MFI are then interpolated from the standard curve to be expressed as pg/mL (*see* **Fig. 1.** and **Note 5**).

3.4.2. Assay Precision

Intraassay variability is determined by assaying multiple replicates of the sample in a single assay; interassay variability is determined by assaying the same sample on different days. Both determinations are expressed as a coefficient of variation (%CV, defined as variability/mean × 100). Because of the generally low levels of endogenous cytokines in human serum, it is recommended that samples be spiked with each target cytokine. In general, CV values of <20% are considered indicative of good assay precision in the multiplexed setting *(11)*.

3.4.3. Assay Accuracy

To determine the accuracy of the assay for determining levels of each target cytokine in clinical samples, samples are spiked with known amounts of the target

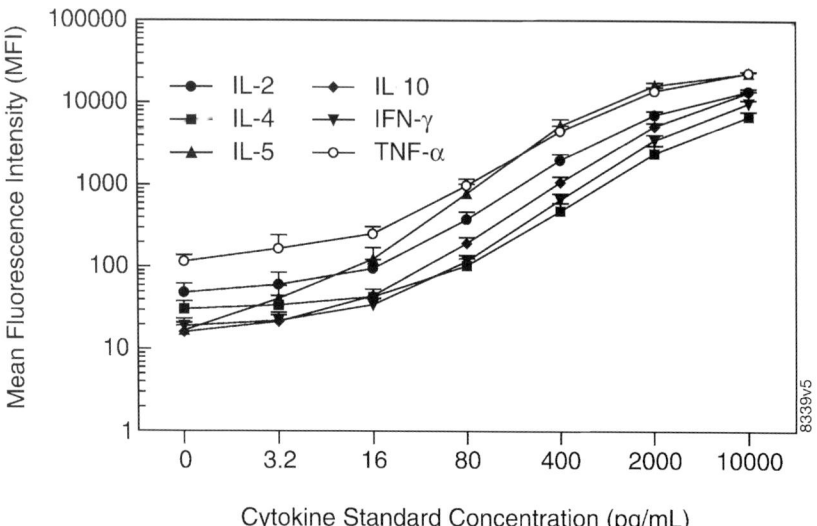

Fig. 1. Standard curves for a six-cytokine panel measured with a multiplexed flow cytometry assay. Data are plotted on a log-linear curve. Data represent mean MFI of eight replicates per determination. Reproducibility (precision) of standard curve from multiple assays was very high; an R^2 value of 1.00 was observed for each cytokine when the backcalculated MFI values were plotted against standard values. IL, interleukin; IFN, interferon; TNF, tumor necrosis factor.

cytokine and percent recovery is determined. Percent recoveries within 25% of the expected (i.e., spiked) concentration are considered acceptable and indicate that the assay is suitable for quantitative measurement of cytokine expression in clinical samples (*see* **Note 6**).

3.5. Data Analysis

Raw data (MFI) are captured using the Bioplex plate reader software (version 2.0). For data analysis, a 5-PL curve fit is applied to each standard curve, and sample concentrations are interpolated from the standard curve.

4. Notes

1. Although the assay methods described in this section are specific to the Bioplex Protein Array System from Bio-Rad, they should also be suitable for use with any other commercial multiplexed flow cytometric kits after verification of assay performance of the other kits.
2. Clinical samples are typically received and stored frozen. In the case of repeat analysis, samples would be subjected to multiple freeze/thaw cycles. Data from our laboratory indicate that serum samples can be frozen and thawed up to three times with no significant change in cytokine concentration (*see* **Fig. 2**). For IL-2, IL-4,

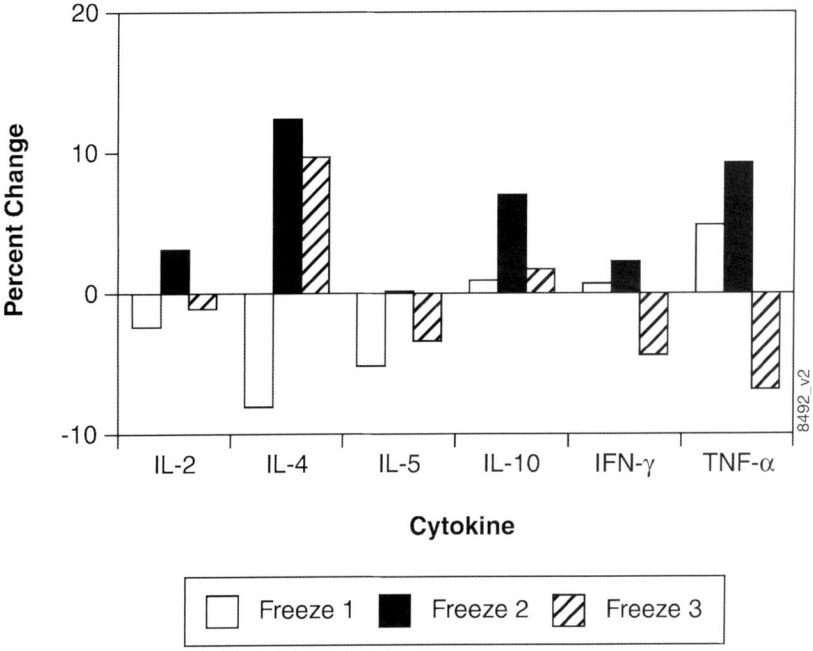

Fig. 2. Percent change in cytokine concentration following repeated freeze/thaw cycles relative to initial analysis. For this analysis, serum samples from four healthy donors were spiked with each of the six cytokines (500 pg/mL) and analyzed fresh and after each of three freeze/thaw cycles. Each value represents average of the percent change values for four subjects. IL, interleukin; IFN, interferon; TNF, tumor necrosis factor.

IL-5, IL-10, IFN-γ, and TNF-α, the percent change values (relative to the initial assay) after each of three freeze/thaw cycles was less than ±20% for each of the four seruma samples tested. Intersubject variability following repeated freeze/thaw cycles was somewhat greater for TNF-α, with percent change values relative to the initial (fresh) recovery ranging from −22.8% to 34.0% in the four samples.

3. Several technical guidelines should be followed to ensure optimal assay performance:

   a. The antibody-immobilized beads are light sensitive and should be shielded from ambient light. The manufacturer indicates that unused mixed antibody-immobilized beads can be stored in the bead mix bottle at 2–8°C for up to 1 mo. The bead mix should not be frozen and should be prepared fresh prior to use.

   b. It is essential that all reagents be brought to room temperature (20–25°C) before use.

   c. During preparation of the cytokine standards, make certain to vortex the higher concentration well before making the next dilution.

   d. Do not store serum samples in glass tubes, but rather use polypropylene tubes to avoid adsorption of cytokines by the tube wall.

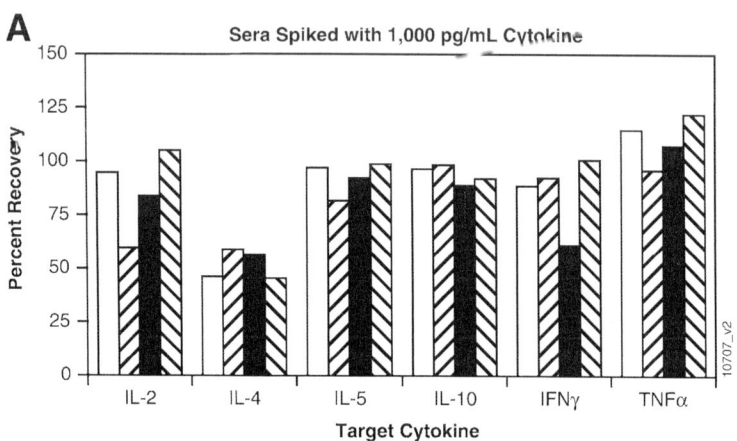

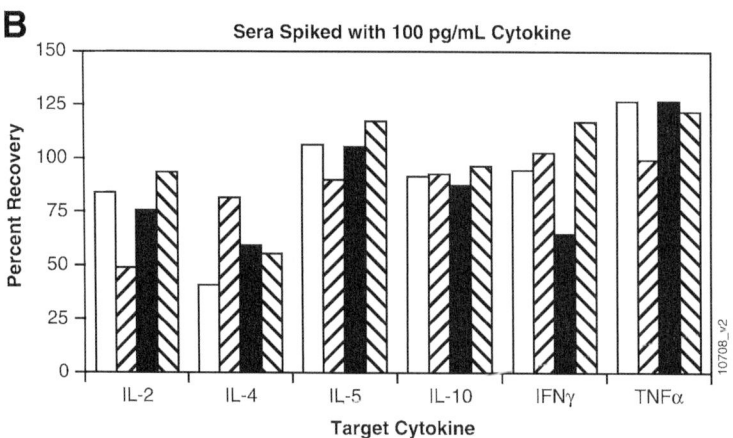

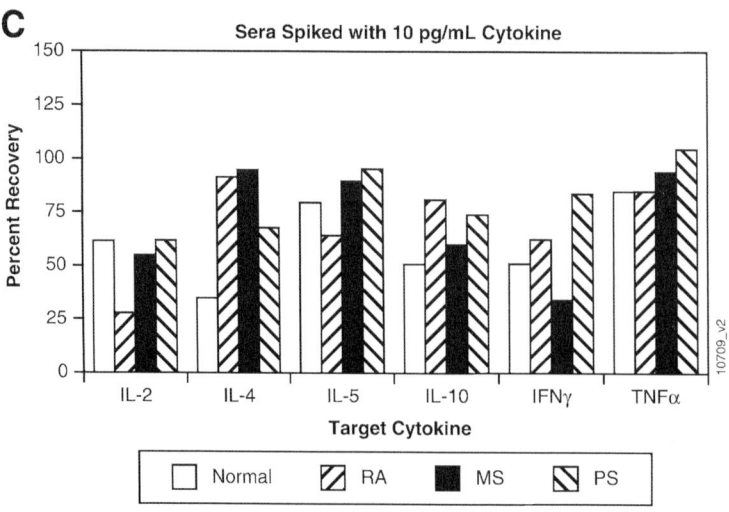

4. Our determination of the number of beads analyzed per sample on the standard curve performance indicated that at least 50 microspheres is necessary to achieve good assay precision (CV < 10%).

5. In assaying serum matrix, we found considerable variability in background MFI values when different lots of the same human cytokine LINCO-plex Kits were used. This prevented calculation of a global LOD/LOQ for the six cytokines using the approach described in **Subheading 3.4.1.** An alternate approach for determining the LOQ was to use the lowest standard that backcalculated to within 25% of the nominal value.

6. Recovery of cytokines from spiked samples is matrix dependent *(4)*. In tissue culture media (PBMC), the recoveries of six cytokines (IL-2, IL-4, IL-5, IL-10, IFN-γ, and TNF-α) were typically within 25% of the expected concentration, confirming the results of other researchers showing that multiplexing can be used for quantitative comparison of cytokine expression in *in* vitro cellular studies *(3–6)*. In serum samples, the recoveries of each of these six cytokines except IL-4 were also within ±25% of the expected concentration when samples were spiked with high (100 and 1000 pg/mL) concentrations of each target cytokine (**Fig. 3A** and **B**). This was true for cytokine recovery in sera from healthy donors and patients with autoimmune diseases (rheumatoid arthritis, multiple sclerosis, or psoriasis). At concentrations more indicative of the physiologic range (10 pg/mL), only the recovery of TNF-α remained high (**Fig. 3C**). For the remaining cytokines, recovery was reduced and showed variability, and this was observed with sera from both normal donors and from patients with autoimmune diseases. de Jager and colleagues *(5)* have also reported greater variance in recovery from 10 pg/mL-spiked cytokine samples.

A further indication of the relationship between cytokine recovery and sample matrix was the observation in our laboratory that recoveries for IL-2 and IL-5 were lower in serum samples from patients with rheumatoid arthritis or multiple sclerosis, respectively, possibly because of the presence of various soluble receptors (e.g., sIL-2R) in the sera from patients with these diseases *(12)*.

These data underscore the point that performance of the multiplexed flow cytometry assay used to simultaneously measure levels of multiple cytokines from clinical samples needs to be thoroughly investigated to permit proper interpretation of the results, particularly when comparing results in patients with different disease

---

Fig. 3. *(Figure on opposite page)* Mean percent recovery of sera spiked with varying concentrations (1000, 100, or 10 pg/mL) of interleukin (IL)-2, IL-4, IL-5, IL-10, interferon-α (IFN-γ), and tumor necrosis factor-α (TNF-α) in from healthy donors (*n* = 2), or patients with rheumatoid arthritis (RA; *n* = 1), multiple sclerosis (MS; *n* = 2), or psoriasis (PS; *n* = 3). Recovery (i.e., concentration) of each cytokine was determined by interpolation from the standard curve. All determinations were made in duplicate. These data underscore the variability in recovery of clinically relevant cytokines from serum samples collected from patients with different disease states. Serum samples for patients with autoimmune diseases were purchased from Bioreclemation (Hicksville, NY).

states or across studies *(9)*. We also recommend that all samples from a single study be analyzed together so that valid comparisons can be made.

References

1. Slifka, M. K., and Whitton, J. L. (2000) Clinical implications dysregulated cytokine production. *J. Mol. Med.* **78,** 74–80.
2. Kellar, K. L., and Iannone, M. A. (2002) Multiplexed microsphere-based flow cytometric assays. *Exp. Hematol.* **30,** 1227–1237.
3. Prabhakar, U., Eirikis, E., and Davis, H. M. (2002) Simultaneous quantification of proinflammatory cytokines in human plasma using the LabMAP assay. *J. Immunol. Methods* **260,** 207–218.
4. Kellar, K. L., Kalwar, R. R., Dubois, K. A., Crouse, D., Chafin, W. D., and Kane, B-E. (2001) Multiplexed fluorescent bead-based immunoassays for quantitation of human cytokines in serum and culture supernatants. *Cytometry* **45,** 27–36.
5. de Jager, W., te Velthuis, H., Prakken, B. J., Kuis, W., and Rijkers, G. T. (2003) Simultaneous detection of 15 human cytokines in a single sample of stimulated peripheral blood mononuclear cells. *Clin. Diagn. Lab. Immunol.* **10,** 133–139.
6. Chen, R., Lowe, L., Wilson, J. D., et al. (1999) Simultaneous quantification of six human cytokines in a single sample using microparticle-based flow cytometric technology. *Clin. Chem.* **45,** 1693–1694.
7. Carson, R. T., and Vignali, D. A. A. (1999) Simultaneous quantitation of 15 cytokines using multiplexed flow cytometric assay. *J. Immunol. Methods* **227,** 41–52.
8. Camilla, C., Mély, L., Magnan, A., et al. (2001) Flow cytometric microsphere-based immunoassay: analysis of secreted cytokines in whole-blood samples from asthmatics. *Clin. Diagn. Lab. Immunol.* **8,** 776–784.
9. Banks, R. E. (2000) Measurement of cytokines in clinical samples using immunoassays: problems and pitfalls. (2000) *Crit. Rev. Clin. Lab. Sci.* **37,** 131–182.
10. Bioanalytical Method Validation, Guidance for Industry. United States Food and Drug Administration. May 2001.
11. Findlay, J. W. A., Smith, W. C., Lee, J. W., et al. (2000) Validation of immunoassays for bioanalysis: a pharmaceutical industry perspective. *J. Pharmaceut. Biomed. Anal.* **21,** 1249–1273.
12. Klimiuk, P. A., Sierakowski, S., Latosiewicz, R., et al. (2003) Interleukin-6, soluble interleukin-2 receptor and soluble interleukin-6 receptor in the sera of patients with different histological patterns of rheumatoid synovitis. *Clin. Exp. Rheumatol.* **21,** 63–69.

14

Validation and Quality Control of Protein Microarray-Based Analytical Methods

Larry J. Kricka and Stephen R. Master

Summary

The microarray has emerged as an important format for simultaneous analysis of tens of thousands of substances present in a sample. Successful adaptation of microarray assays to clinical diagnostics will require particular attention to issues of quality control and quality assurance. Results of an assay can be compromised by a number of preanalytical factors including the quality of the reagents (e.g., the microarray and the detection reagents) and the integrity of the sample. Similarly, numerous factors in the analytical phase of a microarray assay, including changes in the reaction conditions and calibration, can contribute to inaccuracy and imprecision. Furthermore, a microarray combines many reagents or samples in a single device and therefore presents additional issues not usually encountered in discrete testing of a single analyte in a single sample. Various strategies (e.g., replicate analysis, array orientation control features, on-array controls, normalization) have been implemented to control and assess analytical factors that might compromise data generated from a microarray. The current range of measures taken to ensure the analytical accuracy and quality of data generated from protein microarrays is reviewed in the context of the lessons learned from DNA microarrays. The special considerations for protein microarrays as they transition from research into routine clinical analysis and the resulting quality control of clinical test results generated using such devices are discussed.

Key Words: Microarray; quality control; quality assurance; clinical analysis; bioinformatics.

1. Introduction

The microarray has emerged as an important format for simultaneous analysis of tens to hundreds of thousands of substances that may be present in a sample. Arrays of immobilized reagents for testing a liquid sample and arrays of immobilized samples to be tested using a liquid reagent have been developed. Reagents immobilized in an array format include cDNA, oligonucleotides, peptide nucleic acids *(1–8)*, antibodies, and antigens *(9–18)*. The major research

From: *Methods in Molecular Medicine, Vol. 144, Microarrays in Clinical Diagnostics*
Edited by: T. Joos and P. Fortina © Humana Press Inc., Totowa, NJ

applications for DNA arrays have been expression monitoring and sequencing *(19–22)*, and prominent applications for protein arrays include simultaneous cytokine analysis *(23,24)*, allergen testing *(25–28)*, and studies to discover markers for disease diagnosis and management *(18)*.

Two distinct microarray formats have evolved—the 2D microarray and the liquid microarray *(29–32)*. In a 2D microarray format, reagents for individual tests are immobilized as an ordered array or grid on a flat surface such as a microscope slide or in the ends of a fiberoptic bundle. Each location on the array corresponds to a reagent for a particular test. The surface of the array is exposed to the sample and then to the detection reagents. Finally, the surface of the array is scanned, and the presence of a signal (e.g., a fluorescence signal) at particular locations indicates the presence of those particular analytes in the sample.

In a liquid microarray format, the reagents are immobilized on the surface of a set of microbeads. Each member of the microbead set has a unique fluorescence signature that allows it to be distinguished from each of the other members of the set. Additionally, each member of the microbead set is coated with a different reagent. A suspension of the set of microbeads is incubated with the sample and with detection reagents. The fluorescence signal from each individual bead is measured, and this both identifies the microbead and indicates whether it has reacted with a specific analyte present in the sample.

2. Transition of Microarray Tests From Research Into Routine Clinical Analysis

The microarray assay format is ideally suited to the panels of tests that are emerging from the genomic and proteomic initiatives. This assay format is a relatively simple and very convenient way of simultaneously analyzing a sample for many different analytes, and this method of highly parallel testing is more rapid than serial assays. At the forefront of emerging tests that are suited to a microarray format are tests for assessing drug metabolizer status (e.g., CYP450 genotyping) *(33)*, evaluating drug resistance by human immunodeficiency virus (HIV) *(34)*, detecting disease-causing mutations (e.g., cystic fibrosis, p53) *(35,36)*, resequencing *(37,38)*, detecting single-nucleotide polymorphisms (SNPs) *(37–39)*, measuring cytokines *(23,24)*, and assessing allergic responses *(25–28)*.

Regulatory issues aside *(40,41*; www.fda.gov/cdrh/oivd/letter-roche2.html), the transfer of the microarray testing format into the routine clinical laboratory will pose some complex analytical issues. Currently, the clinical laboratory does not use microarrays. Indeed, it is rare to have any simultaneously multianalyte testing using a single analytical device. The closest analogies are a disposable device used for point-of-care allergy testing (e.g., the MASTpette uses an array of allergen-coated cotton threads in a plastic pipete) *(42)* and disposable cartridges for point-of-care testing for drugs of abuse or cardiac markers that use a series of parallel zones of capture reagent printed onto a membrane strip *(43)*. A central issue for the implementation of microarrays is ensuring quality control (QC; activities used to monitor the quality of analytical data and to ensure that it satisfies specified criteria) and quality assurance (QA; the process and practices for assuring the quality of an analytical procedure). This chapter explores the current range of measures taken to assure the analytical accuracy and quality of data generated from protein microarrays and explores special considerations for the QC of clinical test results generated using such devices.

3. Validation of Analytical Performance of Protein Microarray Assays

The principal type of assay performed using a protein microarray is an immunoassay. Competitive assays, sandwich assays, and labeled sample assay designs have been implemented in a microarray format. In the routine clinical laboratory, a well-defined series of QC measures have been widely adopted in order to ensure that the results produced by these analytical methods are accurate and precise. Initially, the performance of the assay is evaluated using an established protocol (e.g., CLSI–www.nccls.org) *(44,45)*. This evaluation includes:

1. Range of linearity.
2. Imprecision (within- and between-batch at low, normal, and high concentrations using specimens that are in an appropriate biological matrix).
3. Analytical specificity (e.g., susceptibility of the assay to common interferents, such as lipids, hemoglobin, bilirubin, common drugs, and drug metabolites).
4. Recovery of pure analyte spiked into the test matrix (e.g., serum, plasma).
5. Limit of detection (defined as the lowest concentration or quantity of an analyte that can be detected with a stated reasonable uncertainty for a given analytical procedure, e.g., the concentration at a signal-to-noise ratio of 2 or the concentration corresponding to a signal 3 standard deviations (SD) above the mean for a calibrator that is free of analyte).
6. Comparability of the results obtained by the method being validated with those obtained by a reference-quality method on 100–200 different samples from patients selected to include a range of values for the analyte likely to be encountered in routine application.

These criteria are now actively advocated by journals in their instructions to authors seeking to publish manuscripts describing the development and evaluation of the performance of new methods and instruments (e.g., *see* http://www.AACC.ORG/ccj/infoauth.stm).

When an immunoassay is set up in a routine clinical laboratory, a further series of established practices is utilized to ensure reliable and accurate results. Calibration standards are assayed to confirm the working range for the assay. Linearity is rechecked periodically using appropriate calibrators or verifiers for immunoassays run on automated analyzers with stored standard curves. In the case of manual immunoassays, the standard curve is included in each batch of assays.

A further measure to monitor assay quality on a daily basis employs QC materials with high, medium, and low levels of the analyte. These are assayed repeatedly in order to establish control ranges that define the performance of the assay over the working analytical range. These controls are then assayed before the analysis of a batch of samples for QC purposes. If analytical results for these controls fall outside of the acceptable range, then the sources of inaccuracy must be determined and rectified until reanalyzed controls are within the defined limits. No clinical analyses are commenced until acceptable control values are obtained, thus indicating that the assay is operating within the desired level of analytical performance. Controls can also be included among the batch of samples to be analyzed. Analytical results on samples are rejected if the values for any of the controls included in the batch of samples fall outside the control ranges.

An additional, but retrospective, level of scrutiny of analytical performance is provided by external QA schemes in which a sample of the same specimen is sent for analysis to different laboratories enrolled in the scheme (*46,47*). The results are collected centrally and subjected to statistical analysis. A mean and SD are calculated from the results returned by each of the laboratories using a particular analytical method or analyzer. Each laboratory can then compare the result that it obtained on the sample with results obtained by all the other laboratories using the same method. Generally, the closer the result is to the all-laboratory method mean, the more likely it is that the analytical method is performing reliably.

In contrast to routine immunoassays, few evaluations of protein microarray immunoassays have conformed to the rigorous evaluation criteria described above (*see* **ref.** *48* for an exemplary study). Likewise, there are no detailed studies of QC measures and strategies for microarray immunoassays, and as yet there is no external QA scheme designed to assess the performance of protein microarray immunoassays. The following sections explore the range of measures taken by different researchers to evaluate and ensure reliable and accurate results from protein microarray experiments. We also discuss some special considerations for using microarrays in clinical analysis and the lessons learned from DNA microarrays.

4. Preanalytical Factors

The preanalytical phase of an analytical procedure includes specimen collection and the preparation of the specimen for analysis. Results of an analytical procedure can be compromised by a number of preanalytical factors including the quality of the reagents (e.g., the microarray and the detection reagents) and the integrity of the sample.

4.1. Microarray Fabrication

Many laboratories print their own arrays, and the technical issues in array fabrication e.g., array surfaces *(49)*, have been considered in detail elsewhere *(11,12)*. Strategies investigated previously in the development of immunoassays, such as optimization of antibody immobilization, have been revisited for microarrays. The influence of random vs specifically orientated antibodies and Fab′ fragments has been studied, and the specific orientation of biotinylated antibodies on a streptavidin surface has been shown to provide up to 10-fold improvement in analyte binding capacity *(50)*. As microarrays are usually produced in small batches, appropriate testing has been performed with controls to evaluate antibody binding prior to using the arrays for analytical purposes *(51)*. Other procedures have assessed printing quality, functionality of printed proteins, and hydration status of the microarray and its influence on reactivity *(52)*.

4.2. Specimen

The preanalytical issues for the specimen to be analyzed using a microarray are no different from those for any other clinical specimen. These include correct specimen collection with an appropriate preservative (if required), along with transport and storage under conditions that will not be detrimental to the integrity of the specimen.

5. Analytical Factors

The analytical phase of a clinical assay comprises all the steps involved in the actual analysis of the sample. Many factors in the analytical procedure can contribute to inaccuracy and imprecision including changes in the reaction conditions and calibration. A microarray combines many reagents or samples in a single device, and this presents additional issues not usually encountered in discrete testing of a single analyte in a single sample. Various strategies have been implemented to control and assess analytical factors that might compromise data generated from a microarray, as outlined below.

5.1. Sample and Dilution

When using a conventional immunoassay, it is common practice to dilute and rerun a sample that is out of range. In contrast, implementing this approach using protein microarrays is currently an expensive and time-consuming option. To address this problem for assays in which the samples are spotted on the array for parallel detection with a single reagent, multiple sample dilutions have been used to that ensure analyte concentration is within the dynamic range, e.g. a 5-point dilution of the sample *(53,54)*.

5.2. Singlicate vs Replicate Analysis

Most conventional immunoassays, especially those performed using automated equipment, have reached a level of reliability such that singlicate analysis is the now the normal practice. For microarrays in their current state of development, many groups have utilized replicate analyses of a given sample. Strategies include duplicate *(51,55)* triplicate *(53)*, quadruplicate *(56,57)*, and up to 10 replicate analyses of the sample on the same array *(58–60)*. Other replication strategies include adjacent duplicates on two different arrays *(61)*. In addition, duplicate spots are sometimes printed in different regions of the same array in order to control for regional differences in reaction conditions *(62)*. For liquid arrays based on microbeads, a greater degree of replication is possible, and up to 500 beads per analyte can be used *(63)*. Commercial arrays have adopted both singlicate (http://www.hypromatrix.com) and duplicate testing strategies (e.g., *see* http://www.panomics.com).

5.3. Controls Included in the Array

Different types of performance controls are included on protein microarrays. Controls include a blank area *(64)*, positive controls such as anti-human IgE or IgE on an allergen microarray *(25,65)*, and negative controls such as buffer or protein *(53,66,67)*. Additionally, controls are sometimes printed on a microarray in great numbers, e.g., 64 positive and 384 negative controls in duplicate on the same array *(62)*. Commercial microarrays for research applications also include positive and negative controls (e.g., *see* http://www.zyomyx.com). Controls provide information on the assay background (e.g., a bovine serum albumin [BSA] control) *(68)*, nonspecific binding (e.g., irrelevant antibodies as controls *[68]* or murine isotype controls for murine antibody microarrays *[66]*), and crossreactivity (e.g., human IgG printed onto allergen microarrays to estimate crossreactivity with the anti-human-IgE conjugate) *(25)*. Controls have been used to test the binding of the conjugate employed to detect analyte bound to individual spots *(68)* and to assess the activity of key assay reagents such as the detection conjugate *(69,70)*.

An interesting control procedure used in bead arrays has been to include beads for detecting sample addition and for the presence of an interfering substance such as rheumatoid factor *(31)*. For example, a serological assay panel included beads with anti-IgG or anti-IgM to verify sample addition. A minimum concentration of the immunoglobulin that should be present was set at 20 mg/dL for IgM and less than 400 mg/dL for IgG. Failure of the beads to register this concentration of analyte in the sample indicates a failure to dispense the sample *(31)*.

5.4. Within-Array and Between-Array Imprecision

In some protein microarray studies, reproducibility has been assessed quali-tatively by demonstrating similar patterns of reactivity for samples from the same specimen tested on two arrays *(51)*. Other investigators have performed quantitative assessments of imprecision by replicate assays for a streptavidin-peroxidase conjugate control and measured a SEM of 2.6–7.6% for a single chip and 19.7% for 17 chips *(25)*. Variation for positives tested on multiple arrays (between-batch precision for controls across four arrays) ranged from 9.2 to 32%, attributed in part to variable development of the array with the detection reagent *(64)*. Analogous studies using a bead microarray-based assay for 15 different cytokines reported that the average intraassay imprecision was 8.7% and the average interassay was 16.5% *(63)*.

5.5. Linearity

Linearity has been assessed in microarray assays using dilutions of standards (bead array) *(63)*, serial dilutions of a positive specimen *(65,71,72)*, or a positive control spotted directly onto the array *(73)*.

5.6. Analytical Specificity

The effects of interfering substances have been evaluated for microarray immunoassays by mixing a positive specimen with specimens containing potential interferents (e.g., hemoglobin, cholesterol) *(65)*. One typical source of nonspecificity in an immunoassay is from the crossreactivity of the antibody used in the assay with other substances in the sample, and usually an antibody with the highest possible specificity is used in order to eliminate or minimize such crossreactions. This consideration is particularly critical for complex protein mixtures in which the range of protein concentrations may span ten orders of magnitude. For example, if a detection antibody has a specificity of 99% and a crossreacting protein is present in a 1000-fold excess, then the assay will be nonspecific for the intended protein analyte *(12)*. Thus, as with conventional immunoassays, great care must be taken with antibody selection for microarray analysis of complex protein mixtures.

5.7. Recovery

Recovery in microarray assays, measured as a percentage of the total analyte, has been assessed by spiking known amounts of the analyte into a set of test sam-ples for analysis and then reassaying the samples. Cytokine recoveries measured using this technique ranged from 91 to 108% for 100 pg/mL and 1000 pg/mL *(63)*.

5.8. Limit of Detection

Detection limits have been determined by standard methods including analysis of standards of known value and serial dilutions of a positive sample spotted onto the array. The detection limit is then determined as the amount of analyte that produces a signal corresponding to the background plus a number of multiples of the SD interpolated from the standard curve *(25,48,56,73–76)*. More extensive detection limit studies have involved the analysis of 20 standard curves on seven arrays for recombinant prostate-specific antigen (PSA) and subsequent use of these data to determine both the limit of detection (background +2 SD: ~10^{-21} mol) and the functional sensitivity of the assay (concentration that can be measured with a coefficient of variation [CV] of 20% within an array; 5×10^{-20}) *(76)*.

5.9. Method Comparison

Several comparisions of microarray and conventional immunoassays have been published. A comparison of antiviral antibody detection on a protein microarray vs a conventional ELISA showed concordant results ($r = 0.836$–0.997) *(58)*. Similarly, a cytokine assay using a liquid microarray showed good correlation with a conventional method ($r = 0.75$–0.99) *(63)*. Comparison of allergen test results generated using a 2D microarray (rolling circle amplification detection of bound IgE) with results generated using a previously validated commercial test also showed good agreement *(65)*. However, a comparison of human cytokine assays performed with both a traditional ELISA and a low-cost membrane protein array showed that this particular type of microarray lacks sensitivity *(64)*.

5.10. Data Acquisition and Data Processing

The magnitude of data generation by microarrays poses new challenges for the laboratory. In particular, data processing has become an important issue with respect to QC and QA.

5.10.1. Bioinformatics

A variety of software packages have been used for spot identification and quantitation from digital images of protein microarray data. Many of these packages have been used for similar analysis of spotted cDNA arrays, and the technical issues relating to spot detection for protein microarrays are fundamentally similar. Publicly available and proprietary packages that have been utilized for analysis of protein microarray images include P-SCAN *(12,77)*, GenePix (Axon Laboratories) *(78)*, ImageQuant (Molecular Dynamics) *(79)*, and GeneTAC (Genomic Solutions) *(62)*.

5.10.2. Array Orientatation and Spot Alignment

Knowing the orientation of an array is vital in order to correctly link the scanned signal to the identity of the spotted reagents or sample. Accordingly, reference features or landmarks can be printed on arrays in order to provide an unambiguous orientation. Materials used to create these features have included Cy3-labeled protein *(80)* and biotinylated BSA *(62)*. Because the analysis of a diverse set of microarray designs requires general-purpose analytical software, manual intervention is usually required in order to ensure proper spot detection and alignment *(12)*.

5.10.3. Quantitation

Once array orientation and spot identification have been confirmed, automated spot quantitation can be undertaken. Intensity is typically estimated using either the total integrated spot intensity or (equivalently) average intensity. Local background has been estimated using the median pixel intensity surrounding the spot *(78)*, and this background value can be subtracted from an average pixel intensity to yield the corrected estimate. Although this approach corrects for general background staining, more substantial defects may require elimination of affected array spots from further analysis and consideration *(62)*. When on-chip replicates are arrayed as adjacent spots, these local defects may affect multiple measurements of the same analyte.

5.10.4. Normalization

A specific manufacturing concern for microarrays is intra- and interarray variation in binding of the analyte owing to differences in the amount of protein spotted in each location on the array and the effect of this on the comparability of data generated from different arrays. One group has recommended validation of data using arrays printed by different methods *(61)*. A specific intraarray normalization strategy requires measurement of the protein concentration in the spots on the array and then using these values to normalize the corresponding signals *(67,81,82)*. For interarray normalization, one strategy has been to print 16 identical control features on each array and then to calculate a median value of the signal from these locations and use this to normalize data between arrays *(58)*.

Another normalization strategy relies on adding a control reagent to the sample. For example, biotinylated BSA has been included with samples and detected via its binding to an anti-BSA spot on the array and a streptavidin-phycoerythrin reagent *(61)*.

5.10.5. Higher Order Bioinformatic Processing

A variety of higher order manipulations (e.g., pattern analysis and classification) of protein abundance data, most of which have been validated for biological research using gene expression datasets, have been used for exploratory data analysis (*see*, e.g., **ref.** *12*). The QC and QA issues associated with such higher order manipulations will be described below (**Subheading 6.4.**).

6. Lessons From DNA Microarrays

Although neither protein nor DNA microarrays (DNA chips) have yet been incorporated into routine clinical diagnostics, the widespread use of DNA microarrays in both basic and (increasingly) clinical research settings has highlighted a number of technical and analytical issues that are directly applicable to protein microarrays. As there is no conceptual difference between patterns of gene expression and patterns of protein abundance, many of the bioinformatics tools in use were originally developed for use with high-throughput gene expression data. In addition, experience with DNA microarrays has provided a framework for discussing a number of important issues in protein microarray analysis, including the importance of control features on the arrays themselves, relative utility of technical and biological replicates, comparability of results across analytical platforms, and standards for data exchange.

6.1. Hybridization Controls

The fundamental differences between protein and DNA microarrays stem from the ability to easily produce oligonucleotides or longer cDNA fragments that specifically hybridize to a sequence of interest. This not only simplifies the reagent development phase of microarray design but also allows for an additional level of control to be incorporated into the assay. For example, Affymetrix oligonucleotide expression arrays measure transcript abundance by measuring binding to specific oligonucleotides that match a sequence of interest. For each of these specific oligonucleotides, a matched control oligonucleotide that incorporates a single base-pair mismatch is used to account for nonspecific binding *(83)*. Such controls, in conjunction with whole-genome sequence information that can be used to minimize the risk of cross-hybridization of closely related sequences, provide a level of validated specificity not currently obtainable using protein microarrays. In this regard, antibody interference and cross-reactivity remain areas of concern for QC/QA of protein microarrays in the clinical laboratory.

6.2. Replicates

Like protein microarrays, DNA microarrays must account for the statistical problems inherent in large-scale analysis. A variety of groups have noted that, for example, a 99% accuracy over 10,000 genes yields an expected 100 false-positive results *(84)*, and various groups have attempted to overcome this problem through whole-array technical replicate assays or within-chip replicate measurements. Whole-array replicates have been shown to markedly decrease the false-positive rate with respect to differential gene expression *(85)*. Similarly, it has recently been shown that greater than 5 (and more often 10–15) samples ("replicate" with respect to, for example, tumor type) are required in order to generate stable analytical results *(86)*. In this latter case, "replicate" samples were obtained from different patients and thus have a direct bearing on the ability to perform robust disease classification on biologically diverse individuals; however, they are not directly germane to the problem of reproducibility of replicate assays performed on an individual patient.

In addition to replicates of the entire assay, two basic types of on-chip replicates have been utilized in DNA microarrays. The first, and most straightforward, type involves simply arraying the same transcript or oligonucleotide in multiple locations on the chip; this approach has been widely adapted for protein microarray work *(12,78)*. The second approach involves creating multiple, independent oligonucleotides that bind to the same transcript. Affymetrix expression microarrays typically detect a given transcript using ≥11 oligonucleotides, and in this way the specificity of the assay is increased *(87)*. Of note, use of data from these individual hybridizations has led to improved quantitation of gene expression as well as improved sensitivity for identifying differential regulation *(88,89)*. Another noteworthy feature of this design is that, by distributing the locations of oligonucleotides detecting a given transcript across an array, the estimate of transcript concentration is less susceptible to spatially restricted defects on the chip. As these oligonucleotides bind to different regions of the transcript, an analogous protein microarray assay would utilize, e.g., multiple antibodies recognizing different epitopes on the analyte. Although the advantage of this approach is significant, its incorporation within protein microarrays will involve substantial additional reagent development.

6.3. Protocols and Data Comparison

The diverse and rapidly evolving set of experimental protocols utilized for DNA microarray construction, sample preparation, and hybridization led to the need for a consensus set of minimal experimental information (the Minimal Information About a Microarray Experiment [MIAME]; *90,91*). A similar format suitable for protein microarrays has begun to emerge (MIAPE; *92,93*), although to date the most detailed work has focused on protein expression data derived from mass spectrometric analysis. These standards provide important information regarding sample origin, location, and processing; it is likely, however, that a different set of standards will be necessary in order to fully capture the variety of patient identifiers that will be necessary for routine clinical diagnostics.

The existence of MIAME, although useful in a research context, does not solve the more fundamental difficulty of comparing results obtained using different assays. In the same way that choosing a given analyte-specific antibody exercises a broad influence on overall test performance, the location and extent of nucleic acid homology that are selected for hybridization with the sample affect both the sensitivity and specificity of the assay. Even within a single array, it has long been recognized that different oligonucleotide sets intended to measure expression of the same gene may give widely varying numerical results; this demonstrates, at the very least, that any absolute (as opposed to relative) quantitation of gene expression requires independent calibration for each transcript. More significantly, cross-platform comparisons of results from reference pools of RNA have reported substantial variability in the relative efficiency of detecting various transcripts *(94)*. Despite these challenges, success in cross-platform comparisons has been reported *(95)*. Furthermore, techniques for correlating cross-platform results using reference samples have recently been described *(96)*, raising the possibility that the systematic correlation of results may become more widely available.

The importance of cross-platform and cross-institutional comparisons will probably be greatest for arrays that survey markers reflecting a broad range of cellular processes. Conversely, protein microarrays intended to assay a restricted range of analytes may be provided in the near term by commercial sources. For the former group, correlation of clinical outcomes will require more than a mere consensus for experimental description; in the same way that adequate RNA standards are required for DNA microarray comparisons *(97)*, protein sets derived from standardized tissues will be necessary. Additionally, the ready exchange of datasets that are more complex than standard laboratory assays by several orders of magnitude (hundreds to tens of thousands of fields) will require further consensus standards. If a standard format or source for protein microarrays were to emerge, it would probably facilitate these exchanges, in the same way that the widespread use of Affymetrix microarrays has recently led to the proposal of a set of consensus "best practices" for sample processing, QC, and data analysis *(98)*.

6.4. Data Analysis

Perhaps the most significant conceptual issue in the use of DNA microarrays for clinical diagnostics is the question of higher order data processing methods used for classification. A variety of groups have undertaken studies utilizing microarray data as a detailed phenotype that can be used for disease classification, and the corresponding analytical tools employed have included bayesian classifiers, threshold correlation coefficients, weighted voting, *k*-nearest-neighbors, support vector machines, neural networks, and hierarchical clustering (*see*, e.g., *99–103*). This overall approach has been extremely promising and appears to provide a substantial improvement in our ability to classify a variety of neoplasms with respect to prognosis.

Similarly, then, a natural application of protein microarray technologies will be for disease classification. From a QC perspective, however, a fundamental question that must be addressed is the relative sensitivity of these approaches to failure of any given measurement. If microarrays are seen as fundamentally similar to previous high-throughput assays within the clinical laboratory, then individual QC measures will be necessary to validate results for each analyte. From this view, each gene or protein measurement is a reportable result. On the other hand, if the overall test is intended to use measurements from a large number of analytes in order to distinguish robustly between disease states in the face of biological variability and experimental noise, then the need for rigorous QC and QA of each individual analyte may be less important than the overall effect that this failure might have on the behavior of a global classifier. As these issues are only now emerging with the movement of DNA microarrays toward clinical use, the approaches taken within the DNA microarray world will probably affect the future use of protein microarrays. In any case, it is crucial that these issues be addressed prior to the introduction of a proposed genomic or proteomic test into clinical practice.

7. Quality Control

There have been a number of concerns about the quality of microarray assays. These range from the activity and reproducibility of spotting of reagents on the array *(104–107)* to the image processing procedure *(108)* and data analysis methods used *(109,110)*. Less attention has been paid to QC of the analytical procedure when it is used as a routine method for analyzing clinical specimens.

In routine practice, the clinical laboratory utilizes QC procedures to ensure that analytical methods are operating within specifications on a day-to-day basis. Structured rules have been devised for interpreting QC data and provide the basis for rejecting or accepting analytical data from specimens analyzed in the same batch or in the same time period as the controls *(111)*.

Routine immunoassays, performed on automated equipment, are usually controlled by analysis of QC materials with high, medium, and low levels of the analyte. These analyses are typically performed at the beginning of each day, and one or more of these control materials are then reanalyzed at the beginning of subsequent 8-h shifts. For immunoassays performed as a single batch, for example, in a microtiter plate, it is usual to include all three controls in wells at both the beginning and the end of the plate. In both cases, examination of the QC data is used to decide on the acceptability of the results obtained on samples from patients. For the assay to be within control, the results for each of the control materials must be within a range of values established by the laboratory for each of the QC materials.

Immunoassays performed on a microtiter plate (e.g., 96 wells) provide the closest analogy to microarray analysis. Usually, only a single reagent is arrayed in the microplate wells, and so the QC problem is relatively simple compared with a microarray with hundreds or thousands of individual reagents. A general strategy for QC of microarray analyses has not been clearly outlined in the literature.

The first requirement for successful QC of a microarray assay is a set of controls with high, medium, and low levels of each of the multitude of analytes to be tested on the array. Formulating controls for some of the current types of protein microarray assays would be complex because of the large number of different analytes that each control must contain in order to test each reagent on the array. Control materials used in the routine clinical laboratory can have more than 80 individual analytes (*see* www.bio-rad.com), but most are stable (e.g., sodium, cholesterol, albumin) and readily obtained in pure form. The stability, purity, and availability of the substances needed to prepare QC materials for protein microarrays may be a significant issue. Also, manufacturing control materials for microarray assays may itself pose a formidable challenge depending on the identity of the individual analytes.

The next requirement is a set of rules for interpreting the QC data and determining the acceptability of the analytical results produced by a given batch of arrays. There is clearly no problem if all the assays on an array with each QC material are within the control limits. However, how does the laboratory deal with a situation in which some of the analytical results for the QC materials are outside the control limits? A situation could arise in which different assays fail with the high, medium, and low controls, as illustrated in the antibody microarrays depicted in **Fig. 1. Figure 1A** shows the results of analyzing a high, medium, and low control with no failures. In contrast, **Fig. 1B** shows failures in the assay of the controls for the analytes that are tested at locations 1, 2, and 3 on the array. **Figure 1C** shows results from three patients analyzed in the same batch as the controls shown in **Fig. 1B**. The issue that now arises is how to interpret the results for the patients in view of the failed controls. There could be two possible courses of action:

1. Rejection of the entire set of data (controls and patients) and a rerun of the controls using new arrays followed by a subsequent reanalysis of the samples from the three patients.
2. Selective acceptance of data from locations on the arrays that passed for each of the three controls. Analysis of samples from patients would proceed, and only data from array locations that passed QC would be accepted. However, in order to complete the intended analysis of the samples from the three patients, it would be necessary to run more arrays to obtain the missing analytical data.

Both courses of action lead to rerunning of the controls or controls and samples using new arrays if all the required assays are to be completed on the three samples. For a liquid array, the further possibility exists of preparing a subset of the array containing the failed reagents and simply retesting the samples using this subset of the array. The chances of a QC failure are expected to increase with

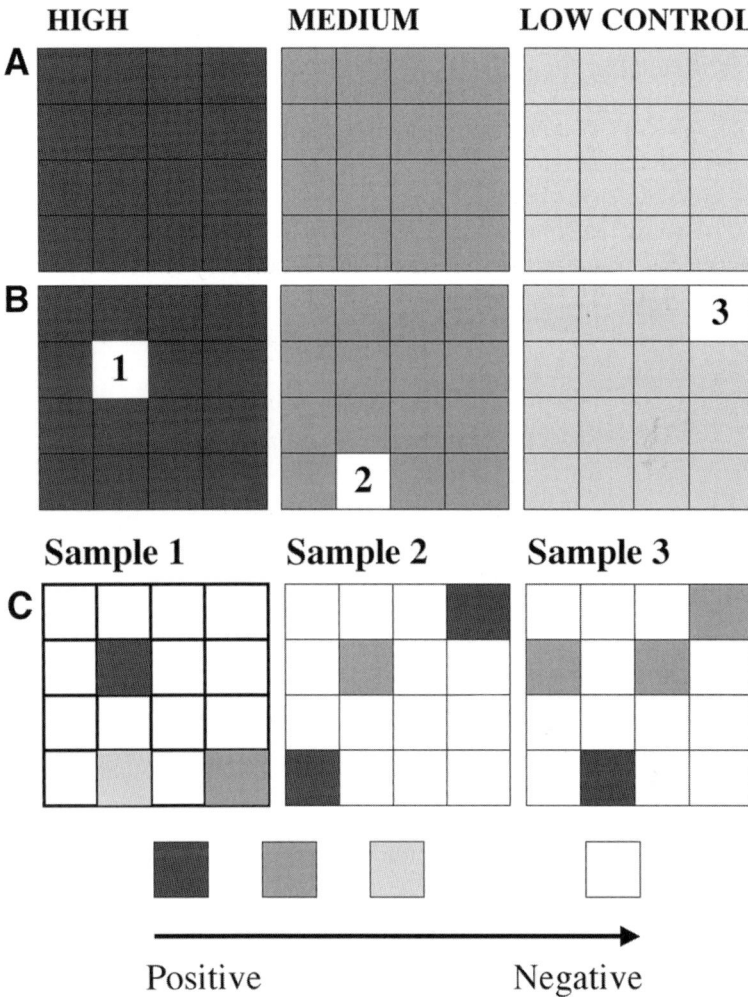

Fig. 1. Testing for multiple analytes using an antibody array. (**A**) Results for three control materials (high, medium, and low concentrations of the various analytes that the array is designed to test for in a sample) analyzed using an antibody array. All control values are within acceptable limits. (**B**) Results for the same three control materials using the antibody array illustrating a failure of the array to produce acceptable results at the locations labeled 1, 2, and 3. All the other assay results from other array locations are within acceptable limits. (**C**) Examples of analytical results generated with samples from three patients analyzed in the same batch as the controls.

an increase in the number of individual reactions performed on the array. Hence, it is imperative that these types of QC issues be resolved if large-scale microarray analysis is to emerge as a routine method for the clinical laboratory.

8. Conclusions

Protein microarray technology is at an early stage of development, and so far its use is confined to the research laboratory. However, this powerful new tool is being used to uncover important diagnostic and prognostic markers that may one day emerge as routine tests in the clinical laboratory. Even during this marker discovery phase, it is important that these methods be subject to adequate control procedures in order to ensure that data of the highest quality are obtained. If quality is not adequately controlled, the assay may yield erroneous results that would mask or confound meaningful diagnostic or prognostic associations. A recent primer on protein microarrays provides an excellent introduction to the many analytical issues in microarray-based assays *(12)*.

This survey of protein microarray methods explores the range of strategies designed to ensure analytical quality, and it also highlights some of the potential pitfalls when one is moving these arrays into routine clinical practice. The ultimate introduction of protein microarray-based testing into wide-scale and routine clinical laboratory practice will be challenging. The larger the number of test spots in an array, the greater and more difficult the challenge in terms of controlling the assay and interpreting the data. It is currently unclear what size of microarray will be necessary or feasible in routine clinical laboratory practice. A logical target for the quality of microarray assays is to match the analytical precision attained by modern-day automated immunoassay systems. These systems routinely achieve an interassay CV of less than 10% and for some assays less than 5% *(112–117)*.

DNA microarrays have been in use longer than protein microarrays, and some of the lessons learned with this type of microarray method have been transferred into protein microarray assays. The first DNA microarray-based in vitro diagnostic test has recently (December, 2004) been approved by the U.S. Food and Drug Administration (Roche Molecular Systems, Amplichip CYP 450 test). What will most likely be the first clinical protein microarray assay system is currently under evaluation (*see* www.randox.com), and it will set the stage for the future use of protein microarrays and other microarray-based assays in routine practice. With the development of appropriate QC and QA measures, we anticipate that these assays will be of substantial benefit in the future practice of laboratory medicine.

References

1. Schena, M., ed. (1999) *DNA Microarrays: A Practical Approach*. Oxford University Press, Oxford.
2. Copland, J. A., Davies, P. J., Shipley, G. L., Wood, C. G., Luxon, B. A., and Urban, R. J. (2003) The use of DNA microarrays to assess clinical samples: the transition from bedside to bench to bedside. *Rec. Prog. Hormone Res.* **58,** 25–53.

3. Geschwind, D. H. (2003) DNA microarrays: translation of the genome from laboratory to clinic. *Lancet Neurol.* **2,** 275–282.

4. Mariadason, J. M., Augenlicht, L. H., and Arango D. (2003) Microarray analysis in the clinical management of cancer. *Hematol. Oncol. Clinics N. Am.* **17,** 377–387.

5. Pusztai, L., Ayers, M., Stec, J., and Hortobagyi, G. N. (2003) Clinical application of cDNA microarrays in oncology. *Oncologist* **8,** 252–258.

6. Sevenet, N. and Cussenot. O. (2003) DNA microarrays in clinical practice: past, present, and future. *Clin. Exp. Med.* **3,** 1–3.

7. Shaughnessy, J. Jr. (2003) Primer on medical genomics. Part IX: scientific and clinical applications of DNA microarrays—multiple myeloma as a disease model. *Mayo Clinic Proc.* **78,** 1098–1109.

8. Kricka, L. J. and Fortina, P. (2001) Microarray technology and applications: an all-language literature survey including books and patents. *Clin. Chem.* **47,** 1479–1482.

9. Albala, J. S. (2001) Array-based proteomics: the latest chip challenge. *Expert Rev. Mol. Diagn.* **1,** 145–152.

10. Cutler, P. (2003) Protein arrays: the current state-of-the-art. *Proteomics.* **3,** 3–18.

11. Schena, M., ed. (2004) *Protein Microarrays.* Jones & Bartlett, Boston.

11a. Eickhoff, H., Konthur, Z., and Lueking, A. (2002) Protein array technology: the tool to bridge genomics and proteomics. *Adv. Biochem. Eng. Biotechnol.* **77,** 103–112.

12. Liotta, L. A., Espina, V., Mehta, A. I., et al. (2003) Protein microarrays: meeting analytical challenges for clinical applications. *Cancer Cell* **3,** 317–325.

13. MacBeath, G. (2002) Protein microarrays and proteomics. *Nat. Gen.* **32(suppl),** 526–532.

14. Schweitzer, B. and Kingsmore, S. F. (2002) Measuring proteins on microarrays. *Curr. Opin. Biotechnol.* **13,** 14–19.

15. Templin, M. F., Stoll, D., Schrenk, M., Traub, P. C., Vohringer, C. F., and Joos, T. O. (2002) Protein microarray technology. *Drug Disc. Today* **7,** 815–822.

16. Zhu, H. and Snyder, M. (2003) Protein chip technology. *Curr. Opin. Chem. Biol.* **7,** 55–63.

17. Kricka, L. J., Joos, T., and Fortina, P. (2003) Protein microarrays: a literature survey. *Clin. Chem.* **49,** 2109.

18. Petricoin, E. F. and Liotta, L. A. (2003) Clinical applications of proteomics. *J. Nutr.* **133(suppl),** 2476S–2484S.

19. Drmanac, R., Drmanac, S., Chui, G., et al. (2002) Sequencing by hybridization (SBH): advantages, achievements, and opportunities. *Adv. Biochem. Eng. Biotechnol.* **77,** 75–101.

20. Prix, L., Uciechowski, P., Bockmann, B., Giesing, M., and Schuetz, A. J. (2002) Diagnostic biochip array for fast and sensitive detection of K-ras mutations in stool. *Clin. Chem.* **48,** 428–435.

21. Schaefer, K. L., Wai, D., Poremba, C., Diallo, R., Boecker, W., and Dockhorn-Dworniczak, B. (2002) Analysis of TP53 germline mutations in pediatric tumor patients using DNA microarray-based sequencing technology. *Med. Pediatr. Oncol.* **38,** 247–253.

22. Warrington, J. A., Shah, N. A., Chen, X., et al. (2002) New developments in high-throughput resequencing and variation detection using high density microarrays. *Hum. Mutat.* **19,** 402–409.

23. Huang, R. P., Huang, R., Fan, Y., and Lin, Y. (2001) Simultaneous detection of multiple cytokines from conditioned media and patient's sera by an antibody-based protein array system. *Anal. Biochem.* **294,** 55–62.

24. Lin, Y., Huang, R., Cao, X., Wang, S. M., Shi, Q., and Huang, R. P. (2003) Detection of multiple cytokines by protein arrays from cell lysate and tissue lysate. *Clin. Chem. Lab. Med.* **41,** 139–145.

25. Fall, B. I., Eberlein-Konig, B., Behrendt, H., Niessner, R., Ring, J., and Weller M. G. (2003) Microarrays for the screening of allergen-specific IgE in human serum. *Anal. Chem.* **75,** 556–562.

26. Harwanegg, C., Laffer, S., Hiller, R., et al. (2003) Microarrayed recombinant allergens for diagnosis of allergy. *Clin. Exp. Allergy* **33,** 7–13.

27. Jahn-Schmid, B., Harwanegg, C., Hiller, R., et al. (2003) Allergen microarray: comparison of microarray using recombinant allergens with conventional diagnostic methods to detect allergen-specific serum immunoglobulin E. *Clin. Exp. Allergy* **33,** 1443–1449.

28. Kim, T. E., Park, S. W., Cho, N. Y., et al. (2002) Quantitative measurement of serum allergen-specific IgE on protein chip. *Exp. Mol. Med.* **34,** 152–158.

29. Lipshutz, R. J., Fodor, S. P., Gingeras, T. R., and Lockhart, D. J. (1999) High density synthetic oligonucleotide arrays. *Nat. Genet.* **21(suppl),** 20–24.

30 Southern, E. M. (2001) DNA microarrays. History and overview. *Methods Mol. Biol.* **170,** 1–15.

31. Martins, T. B. (2002) Development of internal controls for the Luminex instrument as part of a multiplex seven-analyte viral respiratory antibody profile. *Clin. Diagn. Lab. Immunol.* **9,** 41–45.

32. Ekins, R. P. (1998) Ligand assays: from electrophoresis to miniaturized microarrays. *Clin. Chem.* **44,** 2015–2030.

33. Chou, W. H., Yan, F. X., Robbins-Weilert, D. K., et al. (2003) Comparison of two CYP2D6 genotyping methods and assessment of genotype-phenotype relationships. *Clin. Chem.* **49,** 542–551.

34. Wilson, J. W., Bean, P., Robins, T., Graziano, F., and Persing, D. H. (2000) Comparative evaluation of three human immunodeficiency virus genotyping systems: the HIV-GenotypR method, the HIV PRT GeneChip assay, and the HIV-1 RT line probe assay. *J. Clin. Microbiol.* **38,** 3022–3028.

35. Salvado, C. S., Trounson, A. O., and Cram, D. S. (2004) Towards preimplantation diagnosis of cystic fibrosis using microarrays. *Reprod. Biomed.* **8,** 107–114.

36. Ahrendt, S. A., Hu, Y., Buta, M., et al. (2003) p53 mutations and survival in stage I non-small-cell lung cancer: results of a prospective study. *J. Natl. Cancer Inst.* **95,** 961–970.

37. Warrington, J. A., Shah, N. A., Chen, X., et al. (2002) New developments in high-throughput resequencing and variation detection using high density microarrays. *Hum. Mutat.* **19,** 402–409.

38. Hacia, J. G. (1999) Resequencing and mutational analysis using oligonucleotide microarrays. *Nat. Gen.* **21(suppl),** 42–47.

39. Onwuazor, O. N., Wen, X. Y., Wang, D. Y., et al. (2003) Mutation, SNP, and isoform analysis of fibroblast growth factor receptor 3 (FGFR3) in 150 newly diagnosed multiple myeloma patients. *Blood* **102,** 772–773.

40. Kling, J. (2003) Roche's microarray tests. US FDA's diagnostic policy. *Nat. Biotechnol.* **21,** 959–960.

41. Petricoin, E. F. 3rd., Hackett, J. L., Lesko, L. J., et al. (2002) Medical applications of microarray technologies: a regulatory science perspective. *Nat. Genet.* **32(suppl),** 474–479.

42. Brown, C. R., Higgins, K. W., Frazer, K., et al. (1985) Simultaneous determination of total IgE and allergen-specific IgE in serum by the MAST chemiluminescent assay system. *Clin. Chem.* **31,** 1500–1505.

43. Price, C. P. and Hicks, J. M., eds. (1999) *Point-of-Care Testing.* AACC Press, Washington, DC.

44. Koch, D. O. and Peters, T. Jr. (1999) Selection and evaluation methods, in *Tietz Textbook of Clinical Chemistry,* 3rd ed. (Burtis, C.A. and Ashwood, E.R., eds.), WB Saunders, Philadelphia, pp. 320–335.

45. Carey, R. N. and Garber, C. C. (1989) Evaluation of methods, in *Clinical Chemistry. Theory, Practice and Correlation,* 2nd ed. (Kaplan, L. A. and Pesce, A. J., eds.), CV Mosby, St. Louis, pp. 290–310.

46. Hoeltge, G. A. and Duckworth, J. K. (1987) Review of proficiency testing performance of laboratories accredited by the College of American Pathologists. *Arch. Pathol. Lab. Med.* **111,** 1011–1014.

47. Whitehead, T. P. and Woodford, F. P. (1981) External quality assessment of clinical laboratories in the United Kingdom. *J. Clin. Pathol.* **34,** 947–957.

48. Mezzasoma, L., Bacarese-Hamilton, T., Di Cristina, M., Rossi, R., Bistoni, F., and Crisanti, A. (2002) Antigen microarrays for serodiagnosis of infectious diseases. *Clin. Chem.* **48,** 121–130.

49. Angenendt, P., Glokler, J., Murphy, D., Lehrach, H., and Cahill, D. J. (2002) Toward optimized antibody microarrays: a comparison of current microarray support materials. *Anal. Biochem.* **309,** 253–260.

50. Peluso, P., Wilson, D. S., Do, D., et al. (2003) Optimizing antibody immobilization strategies for the construction of protein microarrays. *Anal. Biochem.* **312,** 113–124.

51. Belov, L., Huang, P, Barber, N., Mulligan, S. P., and Christopherson, R. I. (2003) Identification of repertoires of surface antigens on leukemias using an antibody microarray. *Proteomics* **3,** 2146–2154.

52. Kiyonaka, S., Sada, K., Yoshimura, I., Shinkai, S., Kato, N., and Hamachi, I. (2004) Semi-wet peptide/protein array using supramolecular hydrogel. *Nat. Materials* **3,** 58–64.

53. Espina, V., Mehta, A. I., Winters, M. E., et al. (2003) Protein microarrays: molecular profiling technologies for clinical specimens. *Proteomics* **3,** 2091–2100.

54. Grubb, R. L., Calvert, V. S., Wulkuhle, J. D., et al. (2003) Signal pathway profiling of prostate cancer using reverse phase protein arrays. *Proteomics* **3**, 2142–2146.

55. Coleman, M. A., Miller, K. A., Beernink, P. T., Yoshikawa, D. M., and Albala, J. S. (2003) Identification of chromatin-related protein interactions using protein microarrays *Proteomics* **3**, 2101–2107.

56. Kingsmore, S. F. and Patel, D. D. (2003) Multiplexed protein profiling on antibody-based microarrays by rolling circle amplification. *Curr. Opin. Biotechnol.* **14**, 74–81.

57. MacBeath, G. and Schreiber, S. L. (2000) Printing proteins as microarrays for high-throughput function determination. *Science* **289**, 1760–1763.

58. Neuman de Vegvar, H. E., Amara, R. R., Steinman, L., Utz, P. J., Robinson, H. L., and Robinson, W. H. (2003) Microarray profiling of antibody responses against simian-human immunodeficiency virus: postchallenge convergence of reactivities independent of host histocompatibility type and vaccine regimen. *J. Virol.* **77**, 11125–11138.

59. Sreekumar, A., Nyati, M. K., Varambally, S., et al. (2001) Profiling of cancer cells using protein microarrays: discovery of novel radiation-regulated proteins. *Cancer Res.* **61**, 7585–7593.

60. Delehanty, J. B. and Ligler, F. S. (2002) A microarray immunoassay for simultaneous detection of proteins and bacteria. *Anal. Chem.* **74**, 5681–5687.

61. Bouwman, K., Qiu, J. Zhou, H., et al. (2003) Microarrays of tumor cell derived proteins uncover a distinct pattern of prostate cancer serum immunoreactivity. *Proteomics* **3**, 2200–2207.

62. Nam, M. J., Madoz-Gurpide, K. J., Wang, H., et al. (2003) Molecular profiling of the immune response in colon cancer using protein microarrays: occurrence of autoantibodies to ubiquitin *C*-terminal hydrolase L3. *Proteomics* **3**, 2108–2115.

63. de Jager, W., te Velthuis, H., Prakken, B. J., Kuis, W., and Rijkers G. T. (2003) Simultaneous detection of 15 human cytokines in a single sample of stimulated peripheral blood mononuclear cells. *Clin. Diagn. Lab. Immunol.* **10**, 133–139.

64. Copeland, S., Siddiqui, J., and Remick, D. (2004) Direct comparison of traditional ELISAs and membrane protein arrays for detection and quantification of human cytokines. *J. Immunol. Methods* **284**, 99–106.

65. Wiltshire, S., O'Malley, S., Lambert, J., et al. (2000) Detection of multiple allergen-specific IgEs on microarrays by immunoassay with rolling circle amplification. *Clin. Chem.* **46**, 1990–1993.

66. Belov, L., de la Vega, O., dos Remedios, C. G., Mulligan, S. P., and Christopherson, R. I. (2001) Immunophenotyping of leukemias using a cluster of differentiation antibody microarray. *Cancer Res.* **61**, 4483–4489.

67. Ge, H. (2000) UPA, a universal protein array system for quantitative detection of protein-protein, protein-DNA, protein-RNA and protein-ligand interactions. *Nucleic Acids Res.* **28**, e3.

68. Levit-Binnun, N., Lindner, A. B., Zik, O., Eshhar, Z., and Moses, E. (20030 Quantitative detection of protein arrays. *Anal. Chem.* **75**, 1436–1441.

69. Huang, J. X., Mehrens, D., Wiese, R., et al. (2001) High-throughput genomic and proteomic analysis using microarray technology. *Clin. Chem.* **47**, 1912–1916.

70. Lin, Y., Huang, R., Santanam, N., Liu, Y. G., Parthasarathy, S., and Huang, R. P. (2002) Profiling of human cytokines in healthy individuals with vitamin E supplementation by antibody array. *Cancer Lett.* **187**, 17–24.

71. Mullenix, M. C., Wiltshire, S., Shao, W., Kitos, G., and Schweitzer, B. (2001) Allergen-specific IgE detection on microarrays using rolling circle amplification: correlation with in vitro assays for serum IgE. *Clin. Chem.* **47**, 1926–1929.

72. Bacarese-Hamilton, T., Bistoni, F., and Crisanti, A. (2002) Protein microarrays: from serodiagnosis to whole proteome scale analysis of the immune response against pathogenic microorganisms. *Biotechniques* **Suppl.,** 24–29.

73. Feng, Y., Ke, X., Ma, R., Chen, Y., Hu, G., and Liu, F. (2004) Parallel detection of autoantibodies with microarrays in rheumatoid diseases. *Clin. Chem.* **50,** 416–422.

74. Scorilas, A., Bjartell, A., Lilja, H., Moller, C., and Diamandis, E. P. (2000) Streptavidin-polyvinylamine conjugates labeled with a europium chelate: applications in immunoassay, immunohistochemistry, and microarrays. *Clin. Chem.* **46,** 1450–1455.

75. Lueking, A., Horn, M., Eickhoff, H., Bussow, K., Lehrach, H., and Walter, G. (1999) Protein microarrays for gene expression and antibody screening. *Anal. Biochem.* **270,** 103–111.

76. Paweletz, C. P., Charboneau, L., Bichsel, V. E., et al. (2001) Reverse phase protein microarrays which capture disease progression show activation of pro-survival pathways at the cancer invasion front. *Oncogene* **20**, 1981–1989.

77. Carlisle, A. J., Prabhu, V. V., Elkahloun, A., et al. (2000) Development of a prostate cDNA microarray and statistical gene expression analysis package *Mol. Carcinogen.* **28**, 12–22.

78. Haab, B. B., Dunham, M. J., and Brown, P. O. (2001) Protein microarrays for highly parallel detection and quantitation of specific proteins and antibodies in complex solutions. *Genome Biol.* **2**, 1–13.

79. Grubb, R. L., Calvert, V. S., Wulkuhle, J. D., et al. (2003). Signal pathway profiling of prostate cancer using reverse phase protein arrays. *Proteomics* **3,** 2142–2146.

80. Neuman de Vegvar, H. E., Amara, R. R, Steinman, L., Utz, P. J., Robinson, H. L., and Robinson, W. H. (2003). Microarray profiling of antibody responses against simian-human immunodeficiency virus: postchallenge convergence of reactivities independent of host histocompatibility type and vaccine regimen. *J. Viron.* **7,** 11125–11138.

81. Ekins, R. P., Chu, F., and Biggart, E. (1990) Fluorescence spectroscopy and its application to a new generation of high sensitivity, multi-microspot, multianalyte, immunoassay. *Clin. Chim. Acta* **194,** 91–114.

82. Wulfkuhle, J. D., Aquino, J. A., Calvert, V. S., et al. (2003) Signal pathway profiling of ovarian cancer from human tissue specimens using reverse-phase protein microarrays. *Proteomics* **3**, 2085–2090.

83. Lockhart, D. J., Dong, H., Byrne, M. C., et al. (1996) Expression monitoring by hybridization to high-density oligonucleotide arrays. *Nat. Biotechnol.* **14,** 1675–1680.

84. Petricoin, E. F., 3rd, Hackett, J. L., Lesko, L. J., et al. (2002) Medical applications of microarray technologies: a regulatory science perspective. *Nat Genet.* **32(suppl),** 474–479.

85. Lee, M. L., Kuo, F. C., Whitmore, G. A., and Sklar, J. (2000) Importance of replication in microarray gene expression studies: statistical methods and evidence from repetitive cDNA hybridizations. *Proc. Natl. Acad. Sci. USA* **97,** 9834–9839.

86. Pavlidis, P., Qinghong, L., and Noble, W. S. (2003) The effect of replication on gene expression microarray experiments. *Bioinformatics* **19,** 1620–1627.

87. Lipshutz, R. J., Fodor, S. P., Gingeras, T. R., and Lockhart, D. J. (1999) High density synthetic oligonucleotide arrays. *Nat. Genet.* **21(suppl),** 20–24.

88. Cheng Li, C. and Wing Hung Wong, W. H. (2001) Model-based analysis of oligonucleotide arrays: expression index computation and outlier detection. *Proc. Natl. Acad. Sci. USA* **98,** 31–36.

89. Lemon, W. J., Liyanarachchi, S., and You, M. (2003) A high performance test of differential gene expression for oligonucleotide arrays. *Genome Biol.* **4,** R67.

90. Brazma, A., Hingamp, P., Quackenbush, J., et al. (2001) Minimum information about a microarray experiment (MIAME)—toward standards for microarray data. *Nat. Gen.* **29,** 365–371.

91. Ball, C. A., Sherlock, G., Parkinson, H., et al. (2002) Standards for microarray data. *Science*, **298,** 539.

92. Taylor, C. F., Paton, N. W., Garwood, K. L, et al. (2003) A systematic approach to modeling, capturing, and disseminating proteomics experimental data. *Nat. Biotechnol.* **21,** 247–254.

93. Orchard, S., Hermjakob, H., Julian, R. K. Jr, et al. (2004) Common interchange standards for proteomics data: public availability of tools and schema. *Proteomics* **4,** 490–491.

94. Kuo, W. P., Jenssen, T. K., Butte, A. J., Ohno-Machado, L., and Kohane, I. S. (2002) Analysis of matched mRNA measurements from two different microarray technologies. *Bioinformatics* **18,** 405–412.

95. Barczak, A., Rodriguez, M. W., Hanspers, K., et al. (2003) Spotted long oligonucleotide arrays for human gene expression analysis. Genome Res. **13,** 1775-1785.

96. Culhane, A. C., Perrière, G., and Higgins, D. G. (2003) Cross-platform comparison and visualization of gene expression data using co-inertia analysis. *BMC Bioinformatics* **4,** 59.

97. Johnson, K. F. and Lin, S. M. (2001) Critical assessment of microarray data analysis: the 2001 challenge. *Bioinformatics* **17,** 857–858.

98. The Tumor Analysis Best Practices Working Group. (2004) Expression profiling—best practices for data generation and interpretation in clinical trials. *Nat. Rev. Gen.* **5,** 229–237.

99. Wright, G., Tan, B., Rosenwald, A., Hurt, E. H., Wiestner, A., and Staudt, L. M. (2003) A gene expression-based method to diagnose clinically distinct subgroups of diffuse large B cell lymphoma. *Proc. Natl. Acad. Sci. USA* **100,** 9991–9996.

100. van de Vijver, M. J., He, Y. D., van't Veer, L. J., et al. (2002) A gene-expression signature as a predictor of survival in breast cancer. *N. Engl. J. Med.* **347,** 1999–2009.

101. Ramaswamy, S., Ross, K. N., Lander, E. S., and Golub, T. R. (2003) A molecular signature of metastasis in primary solid tumors. *Nat. Gen.* **33,** 49–54.

102. Ross, M. E., Zhou, X., Song, G., et al. 2003. Classification of pediatric acute lymphoblastic leukemia by gene expression profiling. *Blood* **102,** 2951–2159.

103. Sorlie, T., Perou, C. M., Tibshirani, R., et al. (2001) Gene expression patterns of breast carcinomas distinguish tumor subclasses with clinical implications. *Proc. Natl. Acad. Sci. USA* **98,** 10869–10874.

104. Hessner, M. J., Wang X., Khan, S., et al. (2003) Use of a three-color cDNA microarray platform to measure and control support-bound probe for improved data quality and reproducibility. *Nucleic Acids Res.* **31,** e60.

105. Boa, Z., Ma, W. L., Hu, Z. Y., Rong, S., Shi, Y. B., and Zheng, W. L. (2002) A method for evaluation of the quality of DNA microarray spots. *J. Biochem. Mol. Biol.* **35,** 532–535.

106. Shearstone, J. R., Allaire, N. E., Getman, M. E., and Perrin S. (2002) Nondestructive quality control for microarray production. *Biotechniques* **32,** 1051–1052, 1054, 1056–1057.

107. Sengupta, R. and Tompa, M. (2001) Quality control in manufacturing oligo arrays: a combinatorial design approach. *Pacific Symp. Biocomput.* 348–359.

108. Wang, X., Ghosh, S., and Guo, S. W. (2001) Quantitative quality control in microarray image processing and data acquisition. *Nucleic Acids Res.* **29,** E75–5.

109. Johnson, K. and Lin, S. (2003) QA/QC as a pressing need for microarray analysis: meeting report from CAMDA'02. *Biotechniques* **Suppl,** 62–63.

110. Johnson, K. F. and Lin, S. M. (2001) Critical assessment of microarray data analysis: the 2001 challenge. *Bioinformatics* **17,** 857–858.

111. Westgard, J. O., Barry, P. L., Hunt, M. R., and Groth, T. (1981) A multi-rule Shewhart chart for quality control in clinical chemistry. *Clin. Chem.* **27,** 493–501.

112. Ledue T. B., Collins, M. F., and Ritchie, R. F. (2002) Development of immunoturbidimetric assays for fourteen human serum proteins on the Hitachi 912. *Clin. Chem. Lab. Med.* **40,** 520–528.

113. Tello, F. L. and Hernandez, D. M. (2000) Performance evaluation of nine hormone assays on the Immulite 2000 immunoassay system. *Clin. Chem. Lab. Med.* **38,** 1039–1042.

114. Broussard, L. A. and Hanson, L. (1997) Evaluation of DRI enzyme immunoassays for drugs-of-abuse screening on the Cobas Mira. *Clin. Lab. Sci.* **10,** 83–86.

115. Thakkar, H., Newman, D. J., Holownia, P., et al. (1997) Development and validation of a particle-enhanced turbidimetric inhibition assay for urine albumin on the Dade aca analyzer. *Clin. Chem.* **43,** 109–113.

116. Letellier, M., Levesque, A., Daigle, F., and Grant, A. (1996) Performance evaluation of automated immunoassays on the Technicon Immuno 1 system. *Clin. Chem.* **42,** 1695–1701.

117. Sickinger, E., Stieler, M., Kaufman, B., Kapprell, H. P., West, D., Sandridge, A., Devare, S., Schochetman, G., Hunt, J. C., and Daghfal, D. AxSYM Clinical Study Group. (2004) Multicenter evaluation of a new, automated enzyme-linked immunoassay for detection of human immunodeficiency virus-specific antibodies and antigen. *J. Clin. Microbiol.* **42,** 21–29.

15

Tissue Microarrays

Ronald Simon, Martina Mirlacher, and Guido Sauter

Summary

New high-throughput screening technologies have led to the identification of hundreds of genes with potential roles in cancer or other diseases. For evaluation of promising candidate genes, however, *in-situ* analysis of high numbers of clinical tissue samples is mandatory. The tissue microarray (TMA) technology greatly facilitates such analysis. In this method minute tissue samples (0.6 mm in diameter) from up to 1000 different tissues can be analyzed on one microscope glass slide. All *in-situ* methods suitable for histological studies can be applied to TMAs without major changes in protocols, including immunohistochemistry, fluorescence *in-situ* hybridization, or RNA *in-situ* hybridization. Because all tissues are analyzed simultaneously with the same batch of reagents, TMA studies provide an unprecedented degree of standardization, speed, and cost efficiency.

Key Words: Tissue microarrays; high-throughput analysis; *in-situ* methods; immunohistochemistry; translational research; molecular epidemiology.

1. Introduction

To identify the most significant molecular features among all the emerging candidate disease genes, it is often necessary to analyze a large number of genes in a large number of well-characterized tissues. Especially in cancer research, hundreds of tumors must be analyzed for each gene to generate statistically valid results. Using traditional methods, this would lead to an insurmountable workload. Moreover, traditional molecular tissue analyses result in a critical loss of precious material since the number of conventional tissue sections that can be taken from a tumor block does usually not exceed 200–300.

Tissue microarray (TMA) technology significantly facilitates and accelerates tissue analyses by *in situ* technologies *(1,2)*. In this method, minute tissue cylinders (diameter: 0.6 mm) are removed from hundreds of different primary tumor blocks and subsequently brought into one empty "recipient" paraffin block.

From: *Methods in Molecular Medicine, Vol. 144, Microarrays in Clinical Diagnostics*
Edited by: T. Joos and P. Fortina © Humana Press Inc., Totowa, NJ

Sections from such array blocks can then be used for simultaneous *in situ* analysis of hundreds to thousands of primary tumors on the DNA, RNA, and protein level.

The TMA technique has a number of advantages compared with the "sausage" block technique that was introduced more than 10 yr ago *(3)*. The cylindrical shape and the small diameter of the specimen taken out of the donor block maximize the number of samples that can be taken out of one donor block and minimize tissue damage. The latter is important for pathologists since they can now give researchers access to their material and at the same time retain their tissue blocks. Punched tissue blocks remain fully interpretable for all morphological and molecular analyses that may subsequently become necessary, provided that the number of punches is reasonably selected. Dozens of punches can be taken from one tumor without compromising interpretability.

The range of possible TMA applications is broad. One of the best advantages of TMAs is that one set of tissues (which has been reviewed by one pathologist) with available clinical data can be used for an almost unlimited number of studies once these tissues are in a TMA format. The TMA technique is not limited to cancer research, although this has been the predominant application so far. TMAs can easily be made from frozen or paraffin-embedded tissues. They are suited for all kinds of *in situ* analyses including immunohistochemistry (IHC), RNA *in situ* hybridization (RNA-ISH), or fluorescence *in situ* hybridization (FISH). Perhaps the greatest advantage of the technology is that comprehensive studies including up to 20,000 tissue analyses can be done on 1 d under fully standardized conditions *(4)*.

2. Materials

2.1. Sample Collection

1. Standard routine histology microscope for review of tissue sections.
2. Colored pens to mark representative areas on the slides, e.g., red for tumor, blue for normal, and black for premalignant lesions.
3. Sufficient working space, especially for large-scale projects that require extensive sorting of thousands of sections and blocks.

2.2. Preparation of Recipient Blocks

1. PEEL-A-WAY embedding paraffin pellets, melting point 56–58°C (Polysciences, Warrington, PA).
2. Slotted processing/embedding cassettes for routine histology, e.g., EMS cat. no. 70070 (Electron Microscopy Sciences, Fort Washington, PA).
3. Stainless steel base molds for processing/embedding systems, e.g., EMS cat. no. 62510-30 (Electron Microscopy Sciences).
4. Filter papers.
5. Oven for paraffin melting (70°C).

2.3. TMA Fabrication

1. Tissue arrayer (for example, http://www.chemicon.com) and supplies. Many TMA users nowadays use very efficient home-made devices, especially for frozen TMA.
2. Premanufactured empty paraffin recipient blocks.
3. Illuminated magnifying lenses and supplies (e.g., Luxo U wave II/70, cat. no. 27950, Luxo, Switzerland) (optional).

2.4. TMA Sectioning

1. Standard routine histology microtome and supplies (e.g., cat. no. SM2400, Leica Microsystems).
2. Slide label printer (e.g., DAKO Seymour glass slide labeling system, product code S3416; DAKO, Denmark) or special slide marker (e.g., Securline Marker II, Precision Dynamics).
3. Boxes for slide storage.
4. Refrigerator for slide storage.
5. Paraffin Sectioning Aid-System (Instrumedics, cat. no. PSA) containing ultraviolet curing lamp, adhesive-coated PSA slides, TPC solvent, TPC solvent can, hand roller, tape windows (optional).

2.5. TMAs From Frozen Tissues

1. Tissue arrayer and supplies (*see* **Subheading 3.1.3.**)
2. OCT Tissue-Tek compound embedding medium (Sakura, The Netherlands).
3. Dry ice to keep punching needles and recipient block in optimal cooled condition.
4. Freezer for frozen tissue storage (−70°C).

3. Methods
3.1. TMA Fabrication
3.1.1. Sample Collection

Although a device is needed to manufacture TMAs, it must be understood that most of the work (approx 95%) is traditional pathology work that cannot be accelerated by improved (i.e., automated) tissue arrayers. This preparatory work is similar to what is needed for traditional studies involving "large" tissue sections. The major difference is the number of tissues involved, which can be an order of magnitude higher in TMA studies than in traditional projects. The different tasks related to sample collection are described below.

1. Exactly define the TMA to be made. (Often TMA users realize that one critical control tissue has been forgotten only after completion of the TMA block.) Include normal tissues of the organ of interest and—if possible—a selection of other organs as well.
2. Generate a list of potentially suited tissues.

3. Collect all slides from these tumors from the archive.
4. One pathologist must review all sections from all candidate specimens to select the optimal slide. If possible, tumors should be reclassified at that stage according to current classification schemes, and tissue areas suited for subsequent punching should be marked. Different colors are recommended for marking different areas on one section (for example, red for tumor, black for carcinoma *in situ*, blue for normal tissue). It is advisable to have a freshly H & E-stained section if the actual block surface is not well reflected on the available stained section.
5. Collect the tissue blocks that correspond to the selected slides.
6. These blocks and their corresponding marked slides must be matched and sorted in the order of appearance on the TMA.
7. Define the structure (outline) of the TMA, and compose a file that contains the identification numbers of the tissues together with their locations and real coordinates (as they need to be selected on the arraying device). As a distance between the individual samples, 0.2 mm is recommended. To facilitate navigation on the TMA, we recommend arranging the tissues in multiple sections (e.g., quadrants). The distance between the quadrants may be 0.8 mm. For unequivocal identification of individual samples on TMA slides, it is important to avoid a fully symmetrical TMA structure. In most laboratories capital letters define quadrants, whereas small letters and numbers define the coordinates within these quadrants. Examples of a TMA structure (outline) and data file containing the necessary information for making a TMA are given in **Fig. 1** and **Table 1**.

3.1.2. Preparing Recipient Blocks

In contrast to normal paraffin blocks, TMA blocks are cut at room temperature. Therefore, a special type of paraffin is needed with a melting temperature between 55 and 58°C (PEEL-A-WAY paraffin; *see* **Subheading 2.2.**)

1. The paraffin is melted at 60°C, filtered, and poured in a stainless steel mold.
2. A slotted plastic embedding cassette (as is used in every histology lab) is then placed on top of the warm paraffin.
3. Recipient paraffin blocks are then cooled down for 2 h at room temperature and for 2 additional h at 4°C. Blocks are then removed from the mold. It is important not to cool down the paraffin on a cooling plate because of the risk of block damage.
4. Quality check of the recipient blocks is important because they must not contain air bubbles.
5. Large recipient blocks (for example $30 \times 45 \times 10$ mm) are easier to handle than the small blocks (for example, $25-35 \times 5$ mm) that are typically used in routine histology labs.

3.1.3. TMA Performance

Only if all this preparatory work has been done can a tissue-arraying device be employed. Excellent TMAs can be produced in the hands of a talented and experienced person using manually operated devices. However, optimal arrays

Table 1
Example File for TMA Construction

Location	Coordinates	Location	Coordinates	Location	Coordinates
A 1a	0/0	A 2a	0/800	A 3a	0/1600
A 1b	800/0	A 2b	800/800	A 3b	800/1600
A 1c	1600/0	A 2c	1600/800	A 3c	1600/1600
A 1d	2400/0	A 2d	2400/800	A 3d	2400/1600
A 1e	3200/0	A 2e	3200/800	A 3e	3200/1600
A 1f	4000/0	A 2f	4000/800	A 3f	4000/1600
A 1g	4800/0	A 2g	4800/800	A 3g	4800/1600
A 1h	5600/0	A 2h	5600/800	A 3h	5600/1600
A 1i	6400/0	A 2i	6400/800	A 3i	6400/1600
A 1k	7200/0	A 2k	7200/800	A 3k	7200/1600
A 1l	8000/0	A 2l	8000/800	A 3l	8000/1600
A 1m	8800/0	A 2m	8800/800	A 3m	8800/1600
A 1n	9600/0	A 2n	9600/800	A 3n	9600/1600
A 1o	10400/0	A 2o	10400/800	A 3o	10400/1600
A 1p	11200/0	A 2p	11200/800	A 3p	11200/1600
A 1q	12000/0	A 2q	12000/800	A 3q	12000/1600
A 1r	12800/0	A 2r	12800/800	A 3r	12800/1600

can be expected only after a significant training period, mostly including several hundred, if not a few thousand, punches. A patient and enduring personality as well as keen eyesight are important prerequisites for operators of the manual tissue arrayers. Early-generation automated tissue arrayers are available, but these devices are very expensive and do not accelerate the TMA process.

The TMA process consists of four steps that are repeated for each sample placed on the TMA:

1. Punching a hole in an empty (recipient) paraffin block.
2. Removing and discarding the wax cylinder from the needle used for recipient block punching.
3. Removing a cylindrical sample from a donor paraffin block.
4. Placing the cylindrical tissue sample in the premade hole in the recipient block.

Exact positioning of the tip of the tissue cylinder at the level of the recipient block surface is crucial for the quality and the yield of the block. Placing the tissue too deeply into the recipient block results in empty spots in the first sections taken from the block. Positioning the tissue cylinder not deep enough causes empty spots in the last sections taken from this TMA. However, a too superficial location of the tissue cylinder is less problematic than a too deep

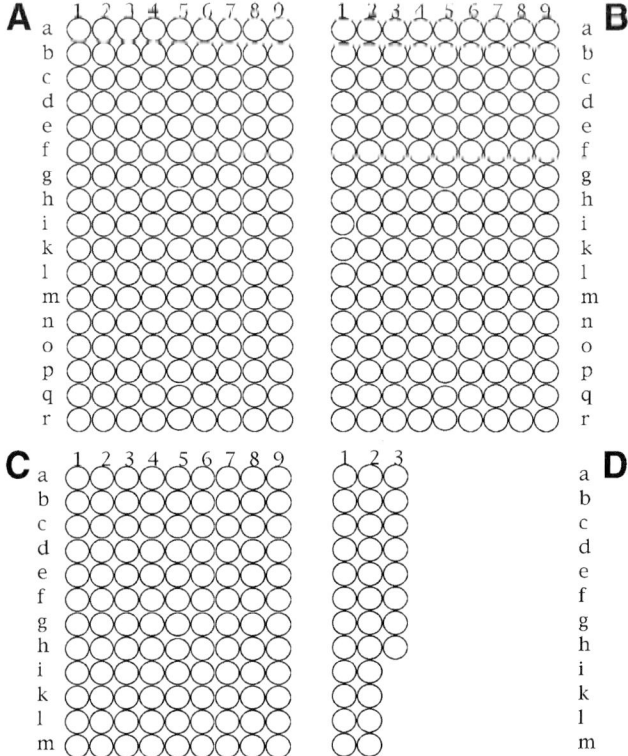

Fig. 1. TMA outline example. The TMA has been divided into four subsections (A–D) to facilitate navigation during microscopy.

position since protruding tissue elements can—to some extent—be leveled out after finishing the punching process. The use of a magnifying lens facilitates precise deposition of samples, especially for beginners.

As soon as all tissue elements are put into the recipient block, the block is heated at 40° for 10 min. Protruding tissue cylinders are then gently pressed deeper into the warmed TMA block using a glass slide.

3.1.4. Array Sectioning

Regular microtome sections may be taken from TMA blocks using standard microtomes. However, the more samples a TMA block contains, the more difficult regular cutting becomes. As a consequence, the number of slides of inadequate quality increases with the size of the TMA, and in turn, fewer sections from the TMA block can be analyzed effectively.

Using a tape sectioning kit (Instrumedics) facilitates cutting and leads to highly regular nondistorted sections (ideal for automated analysis). In addition, the tape system may prevent arrayed samples from floating off the slide, if very harsh pretreatment methods are used. However, the sticky glued slides have the disadvantage of increased background signals between the tissue spots in IHC analyses. The tissue samples themselves do not show increased nonspecific background in IHC. The tape sectioning system is described below.

1. An adhesive tape is placed on the TMA block in the microtome immediately before cutting.
2. A 3–5-μm section is cut. The tissue slice is now adhering to the tape.
3. The tissue slice is placed on a special "glued" slide. (Stretching of the tissue in a water bath or on a heating plate is not necessary.)
4. The slide (tissue on the bottom) is then placed under UV light for 35 s. This leads to polymerization of the glue on the slide and on the tape.
5. Slides are placed into TPC solution (Instrumedics) at room temperature for 5–10 s. The tape can then be gently removed from the glass slide. The tissue remains on the slide.
6. Slides are dried at room temperature.

3.1.5. TMAs From Frozen Tissues

Recently, Fejzo and Slamon *(15)* reported manufacturing TMAs from frozen tissues using a commercially available tissue array device. Commercially available and home-made arrayers arrayers can be utilized for frozen TMA making.

1. Recipient blocks are made by placing OCT into a Tissue-Tek standard cryomold and subsequently mounting on top of a plastic biopsy cassette. As long as the recipient OCT block is sized exactly like a paraffin recipient block (for which the arrayer was constructed), no modifications of the arrayer are necessary to mount the block.
2. The recipient block must be surrounded with dry ice to prevent melting.
3. Tissue biopsies (diameter 0.6 mm; height 4–5 mm) are then punched from OCT-embedded tumor tissues and placed into a recipient OTC array block using a commercial tissue microarrayer. There are four main differences compared with the procedure described for paraffin blocks.

   a. The same needle is utilized for making a hole in the recipient array block and for collecting the core biopsies. Switching to a larger needle is not necessary.
   b. It is important to keep the tissue in the needle frozen during the procedure. This can be done by preecooling the needle with a piece of dry ice before punching and while dispensing the tissue core into the recipient block.
   c. Needles are at a higher risk of bending and breaking. Therefore, punching and coring must be done slowly with minimal pressure to prevent needle breakage.
   d. The frozen TMAs become more irregular and distorted than TMAs from formalin-fixed material. In fact, the commercially available arrayers have not been designed for frozen array making. Therefore a larger space between samples is recommended (e.g. 1 mm).

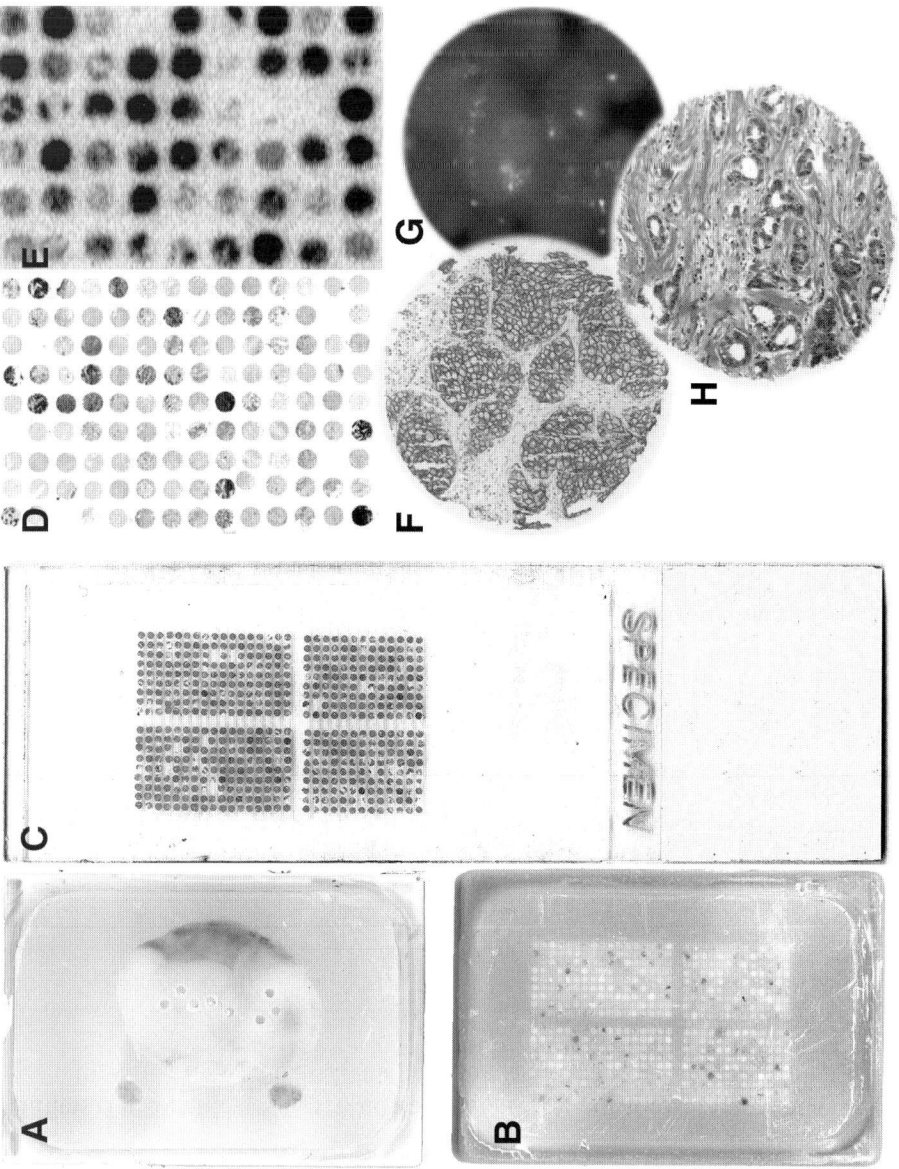

Fig. 2. TMA manufacturing and applications (*see* Color Plate 20 following p. 178). (**A**) Donor block from which several 0.6-mm tissue cores have been removed. Note that the original tissue block remains fully interpretable. Special devices (tissue arrayer) are used for this purpose. (**B**) Completed TMA. Removed tissue cores from several hundred different donor blocks have been assembled in the so-called recipient block. (**C**) Hematoxylin & eosin (H & E)-stained tissue section of the TMA. (**D**) TMA sector analyzed by immunohistochemistry (IHC). (**E**) Image of one TMA sector analyzed by RNA-ISH. The TMA was coated with a photographic emulsion in order to detect the

4. Sections (4–10 µm) of the whole block are cut from the array block. A cryostat microtome (Microm, Germany) can be used with or without the Basic CryoJane Tape Transfer System and slides (Instrumedics).

3.2. TMA Analysis

3.2.1. General Considerations

TMAs are suited for all types of *in situ* analysis methods including IHC, FISH, and RNA-ISH. All protocols that can be used on large sections will also work on TMAs. Examples of stained TMA sections are shown in **Fig. 2**. (*see* Color Plate 20 following p. 178). The most significant difference compared with traditional large section studies is the high level of standardization that can be achieved in TMA experiments. All slides of one TMA study are usually incubated in one set of reagents, ensuring absolutely identical concentrations, temperatures, and incubation times.

Other minor variables that may have an impact on the outcome of *in situ* analyses such as the age of a slide (time between sectioning and use) or section thickness are also fully standardized, as long as all tissues of one study are located on the same TMA section. As a result of this unprecedented standardization within each experiment, surprising interassay variations can occur if experiments are repeated under slightly different conditions. Often, these variations alter the threshold for detection of positivity, thus affecting the overall frequency of positive cases. In contrast, associations between examined parameters and clinicopathological data are usually retained unchanged, because all groups within one TMA (low and high stage, good and bad prognosis) are equally affected by experimental variations. Large numbers of samples on a TMA, however, markedly increase the likelihood of finding significant associations, especially in case of suboptimal IHC or RNA-ISH.

3.2.2. Immunohistochemistry

In general, the same rules apply for IHC analysis on TMA as on large sections. The small size of the arrayed tissues on a TMA facilitates the staining

---

Fig. 2. (*Continued from previous page*) radioactively labeled antisense-RNA probe. **(F)** Magnification of a tissue spot analyzed by IHC. Brown staining indicates expression of the HER2 membrane receptor protein. **(G)** Magnification (630×) of part of a tissue spot analyzed by fluorescence *in situ* hybridization (FISH). Green signals indicate the centromeric region of chromosome 7, and excess of red signal indicates amplification of the EGFR gene. Cell nuclei have been counterstained using a blue fluorescence dye. **(H)** Magnification of an H & E-stained tissue spot of a breast cancer sample. Despite the tiny diameter (0.6 mm), the tissue architecture is fully preserved.

interpretation since predefined criteria can be applied to a well-defined tissue area. This reduces intra- and interobserver variation of IHC interpretation.

For many immunohistochemical tumor analyses, the following information can be recorded:

1. Percentage of positive cells.
2. Staining intensity (0, 1+, 2+, 3+).
3. Subcellular localization of the staining (membraneous, cytoplasmatic, nuclear).
4. Tissue localization of the staining (tumor cells, stroma, vessels).

For statistical analyses, tumors can be classified into three or four groups based on the percentage of positive cells and the staining intensity. These groups include a completely negative, a strongly positive, and two intermediate groups, for example:

Negative	no staining
Weak positivity	1+ in 1–100% or 2+ in ≤20% of cells
Moderate positivity	2+ in 21–79% or 3+ in ≤30% of cells
Strong positivity	2+ in ≥80% or 3+ in >30 of cells

Some of the arrayed tissues may show false-negative or inappropriately weak IHC staining intensity owing to variations in tissue processing (e.g., fixation medium and time). The large number of tissues included in a TMA will often compensate for this phenomenon, which is also encountered in large section IHC analyses. At least a fraction of tissue spots yielding false-negative IHC staining results can be identified in control experiments assessing the antigen integrity of the samples, e.g., IHC detection of tissue type-specific antigens like cytokeratins or vimentin. For tissues with a reasonable proliferative activity, Ki67 (MIB1) is an optimal quality control antibody. MIB1, which leads to strong staining in all mitoses, is often falsely negative in suboptimally processed tissues.

It is highly recommended to use freshly cut sections for IHC analysis. The time span between sectioning and immunostaining should be less than 2 wk. Studies have shown that staining intensity decreases significantly with time for many antibodies (*6*).

3.2.2.1. AUTOMATED IHC ANALYSIS ON TMAS

Systems for automated TMA analysis are increasingly offered from commercial vendors. Although the most uncomplicated and inexpensive systems have great value for documentation purposes (image database generation) and can also efficiently quantify excellent immunostaining, there are several unsurmountable obstacles to automated TMA analysis. Many immunostaining protocols cannot prevent background staining and artifacts in certain tissues, which cannot be identified even by sophisticated automated analysis systems. Also there are often unexpected findings such as staining in stromal compartments like blood vessels or certain normal epithelial cell types. All these problems can be handled automatically by an experienced pathologist. In fact, perhaps the method described in this chapter should not have been named "tissue microarray," as investigators coming from the "array field" are mislead into believing that issues that have great importance in DNA or protein array techniques like automated manufacturing or automated analysis must also have importance in TMAs. As opposed to than classical array methods, the TMA method represents just a miniaturization of pathology analysis. As such, it is still subject to the traditional strengths and weaknesses of pathology and is dependent on the skill of the examiner.

3.2.3. FISH

Because biopsies are all treated individually at the time when they are removed, fixed, and subsequently paraffin-embedded, one must expect a certain degree of heterogeneity with respect to protein and nuclear acid preservation. That this is indeed true is best illustrated in the outcome of FISH analyses. As in large section studies, TMA FISH analyses yield interpretable results in only about 60–90% of the analyzed tumors (depending on the quality and size of the FISH probe) at the first attempt. Again, as in large section studies, it is possible to achieve interpretability in a fraction of initially noninformative cases by changing experimental conditions. For example, an increased proteinase concentration or a longer exposure time to proteinase for slide pretreatment will result in interpretable signals in some initially noninformative cases at the cost of overdigestion of some previously interpretable samples. In general, we do not attempt to improve the fraction of FISH-informative cases by changing experimental conditions. Because of the high number of tumors on our TMAs (usually >500), we tolerate a fraction of noninterpretable tumors rather than using too many precious TMA sections for additional experiments.

3.2.4. RNA-ISH

The question of whether RNA-ISH analysis can be reliably done on sections from archival tissue is disputed. Laboratories that feel confident in doing RNA-ISH analyses on formalin-fixed sections will also be able to execute such analyses on TMAs using the same protocols that are successful on large sections. As in IHC analysis, control experiments to detect expression of housekeeping genes (e.g., β-actin, GAPDH) may be performed to estimate the degree of RNA preservation in the different spots of a TMA, thus allowing exclusion of critical tissues from analysis. Alternatively, TMAs from frozen tissues may be utilized, especially for RNA quantification experiments. RNA-ISH yields reliable results because of the superior RNA preservation compared with formalin-fixed tissues.

References

1. Kononen, J., Bubendorf, L., Kallioniemi, A., et al. (1998). Tissue microarrays for high-throughput molecular profiling of tumor specimens. *Nat. Med.* **4,** 844–847.
2. Simon, R., Mirlacher, M., and Sauter, G. (2004) Tissue microarrays. *Biotechniques* **36,** 98–105.
3. Battifora, H. (1986) The multitumor (sausage) tissue block: novel method for immunohistochemical antibody testing. *Lab. Invest.* **55,** 244–248.
4. Sauter, G., Simon, R., and Hillan, K. (2003) Tissue microarrays in drug discovery. *Nat. Rev. Drug. Discov.* **2,** 962–972.
5. Fejzo, M. S. and Slamon, D. J. (2001) Frozen tumor tissue microarray technology for analysis of tumor RNA, DNA, and proteins. *Am. J. Pathol.* **159,** 1645–1650.
6. Mirlacher, M., Kasper, M., Storz, M., et al. (2004) Influence of slide aging on results of translational research studies using immunohistochemistry. *Mod. Pathol.* **17,** 1414–1420.

Index